Lecture Notes in Computer S

T0237880

Commenced Publication in 1973
Founding and Former Series Editors:
Gerhard Goos, Juris Hartmanis, and Jan van Leeuwen

Editorial Board

Leonardo Vanneschi   Steven Gustafson
Alberto Moraglio   Ivanoe De Falco
Marc Ebner (Eds.)

Genetic Programming

12th European Conference, EuroGP 2009
Tübingen, Germany, April 15-17, 2009
Proceedings

 Springer

Volume Editors

Leonardo Vanneschi
University of Milano-Bicocca
Department of Informatics, Systems and Communication (D.I.S.Co.)
viale Sarca 336-U14, 20126 Milano, Italy
E-mail: vanneschi@disco.unimib.it

Steven Gustafson
GE Global Research
Niskayuna, NY 12309, USA
E-mail: steven.gustafson@research.ge.com

Alberto Moraglio
University of Coimbra
Department of Computer Engineering
Polo II - Pinhal de Marrocos, 3030 Coimbra, Portugal
E-mail: moraglio@dei.uc.pt

Ivanoe De Falco
Institute of High Performance Computing and Networking
National Research Council of Italy (ICAR - CNR)
Via P. Castellino 111, 80131 Napoli, Italy
E-mail: ivanoe.defalco@na.icar.cnr.it

Marc Ebner
Eberhard Karls Universität Tübingen
Wilhelm Schickard Institut für Informatik, Abt. Rechnerarchitektur
Sand 1, 72076 Tübingen, Germany
E-mail: marc.ebner@wsii.uni-tuebingen.de

Library of Congress Control Number: Applied for

CR Subject Classification (1998): D.1, F.1, F.2, I.5, I.2, J.3

LNCS Sublibrary: SL 1 – Theoretical Computer Science and General Issues

ISSN   0302-9743
ISBN-10   3-642-01180-2 Springer Berlin Heidelberg New York
ISBN-13   978-3-642-01180-1 Springer Berlin Heidelberg New York

springer.com

© Springer-Verlag Berlin Heidelberg 2009
Printed in Germany

Typesetting: Camera-ready by author, data conversion by Scientific Publishing Services, Chennai, India
Printed on acid-free paper   SPIN: 12652358   06/3180   5 4 3 2 1 0

Preface

The 12th European Conference on Genetic Programming, EuroGP 2009, took place in Tübingen, Germany during April 15–17 at one of the oldest universities in Germany, the Eberhard Karls Universität Tübingen. This volume contains manuscripts of the 21 oral presentations held during the day, and the nine posters that were presented during a dedicated evening session and reception. The topics covered in this volume reflect the current state of the art of genetic programming, including representations, theory, operators and analysis, feature selection, generalization, coevolution, and numerous applications.

A rigorous, double-blind peer-review process was used, with each submission reviewed by at least three members of the international Program Committee. In total, 57 papers were submitted with an acceptance rate of 36% for full papers and an overall acceptance rate of 52% including posters. The MyReview management software originally developed by Philippe Rigaux, Bertrand Chardon, and other colleagues from the Université Paris-Sud Orsay, France was used for the reviewing process. We are sincerely grateful to Marc Schoenauer from INRIA, France for his continued assistance in hosting and managing the software. Paper review assigments were largely done by an optimization process matching paper keywords to keywords of expertise submitted by reviewers.

EuroGP 2009 was part of the larger Evo* 2009 event, which also included three other co-located events: EvoCOP 2009, EvoBIO 2009, and EvoWorkshops 2009. We would like to thank the many people who made EuroGP and Evo* a success. The great community of researchers and practitioners who submit and present their work, as well as serve the vital task of making timely and constructive reviews, are the foundation for our success. We are indebted to the local organizer, Marc Ebner from the Wilhelm Schickard Institute for Computer Science at the University of Tübingen for his hard and invaluable work that really made Evo* 2009 an enjoyable and unforgettable event. We extend our thanks to Andreas Zell, Chair of Computer Architecture at the Wilhelm Schickard Institute for Computer Science at the University of Tübingen and Peter Weit, Vice Director of the Seminar for Rhetorics at the New Philology Department at the University of Tübingen for local support. Our acknowledgments also go to the Tourist Information Center of Tübingen, especially Marco Schubert, and to the German Research Foundation (DFG) for financial support. We express our gratitude to the Evo* Publicity Chair Ivanoe De Falco, from ICAR, National Research Council of Italy and to Antonio Della Cioppa from the University of Salerno, Italy for his collaboration in maintaining the Evo* Web page.

We would like to thank our two internationally renowned invited keynote speakers: Peter Schuster, President of the Austrian Academy of Sciences and Stuart R. Hameroff, Professor Emeritus of the Departments of Anesthesiology and Psychology and Director of the Center for Consciousness Studies at the University of Arizona, Tucson, USA.

Last but not least, also this year, like every year since the first EuroGP edition in 1998, the work of organization has been made efficient, pleasant and enjoyable by the continued and invaluable coordination and assistance of Jennifer Willies and we are wholeheartedly grateful to her. Without her dedicated effort and the support from the Centre for Emergent Computing at Edinburgh Napier University, UK, these events would not be possible.

April 2009

Leonardo Vanneschi
Steven Gustafson
Alberto Moraglio
Ivanoe De Falco
Marc Ebner

Organization

Administrative details were handled by Jennifer Willies, Centre for Emergent Computing at Edinburgh Napier University, UK.

Organizing Committee

Program Co-chairs	Leonardo Vanneschi (University of Milano-Bicocca, Italy)
	Steven Gustafson (GE Global Research, USA)
Publication Chair	Alberto Moraglio (University of Coimbra, Portugal)
Publicity Chair	Ivanoe De Falco (ICAR, National Research Council of Italy)
Local Chair	Marc Ebner (University of Tübingen, Germany)

Program Committee

Hussein Abbass	UNSW@ADFA, Australia
Lee Altenberg	University of Hawaii at Manoa, USA
R. Muhammad Atif Azad	University of Limerick, Ireland
Wolfgang Banzhaf	Memorial University of Newfoundland, Canada
Anthony Brabazon	University College Dublin, Ireland
Nicolas Bredeche	Université Paris-Sud, France
Edmund Burke	University of Nottingham, UK
Stefano Cagnoni	Università degli Studi di Parma, Italy
Philippe Collard	Laboratoire I3S (UNSA-CNRS), France
Pierre Collet	LSIIT-FDBT, France
Ernesto Costa	Universidade de Coimbra, Portugal
Ivanoe De Falco	ICAR, National Research Council of Italy, Italy
Michael Defoin Platel	University of Auckland, New Zealand
Edwin DeJong	Universiteit Utrecht The Netherlands
Antonio Della Cioppa	Università di Salerno, Italy
Ian Dempsey	Pipeline Financial Group, Inc., USA
Federico Divina	Universidad Pablo de Olavide, Spain
Marc Ebner	Universität Tübingen, Germany
Anikò Ekàrt	Aston University, UK
Anna Esparcia-Alczar	ITI Valencia, Spain
Daryl Essam	UNSW@ADFA, Australia
Francisco Fernàndez de Vega	Universidad de Extremadura, Spain
Christian Gagné	MDA, Canada

Table of Contents

Oral Presentations

Posters

One-Class Genetic Programming

Robert Curry and Malcolm I. Heywood

Dalhousie University
6050 University Avenue
Halifax, NS, Canada
B3H 1W5
{rcurry,mheywood}@cs.dal.ca

Abstract. One-class classification naturally only provides one-class of exemplars, the target class, from which to construct the classification model. The one-class approach is constructed from artificial data combined with the known in-class exemplars. A multi-objective fitness function in combination with a local membership function is then used to encourage a co-operative coevolutionary decomposition of the original problem under a novelty detection model of classification. Learners are therefore associated with different subsets of the target class data and encouraged to tradeoff detection versus false positive performance; where this is equivalent to assessing the misclassification of artificial exemplars versus detection of subsets of the target class. Finally, the architecture makes extensive use of active learning to reinforce the scalability of the overall approach.

Keywords: One-Class Classification, Coevolution, Active Learning, Problem Decomposition.

1 Introduction

The ability to learn from a single class of exemplars is of importance under domains where it is not feasible to collect exemplars representative of all scenarios e.g., fault or intrusion detection; or possibly when it is desirable to encourage fail safe behaviors in the resulting classifiers. As such, the problem of one-class learning or 'novelty detection' presents a unique set of requirements from that typically encountered in the classification domain. For example, the discriminatory models of classification most generally employed might formulate the credit assignment goal in terms of maximizing the separation between the in- and out-class exemplars. Clearly this is not feasible under the one-class scenario. Moreover, the one-class case often places more emphasis on requiring fail safe behaviors that explicitly identify when data differs from the target class or 'novelty detection'.

Machine learning algorithms employed under the one-class domain therefore need to address the discrimination/ novelty detection problem directly. Specific earlier works include Support Vector Machines (SVM) [1,2,3], bottleneck neural

L. Vanneschi et al. (Eds.): EuroGP 2009, LNCS 5481, pp. 1–12, 2009.

networks [4] and a recent coevolutionary genetic algorithm approach based on an artificial immune system [5] (for a wider survey see [6,7]).

In particular the one-class SVM model of Schölkopf relies on the correct identification of "relaxation parameters" to separate exemplars from the origin (representing the second unseen class) [1]. Unfortunately, the values for such parameters vary as a function of the data set. However, a recent work proposed a kernel autoassociator for one-class classification [3]. In this case the kernel feature space is used to provide the required non-linear encoding, this time in a very high dimensional space (as opposed to the MLP approach to the encoding problem). A linear mapping is then performed to reconstruct the original attributes as the output. Finally, the work of Tax again uses a kernel based one-class classifier. This approach is distinct in that data is artificially generated to aid the identification of the most concise hypersphere describing the in-class data [2]. Such a framework builds on the original support vector data description model, whilst reducing the significance of specific parameter selections. The principle drawback, however, is that tens or even hundreds of thousands of artificially generated training exemplars are required to build a suitably accurate model [2]. The work proposed in this paper uses the artificial data generation model of Tax, but specifically addresses the training overhead by employing an active learning algorithm. Moreover, the Genetic Programming (GP) paradigm provides the opportunity to solve the problem using an explicitly multiple objective model, where this provides the basis for cooperative coevolutionary problem decomposition.

2   Methodology

The general principles of the one-class GP (OCGP) methodology, as originally suggested in [8], is comprised of four main components:

(1) **Local membership function:** Conventionally, GP programs provide a mapping between the multi-dimensional attribute space and a real-valued one-dimensional number line called the *gpOut* axis. A binary switching function (BSF), as popularized by Koza, is then used to map the one-dimensional 'raw' *gpOut* to one of two classes, as shown in Fig. 1(a) [9]. However, a BSF assumes that the two classes can be separated at the origin. Moreover, under a one-class model of learning – given that we have no information on the distribution of out-class data – exemplars belonging to the unseen classes are just as likely to appear on either side of the origin, resulting in high false positive rates. Therefore, instead of using the 'global' BSF, GP individuals utilize a Gaussian or 'local' membership function (LMF), Fig. 1(b). A small region of the *gpOut* axis is therefore evolved for expressing in-class behavior, where this region is associated with a subset of the target distribution encountered during training. In this way GP individuals act as novelty detectors, as any region on the *gpOut* axis other than that of the LMF is associated with the out-class conditions; thus supporting conservative generalization properties when deployed. Moreover, instead of a single classifier providing a single mapping for all in-class exemplars,

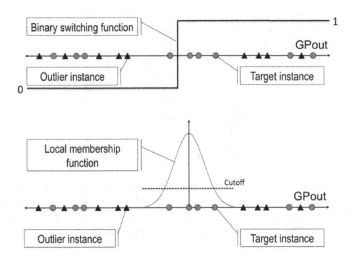

Fig. 1. (a) 'Global' binary switching function vs. (b) Gaussian 'local' membership function

our goal will be to encourage the coevolution of multiple GP novelty detectors to map unique subsets of the in-class exemplars to their respective LMF.

(2) **Artificial outlier generation:** In a conventional binary classification problem the decision boundaries between classes are supported by exemplars from each class. However, in a one-class scenario it is much more difficult to establish appropriate and concise decision boundaries since they are only supported by the target class. Therefore, the approach we adopt for developing one-class classifiers is to build the outlier class data artificially and train using a two class model. The 'trick' is to articulate this goal in terms of finding an optimal trade-off between detection and false positive rates as opposed to explicitly seeking solutions with 'zero error'.

(3) **Balanced block algorithm (BBA):** When generating artificial outlier data it is necessary to have a wider range of possible values than the target data and also to ensure that the target data is surrounded in all attribute directions. Therefore, the resulting 'two-class' training data set tends to be unbalanced and large; implying artificial data partitions in the order of tens of thousands of exemplars (as per Tax [2]). The increased size of the training data set has the potential to significantly increase the training overhead of GP. To address this overhead the BBA active learning algorithm is used [10]; thus fitness evaluation is conducted over a much smaller subset of training exemplars, dynamically identified under feedback from individuals in the population, Fig. 2.

(4) **Evolutionary multi-objective optimization (EMO):** EMO allows multiple objectives to be specified, thereby providing a more effective way to express the quality of GP programs. Moreover, EMO provides a means of comparing individuals under multiple objectives without resorting to *a priori* scalar weighting

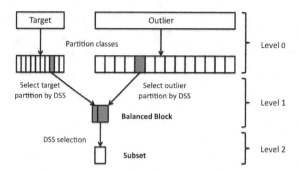

Fig. 2. The balanced block algorithm first partitions the target and outlier classes (Level 0) [10]. Balanced blocks of training data are then formed by selecting a partition from each class by dynamic subset selection (Level 1). For each block multiple subsets are then chosen for GP, training again performed by DSS (Level 2).

functions. In this way the overall OCGP classifier is identified through a cooperative approach that supports the simultaneous development of multiple programs from a *single* population.

The generation of artificial data in and around the target data means that outlier data lying within the actual target distribution cannot be avoided. Thus, when attempting to classify the training data it is necessary to cover as much of the target data as possible (i.e., maximize detection rate), while also minimizing the amount of outlier data covered (i.e., minimize false positive rate); the first two objectives. Furthermore, it is desirable to have an objective to encourage diversity among solutions by actively rewarding non-overlapping behavior between the coverage of different classifiers as evolved from the same population. Finally, the fourth objective encourages solution simplicity, thus reducing the likelihood of overfitting and promoting solution transparency.

GP programs are compared by their objectives using the notion of *dominance*, where a classifier A is said to *dominate* classifier B if it performs at least as well as B in all objectives and better than B in at least one objective. *Pareto ranking* then combines the objectives into a scalar fitness by assigning each classifier a *rank* based on the number of classifiers by which it is dominated [11,12]. A classifier is said to be *non-dominated* if it is not dominated by any other classifier in the population and has a rank of zero. The set of all non-dominated classifiers is referred to as the *Pareto front*. The Pareto front programs represent the current best trade-offs of the multi-objective criteria providing a range of candidate solutions. Pareto front programs influence OCGP by being favored for reproduction and update the archives of best programs which determine the final OCGP classifiers.

2.1   OCGP Algorithm

The general framework of our algorithm is described by the flowchart in Fig. 3. The first step is to initialize the random GP population of programs and the

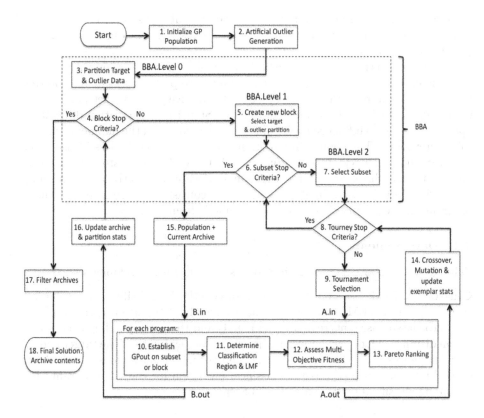

Fig. 3. Framework for OCGP assuming target data is provided as input

necessary data structures, Step 1. OCGP is not restricted to a specific form of GP but in this work a linear representation is assumed. Artificial outlier data is then generated in and around the provided target data, Step 2.

The next stage outlines the three levels of the balanced block algorithm (BBA), Fig. 2, within the dashed box. Level 0, Step 3, partitions the target and outlier data. The second level, Step 5, selects a target and outlier partition to create the Level 1 balanced block. At Level 2, Step 7, subsets are selected from the block to form the subset of exemplars over which fitness evaluation is performed (steady state tournament).

The next box outlines how individuals are evaluated. First programs are evaluated on the current data subset to establish the corresponding *gpOut* distribution, Step 10. The classification region is then determined to parameterize the LMF, Step 11, and the multi-objective fitness is established, Step 12. Once all programs have their multi-objective fitness the programs can be Pareto ranked, Step 13.

The Pareto ranking determines the tournament winners, or parents, from which genetic operators can be applied to create children, Step 14. In addition parent programs update the difficulty of exemplars in order to influence future subset selections. That is to say, previous performance (error) on artificial–target class data is used to guide the number of training subsets sampled from level 1 blocks. As such, difficulty values are averaged across the data in level 1 and 2, Step 6 and 8 respectively.

Once training is complete at level 2, the population and the archives associated with the current target partitions are combined, Step 15, and evaluated on the Level 1 block (Step 10 through Step 13). The archive of the target partition is then updated with the resulting Pareto front, Step 16, and partition difficulties updated in order to influence future partition selections and the number of future subset iterations. The change in partition error rates for each class is also used to determine the block stop criteria, Step 4. Once the block stop criteria has been met the archives are filtered with any duplicates across the archives being removed, Step 17, and the final OCGP classifier consists of the remaining archive programs. More details of the BBA are available from [10].

2.2   Developments

Relative to the above, this work introduces the following developments:

Clustering. In the original formulation of OCGP the classification region for each program (i.e., LMF location) was determined by dividing a program's *gpOut* axis into class-consistent 'regions' (see Fig. 4) and the corresponding class separation distance (1), or *csd*, between adjacent regions estimated. The target region that maximizes *csd* with respect to the neighboring outlier regions has the best separability and is chosen as the classification region for the GP program (determining classification region at Fig. 3 Step 11).

$$csd_{\,0/1} = \frac{|\mu_0 - \mu_1|}{\sqrt{\sigma_0^2 + \sigma_1^2}} \tag{1}$$

In this work the LMF is associated with the most dense region of the *gpOut* axis i.e., independent of the label information. Any artificial exemplars lying within the LMF might either result in the cluster being penalized once the fitness criteria is applied or be considered as outliers generated by the artificial

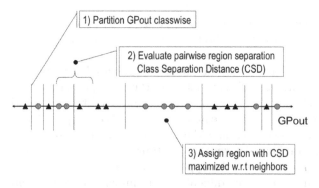

Fig. 4. Determining GP classification region by class separation distance

data generation model. In effect we establish a 'soft' model of clustering (may contain some artificial data points) in place of the previous 'hard' identification of clusters (no clusters permitted with any artificial data points).

To this end, subtractive clustering [13] is used to cluster the one-dimensional *gpOut* axis and has the benefit of not requiring *a priori* specification of the number of clusters. The mixed content of a cluster precludes the use of a class separation distance metric. Instead the sum squared error (SSE) of the exemplars within the LMF are employed. The current GP program's associated LMF returns confidence values for each of the exemplars in the classification region. Therefore, the error for each type of exemplar can be determined by subtracting their confidence value from their actual class label (i.e., reward for target exemplars in classification region and penalize for outliers). The OCGP algorithm using *gpOut* clustering will be referred to as OCGPC.

Caching. The use of the clustering algorithm caused the OCGPC algorithm to run much more slowly than OCGP. The source was identified to be the clustering of the entire GP population and the current archive on the entire Level 1 block (Fig. 3 Step 15). Therefore, instead of evaluating the entire population and archive on the much larger Level 1 block, the mean and standard deviation is cached from the subset evaluation step. Caching was introduced in both the OCGP and OCGPC algorithms and was found to speed up training time without negatively impacting classification performance.

Overlap. The overlap objective has been updated from the previous work to compare tournament programs against the current archive instead of comparing against the other tournament programs (assessing multi-objective fitness at Step 12). Individuals losing a tournament are destined to be replaced by the search operators, thus should not contribute to the overlap evaluation. Moreover, comparison against the archive programs is more relevant, as they represent the current best solution to the current target partition (i.e., target exemplars already covered) and thus encourages tournament programs to classify the remaining uncovered target exemplars.

Artificial outlier generation. Modifications have been made in order to improve the quality of the outlier data (Fig. 3 Step 2). Previously a single radius, R, was determined by the attribute of the target data having the largest range and was then used as the radius for all attribute dimensions when creating outliers. If a large disparity exists between attribute ranges, this can lead to large volumes of the outlier distribution with little to no relevance to the target data. Alternatively, a vector $\overline{R}$ of radii is used consisting of a radius for each attribute. Additionally, when the target data consists of only non-negative values, negative outlier attribute values are restricted to within a close proximity to zero.

3   Experiments

In contrast to the previous performance evaluation [8], we concentrate on benchmarking against data sets that have large unbalanced distributions in the

Table 1. Binary classification datasets. The larger in-class partition of Adult, Census and Letter-vowel was matched with a larger artificial exemplar training partition.

Dataset	Adult		Census		Letter-vowel	
Features	14		40		16	
Class	Train	Test	Train	Test	Train	Test
0	50,000	11,355	50,000	34,947	50,000	4,031
1	7,506	3,700	5,472	2,683	2,660	969
Total	57,506	15,055	55,472	37,630	52,660	5,000
Dataset	Letter-a		Letter-e		Mushroom	
Features	16		16		22	
Class	Train	Test	Train	Test	Train	Test
0	10,000	4,803	10,000	4,808	10,000	872
1	574	197	545	192	1,617	539
Total	10,574	5,000	10,545	5,000	11,617	1,411

underlying exemplar distribution, thus are known to be difficult to classify under binary classification methods. Specifically, the Adult, Census-Income (KDD), Mushroom and Letter Recognition data sets from the UCI machine learning repository [14] were utilized (Table 1). The Letter Recognition data set was used to create three one-class classification data sets where the target data was alternately all vowels (Letter-vowel), the letter 'a' (Letter-a) and the letter 'e' (Letter-e). For the Adult and Census data sets a predefined training and test partition exists. For the Letter Recognition and Mushroom data sets the data was first divided into a 75% training and 25% test data set while maintaining the class distributions of the original data. The class 0 exemplars were removed from the training data set to form the one-class target data set.

Training was performed on a dual G4 1.33 GHz Mac Server with 2 GB of RAM. All experiments are based on 50 GP runs where runs differ only in their choice of random seeds for initializing the population while all other parameters remain unchanged. Table 2 lists the common parameter settings for all runs.

The OCGP algorithm results are compared to results found by a one-class support vector machine (OC ν-SVM) [1] and a one-class or bottleneck neural network (BNN)[1], where both algorithms are trained on the target data alone. Additionally, the OCGP results will be compared to a two-class support vector machine (ν-SVM) which use both the artificial outlier and the target data. The two-class SVM is used as a baseline binary classification algorithm in order to assess to what degree the OCGP algorithms are able to provide a better characterization of the problem i.e., both algorithms are trained on the target and artificial outlier data. Comparison against Tax's SVM was not possible as the current implementation does not scale to large data sets.

---

[1] Unlike the original reference ([4]) the BNN was trained using the efficient second Conjugate Gradient weight update rule with tansig activation functions; both of which make a significant improvement over first order error back-propagation.

Table 2. Parameter Settings

Dynamic Page-Based Linear GP			
Population size	125	Tournament Size	4
Max # of pages	32	Number of registers	8
Page size	8 instructions	Instr. prob. 1, 2 or 3	0/5, 4/5, 1/5
Max page size	8 instructions	Function set	$\{+, -, \times, \div\}$
Prob. Xover, Mut., Swap	0.9, 0.5, 0.9	Terminal set	$\{\#$ of attributes$\}$
Balanced Block Algorithm Parameters			
Target Partition Size	$\approx \frac{\#Patterns}{\#Archives}$	Max subset iterations	10
Outlier Partition Size	500	Tourneys per subset	6
Max block selections	2000	Level 2 subset size	100
Archive Parameters			
Number of Archives	Adult = 15, Census = 11, Letter-vowel = 10, Letter-a = 6, Letter-e = 6, Mushroom = 4		
Archive Size	10		

The algorithms are compared in terms of ROC curves of (FPR, DR) pairs on test data sets (Fig. 5). Due to the large number of runs of the OCGP algorithms and the multiple levels of voting possible, plotting all of the OCGP solutions becomes too convoluted. Alternatively, only non-dominated solutions will be shown, where these represent the best-case OCGP (FPR, DR) pairs over the 50 runs. Similarly only the non-dominated solutions of the bottle-neck neural networks will be shown, while for the other algorithms only a small number of solutions are present and so all results will be plotted. Comments will be made as follows on an algorithm-by-algorithm basis:

OCGPC. Of the two one-class GP models the cluster variant for establishing the region of interest on the *gpOut* axis described in Sect. 2.2 appeared to be the most consistent. Specifically, OCGPC provided the best performing curves under the Adult and Vowel data sets (Fig. 5(a) and (c)) and consistently the runner up under all but the Census data set. In each of the three cases where it appears as a runner up it was second to the BNN. However, GP retains the advantage of indexing a subset of the attributes; whereas multi-layer neural networks index all features – a bias of NN methods in general and the autoassociator method of deployment in particular.

OCGP. When assuming the original region based partitioning of *gpOut*, ROC performance decreases significantly with strongest performance on the Census data set; and runner up performance under Adult and Mushroom. Performance on the remaining data sets might match or better the one-class ν-SVM, but generally worse than BNN or OCGPC.

BNN. As remarked above the BNN was the strongest one-class model, with best ROC curves on Letter 'a', 'e', and mushroom; and joint best on Census (with OCGP). Moreover, the approach was always better than the SVM methods benchmarked in this work.

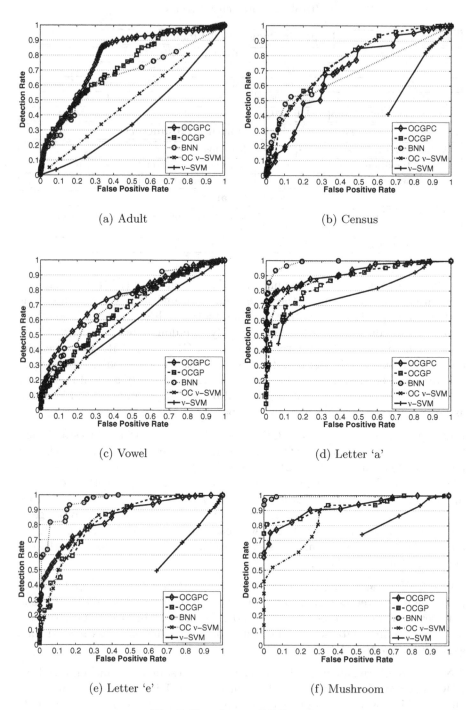

(a) Adult

(b) Census

(c) Vowel

(d) Letter 'a'

(e) Letter 'e'

(f) Mushroom

Fig. 5. Test dataset ROC curves

SVM methods. Neither SVM method – one-class or binary SVM trained on the artificial versus target class – performed well. Indeed the binary SVM was at best degenerate on all but the Letter 'a' data set; resulting in it being ranked worst in all cases. This indicates that merely combining artificial data with target class data is certainly not sufficient for constructing one-class learners as the learning objective is not correctly expressed. The one-class SVM model generally avoided degenerate solutions, but never performed better than OCGPC or BNN.

4   Conclusion

A Genetic Programming (GP) classifier has been proposed in this work for the one-class classification domain. Four main components of the algorithm have been identified. Artificial outlier generation is used to establish more concise boundaries and to enable the use of a two-class classifier. The active learning algorithm BBA tackles the class imbalance and GP scalability issues introduced by the large number of artificial outliers required. Gaussian 'local' membership functions allow GP programs to respond to specific regions of the target distribution and act as novelty detectors. Evolutionary multi-objective optimization drives the search for improved target regions, while allowing for the simultaneous development of multiple programs that cooperate towards the overall problem decomposition through the objective for minimizing overlapping coverage.

A second version of the OCGP algorithm is introduced, namely OCGPC, which determines classification regions by clustering the one-dimensional *gpOut* axis. In addition, 'caching' of the classification regions is introduced to both algorithms, in order to eliminate the need to redetermine classification regions over training blocks. Caching reduces training times without negatively impacting classification performance. Modifications were also made to improve the quality of generated artificial outliers and to the objectives used to determine classification regions, including the use of the sum-squared error and improving the overlap objective by comparing to only the current best archive solutions.

The OCGP and OCGPC algorithms were evaluated on six data sets larger than previously examined. The results were compared against two one-class classifiers trained on target data alone, namely one-class ν-SVM and a bottleneck neural network (BNN). An additional comparison was made with a two-class SVM trained on target data and the generated artificial outlier data. The OCGPC and BNN models were the most consistent performers overall; thereafter model preference might be guided by the desirability for solution simplicity. In this case the OCGPC model makes additional contributions as it operates as a classifier as opposed to an autoassociator i.e., autoassociators are unable to simplify solutions in terms of attributes indexed. Future work will concentrate in this direction and to applying OCGPC to learning without artificial data.

Acknowledgements

BNN and SVM models were built in MATLAB and LIBSVM respectively. The authors acknowledge MITACS, NSERC, CFI and SwissCom Innovations.

References

1. Scholköpf, B., Platt, J., Shawe-Taylor, J., Smola, A., Williamson, R.: Estimating the support of a high-dimensional distribution. Neural Computation 13, 1443–1471 (2001)
2. Tax, D., Duin, R.: Uniform object generation for optimizing one-class classifiers. Journal of Machine Learning Research 2, 155–173 (2001)
3. Zhang, H., Huang, W., Huang, Z., Zhang, B.: A kernel autoassociator approach to pattern classification. IEEE Transactions on Systems, Man and Cybernetics - Part B 35(3), 593–606 (2005)
4. Manevitz, L., Yousef, M.: One-class document classification via neural networks. Neurocomputing 70(7-9), 1466–1481 (2007)
5. Wu, S., Banzhaf, W.: Combatting financial fraud: A coevolutionary anomaly detection approach. In: Genetic and Evolutionary Computation Conference (GECCO), pp. 1673–1680 (2008)
6. Markou, M., Singh, S.: Novelty detection: A review – part 1: Statistical approaches. Signal Processing 83, 2481–2497 (2003)
7. Markou, M., Singh, S.: Novelty detection: A review – part 2: Neural network based approaches. Signal Processing 83, 2499–2521 (2003)
8. Curry, R., Heywood, M.: One-class learning with multi-objective Genetic Programming. In: Proceedings of the IEEE Systems, Man and Cybernetics Conference, SMC, pp. 1938–1945 (2007)
9. Koza, J.: Genetic programming: On the programming of computers by means of natural selection. Statistics and Computing 4(2), 87–112 (1994)
10. Curry, R., Lichodzijewski, P., Heywood, M.: Scaling genetic programming to large datasets using hierarchical dynamic subset selection. IEEE Transactions on Systems, Man and Cybernetics - Part B 37(4), 1065–1073 (2007)
11. Kumar, R., Rockett, P.: Improved sampling of the pareto-front in multiobjective genetic optimizations by steady-state evolution: A pareto converging genetic algorithm. Evolutionary Computation 10(3), 283–314 (2002)
12. Zitzler, E., Thiele, T.: Multiobjective evolutionary algorithms: A comparitive case study and the strength pareto approach. IEEE Transactions on Evolutionary Computation 3(4), 257–271 (1999)
13. Chiu, S.: 9. In: Fuzzy Information Engineering: A Guided Tour of Applications. John Wiley & Sons, Chichester (1997)
14. Asuncion, A., Newman, D.J.: UCI Repository of Machine Learning Databases. Dept. of Information and Comp. Science. University of California, Irvine (2008), http://www.ics.uci.edu/~mlearn/mlrepository.html

Genetic Programming Based Approach for Synchronization with Parameter Mismatches in EEG

Dilip P. Ahalpara[1], Siddharth Arora[2], and M.S. Santhanam[2,3]

[1] Institute for Plasma Research, Near Indira Bridge, Bhat, Gandhinagar-382428, India
[2] Physical Research Laboratory, Navrangpura, Ahmedabad-380009, India
[3] Indian Institute of Science Education and Research, Pashan, Pune-411021, India
dilip@ipr.res.in, sidarora007@gmail.com, santh@prl.res.in

Abstract. Effects of parameter mismatches in synchronized time series are studied first for an analytical non-linear dynamical system (coupled logistic map, CLM) and then in a real system (Electroencephalograph (EEG) signals). The internal system parameters derived from GP analysis are shown to be quite effective in understanding aspects of synchronization and non-synchronization in the two systems considered. In particular, GP is also successful in generating the CLM coupled equations to a very good accuracy with reasonable multi-step predictions. It is shown that synchronization in the above two systems is well understood in terms of parameter mismatches in the system equations derived by GP approach.

1 Introduction

Synchronized time series with errors are known to exist in nature as well as in many nonlinear dynamical systems [1,2,3,4]. For instance, consider two dynamical systems $X(\mu)$ and $Y(\mu)$ which are coupled to one another and μ is an internal parameter. If $x(t)$ and $y(t)$ are the measured output of some variable from X and Y systems respectively, then two possible scenarios can arise within the context of synchronization. This could lead to complete synchronization, implying $x(t)=y(t)$ for all t after certain transient time. Another possibility is that $|x(t)-y(t)|<\epsilon$ for all t, where $0<\epsilon\ll1$. This means that instead of complete synchronization, it is possible to achieve nearly synchronized state, where the error represented by ϵ is the upper bound. For the case of complete synchronization, $\epsilon=0$. In the present study we start with a premise that synchronization with error arises from internal parameter mismatches between coupled systems. In the present discussion, the two systems $X(\mu_1)$ and $Y(\mu_2)$ with parameters μ_1 and μ_2 can be expected to display synchronization with error under certain conditions. We consider two dynamical non-linear systems for generating synchronized series, namely 1) analytical system of coupled logistic map (CLM) and 2) Electroencephalograph (EEG) signals [5] (we have used publicly available

L. Vanneschi et al. (Eds.): EuroGP 2009, LNCS 5481, pp. 13–24, 2009.

EEG datasets from University of Bonn). The analytical system CLM is used to prove the concept and this is then tested on experimental EEG data. Further discussion on parameter mismatch for the above two systems is given in sections 2 and 3 respectively.

A generic way to demonstrate synchronization with error is to consider coupled logistic maps. This has the advantage of being amenable to analytical treatment to some extent. The logistic map given by $x_{n+1}=ax_n(1-x_n)$, with a being the chaos parameter has been widely studied as a paradigm for classical chaos. Further it is also widely applied to study various synchronization scenarios in the presence of chaos. Here we consider two logistic maps whose outputs are coupled to one another, referred later as coupled logistic map (CLM). In section 2 we consider CLM corresponding to high, moderate and low coupling strengths giving rise to synchronized and non-synchronized series. We then analyze these series using Genetic Programming approach [6,7,8,9,10], which is well suited in the present analysis because given the time series, it allows us to evolve and generate its underlying dynamical map equations. Here we extract dynamical system equations with a view to understand effects of parameter mismatch on synchronization. We use time delayed vectors generated in the reconstructed phase space [11] to generate a map of a given dynamical variable based on the components of its immediate past. It may be emphasized that in an earlier study [12] we have essentially carried out GP analysis quite successfully on logistic map system involving only a single map equation; whereas in the present study we have extended this approach to allow GP analysis on a coupled pair of equations as required for CLM system. Using standard time delayed vectors for generating symbolic chromosome structure that undergo successive refinement in an iterative manner by GP approach (described by Szpiro [10] and suitably modified by Ahalpara et al in [12,13]), we carry out the optimization analysis that help us generate dynamical map equations for the coupled system. It is shown that GP analysis is quite successful in generating dynamical map equations (section 2.1) that give excellent results both in terms of correctness of system equations (in their reduced form, as described later) as well as multi-step dynamic predictions within in-sample and out-of-sample regimes. It is shown that the aspects of parameter mismatch typically occurring in non-linear dynamical systems are well brought out by GP analysis. Some more useful properties of the system equations are also investigated in this section with a view to test the GP map equations.

Having demonstrated the efficacy of the methods used for CLM system, we apply this approach on EEG data [5] in section 3. The synchronized series identified using a correlation matrix formalism [14] are first smoothened [11] using a discrete wavelet transform based on Daubechies (Db-4) wavelet [15], where we have used the approach developed by Manimaran et al [16] and subsequently used by Ahalpara et al [17]. This approach essentially helps separate trend from fluctuations of real time series. The smoothened series are further analyzed to extract underlying map equations using GP approach in section 3.1. The crisp map equations generated by GP analysis have dominant linear terms that help in understanding the underlying role of parameter mismatches for the EEG data. Finally concluding remarks are presented in section 4.

2   Synchronization in Coupled Logistic Map (CLM)

The coupled logistic map is an analytical system that provides us a useful prototype for studying cooperative behavior and synchronization within coupled dynamical system. The following pair of logistic maps constitute a coupled logistic map (CLM) system,

$$X_{n+1} = (1 - \epsilon_1)a_1 X_n(1 - X_n) + \epsilon_1 a_2 Y_n(1 - Y_n) \tag{1}$$
$$Y_{n+1} = \epsilon_2 a_1 X_n(1 - X_n) + (1 - \epsilon_2)a_2 Y_n(1 - Y_n),$$

where X_0, Y_0 are initial conditions for the two maps, a_1, a_2 are system parameters and ϵ_1, ϵ_2 are coupling strengths between the two logistic maps for X_n and Y_n. To attain synchronization with small error between the two maps, we choose a sufficiently high coupling strength (ϵ_1, $\epsilon_2 > 0.25$ and close to 1). It can be shown that error due to internal parameter mismatches for the system in Eq. 1 is of the order of δ_a, where $\delta_a = a_1 - a_2$. Thus for $\delta_a = 0$, there is complete synchronization. If the internal parameters a_1 and a_2 mismatch by δ_a, then the upper bound for synchronization error is $O(\delta_a)$. If the coupling between the two systems is small (ϵ_1, $\epsilon_2 < 0.25$), then synchronization is not achieved [3,4]. We consider time series of 500 data points after bypassing initial 5000 data points when synchronization with errors is observed.

Since we are studying nearly synchronized states after the transients have died out, for later purposes, it is convenient to rewrite Eq. 1, under the approximation $X_n = Y_n$, to obtain reduced set of equations,

$$X_{n+1} = (a_1 - \epsilon_1 \delta_a)X_n(1 - X_n) \tag{2}$$
$$Y_{n+1} = (a_2 + \epsilon_2 \delta_a)Y_n(1 - Y_n).$$

The results from GP approach can be compared with the coefficients of $X_n(1 - X_n)$ and $Y_n(1 - Y_n)$.

We report here results for synchronized series in coupled logistic maps with parameter mismatch in map parameters a_1 and a_2. We may note that calculations have also been made for parameter mismatches in coupling parameters ϵ_1 and ϵ_2 as well as initial conditions. These results are beyond the scope of the present paper and hence will be reported elsewhere.

We consider 2 CLM maps giving synchronized series and one map giving non-synchronized series. The parameters of CLM maps are given in Table 1. As the table shows, the two synchronized series correspond to high coupling with $\epsilon_1 = \epsilon_2 = 0.7$ (map no. 1) and moderate coupling with $\epsilon_1 = \epsilon_2 = 0.3$ (map no. 2). Non-synchronized series are generated by considering relatively low coupling with $\epsilon 1 = 0.02$, $\epsilon 2 = 0.06$ (map no. 3), that is below the threshold limit for synchronization.

Table 1. Parameters of 3 CLM maps considered that give rise to synchronized series X_n and Y_n (for map no. 1 and 2) and non-synchronized series (for map no. 3)

Map no.	Map Type	Coupling Parameters		Logistic Map Parameters		Initial Condition	
		ϵ_1	ϵ_2	a_1	a_2	X_0	Y_0
1	Synchronization with error	0.7	0.7	3.7	3.8	0.3	0.3
2	Synchronization with error	0.3	0.3	3.9	4.0	0.3	0.3
3	Non-Synchronization	0.02	0.06	3.7	3.8	0.3	0.8

2.1   GP Analysis on Coupled Logistic Map

Since CLM equations are coupled, we find it useful to carry out GP optimization separately for X_n and Y_n maps where the chromosome structures are allowed to have coupling between X_n and Y_n. While carrying out GP analysis on these series we find that due to coupling between X_n and Y_n series, the template such as

$$((A \otimes B) \otimes (C \otimes D)) \tag{3}$$

(where A, B, C, D being time lagged vector components or real numbers and $\otimes$ being one of the arithmetic operators +, -, × or ÷) proves too general in the present context with the result that GP optimization turns out to be very slow in obtaining a reasonably good fitness value. This is because the structure of the general template does not directly incorporate any coupling between X_n and Y_n maps. In order to overcome this problem of slow convergence, we have therefore found it quite useful to also consider a revised template having a general structure with built-in coupling. The use of coupled templates is found to speed up GP optimization process considerably. We thus impose a revised template with built-in coupling of additive or multiplicative type as follows,

$$(A \otimes f(X_n)) \otimes (B \otimes g(Y_n)) \tag{4}$$

where f and g are polynomial functions of given degree and other terms in the equation have similar meaning as in the previous template (Eq. 3). We use up to 2nd degree polynomials for both X_n and Y_n. It may be noted that during GP evolution the chromosomes are allowed to evolve to polynomials of higher degrees and hence the structure of revised template chosen does not lead to any loss of generality. In fact we use a mix of both the templates in which majority of chromosomes in the pool have general structure as per template in Eq. 3, and additionally some chromosomes are included as per the revised structure of template in Eq. 4.

Thus, in the GP optimization we use map equation to have the form involving time lagged components,

$$X_{t+1} = f(X_t, X_{t-\tau}, X_{t-2\tau}, X_{t-3\tau}, ...X_{t-(d-1)\tau}) \tag{5}$$

where f represents a function involving time series values X_t in the immediate past, arithmetic operators (+, -, × and ÷) and numbers bound between -10 and

10 with a precision of 1 digit after decimal point; d represents dimension of time delayed vector, i.e. number of previous time lagged components that may appear in the function and τ represents a time lag (as described in [11]). The iterative procedure of GP thus considers sum of squared errors (SSE) as a measure of fitness that is minimized,

$$\triangle^2 = \sum_{i=1}^{i=N}(X_i^{calc} - X_i^{given})^2 \tag{6}$$

where N represents number of X_t values (Eq. 5) that are fitted during the GP optimization. The fitness measure derived from $\triangle^2$ is defined as [10], $R^2 = 1 - \triangle^2 / \sum_{i=1}^{i=N}(X_i^{given} - \overline{X_i^{given}})^2$, where $\overline{X_i^{given}}$ is the average of all X_t (Eq. 5) to be fitted. However, in order to reduce bloating effect in some sense [10], a modified fitness measure is defined as follows,

$$r = 1 - (1 - R^2)\frac{N-1}{N-k} \tag{7}$$

where N is the number of equations to be fitted in the training set and k is the total number of time lagged variables of the form $X_t, X_{t-\tau}, X_{t-2\tau}, \ldots$ etc (including repetitions) occurring in the given chromosome. The modified fitness measure prefers a parsimonious model and is henceforth used to refer to fitness value.

With this modified initial pool of chromosomes, the GP optimization scheme has been able to generate quite good fits ($\triangle^2$ values close to 0 and fitness close to 1) as well as good structures for coupled logistic maps. It may be noted that we have developed a code in C++ for carrying out Genetic Programming in which the basic scheme given by Szpiro [10] has been used and also it has been improved in some sense [12]. A binary tree representation has been used for the structure of chromosomes that helps storage, growth and evaluation of symbolic expressions in an effective manner. Further we have extended GP optimization process to incorporate coupling of equations for X_n and Y_n as required by CLM. The GP experiments have been carried out using following parameters: (numbers and time delayed components $X_t, X_{t-\tau}, X_{t-2\tau}, \ldots X_{t-(d-1)\tau}$, where d is vector dimension) as terminal set, $(+, -, *, /)$ as function set, 200 as population size, 300 generations, 0.5 as probability to select either a time delayed component or a number, 0.2 as mutation probability and 30 fitness runs. The GP results are shown in Table 2 and the map equations evolved by GP for the 6 series corresponding to 3 maps are shown in Eqn. 8 to 13.

$$X_{n+1} = T_1 X_n (1 - X_n) + 3Y_n (1 - Y_n) \tag{8}$$
$$Y_{n+1} = 3X_n (1 - X_n) + 0.733Y_n (1 - Y_n) + T_2 \tag{9}$$
$$X_{n+1} = 2.93X_n (1 - X_n) + T_3 Y_n (1 - Y_n) \tag{10}$$
$$Y_{n+1} = T_4 X_n (1 - X_n) + 3Y_n (1 - Y_n) \tag{11}$$
$$X_{n+1} = 3.5X_n (1 - X_n) + T_5 Y_n (1 - Y_n) \tag{12}$$
$$Y_{n+1} = T_6 X_n (1 - X_n) + 3.51Y_n (1 - Y_n) \tag{13}$$

Table 2. GP fitness obtained for various coupled Logistic maps of Table 1 having 500 points and $dim=2$, $tau=1$

Coupled Logistic Map			GP Fitness		
Map no.	Map Type	Series	$\triangle^2$	Fitness	Eqn. no.
1	Synchronized (high coupling)	X_n	0.00444	0.999833	8
		Y_n	8.446e-5	0.999997	9
2	Synchronized (medium coupling)	X_n	2.099e-5	1.0	10
		Y_n	0.00027	0.999995	11
3	Non-Synchronized (low coupling)	X_n	0.00125	0.999946	12
		Y_n	0.00215	0.99991	13

Here each term represented by T_1 to T_6 in the GP solutions have a structure of polynomial expression involving Padé type terms [18], which are known to have good mathematical properties (as against a power series) including their capability to model functions with singularity. We note that the general structure of coupling of X_n and Y_n maps in CLM system is well brought out by GP analysis. The explicit expressions for terms T_1 to T_6 are rather involved, but when evaluated, they show rather small fluctuations around their average values 0.746, 0.0003, 1.005, 0.978, 0.191 and 0.198 respectively, and therefore they serve as corrective terms in the GP equations 8 to 13. For the synchronized series if we approximate $Y_n \approx X_n$, then Eq. 8 to 11 assume their reduced forms (i.e. having X_n or Y_n terms only) that can be compared with the corresponding reduced map equations (Eq. 2) for the original logistic map equations as per the systems parameters of table 1. The two sets of reduced map equations compared in table 3 show excellent results. Thus, for the synchronized series, what contributes more in GP coupled equations is the total component from maps $X_n(1-X_n)$ and $Y_n(1-Y_n)$ together rather than their individual contributions. On the other hand, for the non-synchronized series (map no. 3 in Table 1 and 2), the relative contributions from X_n and Y_n are important, as X_n and Y_n series are no longer synchronized, and hence it is not meaningful to cast them to reduced forms.

Table 3. Comparison of reduced equations for coupled logistic map obtained by setting $Y_n=X_n$ for synchronized series. It is seen that the reduced equations generated by GP compare well with the corresponding original CLM equations as given by Eq. (2).

Coupled logistic map			Reduced Map equations	
Map no.	Map Type	Series	Original CLM Equations	GP map equation
1	Synchronized (high coupling)	X_n	$X_{n+1} = 3.77X_n(1-X_n)$	$X_{n+1} = 3.746X_n(1-X_n)$
		Y_n	$Y_{n+1} = 3.73Y_n(1-Y_n)$	$Y_{n+1} = 3.733Y_n(1-Y_n)$
2	Synchronized (medium coupling)	X_n	$X_{n+1} = 3.93X_n(1-X_n)$	$X_{n+1} = 3.935X_n(1-X_n)$
		Y_n	$Y_{n+1} = 3.97Y_n(1-Y_n)$	$Y_{n+1} = 3.978Y_n(1-Y_n)$

Having obtained explicit forms of GP map equations for the CLM system, we further analyze them to verify the goodness of the equations. First, we carry out an up-down analysis. Here, starting at successive datapoint in the fitted set, it is checked whether the up-down trend for the next given point to the time lagged vector is correctly reproduced by GP map equations in terms of increasing or decreasing values. For the total of 500 datapoints used for fitting purpose, we consider pairs of successive points and find the percentage of the correct reproduction of up-down trend. The percentage of correct signs for the 6 series shown in Table 2 are as follows: 98.8 for the first series X_n, and 100.0 each for the remaining 5 series. Thus, up-down trend by GP maps is in excellent agreement.

On the predictive capabilities of GP equations, we checked that 1-step predictions for X_n and Y_n series individually is good (as reflected in the fitness values). However, a rigorous test of the coupled equations is to carry out multi-step dynamic predictions. Starting from given X_0, Y_0 values, we iterate the coupled GP equations for successive steps and compare the resulting X_n, Y_n values after a given number of steps with the given values. First we carry out in-sample multistep dynamic predictions starting from 1st given datapoint and the results for 16 multistep predictions are shown in Fig. 1. We observe that the dynamic predictions hold good up to around 10 points, after which it begins to deteriorate.

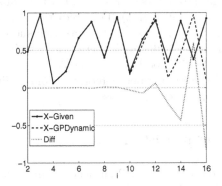

Fig. 1. Multi-step dynamic predictions up to 16 points for X_n series starting from the 2nd point onwards. Coupled logistic map equations from GP analysis have been used for making the predictions.

To understand the results of dynamic predictions, we have calculated Lyapunov exponent for the coupled Logistic map using the programs given in Kantz et al [19]. As is known, Lyapunov exponent provides us an estimate for the sensitivity measure of the initial condition on its growth in time. The Lyapunov exponents for coupled Logistic map for the above time series of X_n is 0.643. Based on this it is estimated that a typical initial error of 0.0001 would grow to 0.1 in around 10 steps and rapidly to higher values. Thus, the dynamic predictions given by GP equations are consistent with the theoretical limits

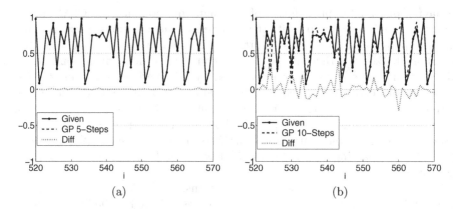

Fig. 2. Out-of-sample multi-step predictions for X-series using equations obtained by GP analysis. The multi-step dynamic predictions are made for 5 steps (a) and 10 steps (b) starting at successive datapoints in the X_n series.

provided by Lyapunov exponent and therefore we understand the inherent limitation of predictability for the coupled Logistic map equations due to its chaotic nature.

Further we also carry out these multi-step predictions for fixed number of steps starting at different datapoints corresponding to out-of-sample predictions (i.e. outside the fitted set of datapoints). The predictions are made for 5 and 10 multi-steps for X_n as well as Y_n series starting at each successive data point in the series and we present here results for X_n series (Fig. 2). It is observed that the dynamic predictions hold quite good up to around 5-steps; and with increasing number of multi-steps, the predictions deteorate rapidly. Similar results have been obtained for in-sample predictions also.

3   Synchronization in EEG Data

We consider the EEG data set [5] for 2 subjects F and S as is made publicly available from University of Bonn. Using correlation matrix formalism [14] we first identify the synchronized series from an ensemble of EEG series. Thus, for subject F we consider channel no. 17 and 35 (synchronized) and channel no. 65 (non-synchronized); whereas for subject S we consider channel no. 42 and 94 (synchronized) and channel no. 75 (non-synchronized). These are experimentally measured series, and therefore we first use wavelet method to smoothen them using a discrete wavelet transform based on Daubechies (Db-4) wavelet [15]. Here we have used the approach developed by Manimaran et al [16] and subsequently used by Ahalpara et al [17]. A standard time delay embedding approach has been used on the smoothened series to reconstruct the phase space akin to the original state space of the system. The time delay τ and dimension d are calculated using standard techniques of mutual information analysis and false nearest neighbors analysis respectively [11].

Table 4. GP model equations obtained for subject F (channels 17 and 35 are synchronized and channel 65 is out of synchronization) and for subject S (channels 42 and 94 are synchronized and channel 75 is out of synchronization)

EEG series		Db4 smoothened series			GP model	
Subject	Channel	Db4-Level	Dimension	Tau	Fitness	Equation No.
F	17	5	4	2	0.999836	14
F	35	5	4	2	0.999835	15
F	65	5	3	2	0.999330	16
S	42	5	4	2	0.999680	17
S	94	5	4	1	0.999636	18
S	75	5	4	2	0.998700	19

3.1   Analysis of Synchronized Series in EEG Using Genetic Programming

Following the methods presented for CLM analysis, we apply GP on smoothened EEG datasets for synchronized and non-synchronized channels for subjects F and S. The results are shown in Table 4. The GP model equations corresponding to level=5 for these channels are shown in Eq. 14 to 19, where T is a relatively small correction.

$$X_{t+1}^{Ch17} = 1.4995X_t - 0.508X_{t-1} + 0.0031X_{t-3} \tag{14}$$

$$X_{t+1}^{Ch35} = 1.476X_t - 0.481X_{t-1} + 0.00869X_{t-3} \tag{15}$$

$$X_{t+1}^{Ch65} = 1.5\left(\frac{X_t^2}{X_{t-1} + \frac{X_t^2}{X_{t-1}+X_t}}\right) \tag{16}$$

$$X_{t+1}^{Ch42} = 1.606X_t - 0.716X_{t-1} + 0.0769X_{t-2} + 0.029X_{t-3} + 0.0359 + T \tag{17}$$

$$X_{t+1}^{Ch94} = 1.557X_t - 0.322X_{t-1} - 0.0568X_{t-2} - 0.188X_{t-3} - 0.0824 \tag{18}$$

$$X_{t+1}^{Ch75} = 1.102X_t - 0.153X_{t-3} - 1.653 + 0.153\left(\frac{X_t^2}{7.5X_t + 1.5X_{t-1}}\right) \tag{19}$$

It is interesting to note that the equations are mostly linear with some additive Padé terms. In the first two terms (for X_t and X_{t-1}), the pair of synchronized channels have a minor parameter mismatch; whereas the corresponding non-synchronized channels have much more variation in the parameters. Thus, we see that synchronization (or non-synchronization) is related to mismatch in the parameters. This feature is clearly seen in the coefficients of successive terms in GP equations for (a) channels 17 and 35 and (b) for channels 42, 94 and 75, as shown in Fig. 3.

The synchronized channels have parameters showing similar trend. But for non-synchronized channel a different trend for parameters is well brought out in Fig. 3(b). It is observed that X_t and X_{t-1} are major terms in GP equation and therefore they play a dominant role in generating synchronization for channels

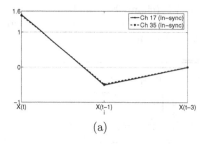 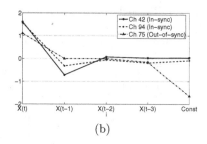

(a)   (b)

Fig. 3. Coefficients of X_t, X_{t-1}, ... in the GP map equations for EEG data (a) for subject F, and (b) for subject S. Matching and mismatch of parameters for synchronized and non-synchronized channels respectively are clearly seen.

42 and 94 (whose coefficients for X_t and X_{t-1} are closely similar) as against non-synchronized channel 75 (whose coefficients for X_t and X_{t-1} are quite different). For the non-synchronized channel 65, the GP map equation is in Padé form, and hence not included in Fig. 3 (a).

We further analyze the GP map equations by inspecting their various properties. First we note that for synchronized channels (say channel 17 and 35 for subject F), we would expect that map equation of channel 17 should approximately hold good for channel 35 and vice versa. Thus, Table 5 shows the effect on sum squared error $\triangle^2$ (Eq. 6) when the equations for synchronized channels are swapped with one another; and interestingly it is seen that the effect is indeed marginal (e.g. for channel 17, $\triangle^2=290.27$ (shown in bold) change to 294.45 when map equation of its synchronized channel 35 is used). On the other hand when we use map equation of non-synchronized channel 65 for channel 17, as we would expect, the corresponding value changes drastically to $\triangle^2=1704.06$. For subject S similar result is obtained for channel 42. However it may be noted that for channel 94 improved fitness is obtained when Eq. 42 is used, and this is attributed to different τ values for the two channels (Table 4).

Next we consider relative contributions of individual terms in GP map equations. For example for GP map equation for channel 17 (Eq. 14) that has 3 terms, we have $\triangle^2=290.27$, which increases to 4130351, 474845 and 309.4 when

Table 5. Interchange of GP model equations to see its effect on sum squared error $\triangle^2$ for both synchronized and non-synchronized channels of EEG data

Subject	Channel	$\triangle^2$		
		Using Eq. for Ch. 17	Using Eq. for Ch. 35	Using Eq. for Ch. 65
F	17	**290.27**	294.45	1704.06
F	35	348.39	**345.04**	2030.24
		Using Eq. for Ch. 42	Using Eq. for Ch. 94	Using Eq. for Ch. 75
S	42	**4032.24**	4959.69	24311.40
S	94	3070.28	**3755.06**	19812.90

we remove terms 1, 2 and 3 respectively, implying that terms 1 and 2 are most important in the map equation. In order to test the goodness of GP map equations, we further carry out an up-down analysis (as described earlier) for CLM system. The percentage of correct sign for the up-down analysis for channels 17, 35, 65 (for subject F) are 95.33, 95.43 and 94.64 respectively; and for channels 42, 94 and 75 (for subject S) the percentage of correct sign are 94.32, 95.43 and 96.53 respectively. It is seen that GP map equations reproduce the up-down trend reasonably well.

4 Conclusion

The present study focuses on synchronization with errors in an analytical system (coupled logistic map) and a real system (EEG data). In both these cases, we identify the regimes or channels that display synchronization with error and show using GP methods that internal parameter mismatches give rise to such synchronization with error. It must be pointed out that in an analytical system like the CLM, it is fairly easy to realize and study cases of synchronization with error and relate it to parameter mismatches. However, this is a challenging task in a real and complex system like the EEG data. This is even more daunting considering that we do not know the internal parameters in the brain from which the EEG signals are generated and therefore it is not clear how mismatches in parameters can possibly be understood. Use of the GP technique allows us to circumvent this problem through a gross modeling approach. Hence, GP allows to detect parameter mismatches in systems as complex as EEG. However we stress that the parameters we obtain from the GP equations cannot be easily related to the real parameters inside the brain. While synchronization mechanism in EEG is not yet completely understood, our work points to a different scenario that can lead to break down of synchronization. It is known for long that under certain conditions if the parameters do not match the synchrony breaks down. We believe we have used a novel approach to demonstrate this in a complex system for which we have no access to the internal parameters.

In order to apply Genetic Programming (GP) approach to extract underlying system equations using the time series, we have extended it in some way to incorporate coupling of map equations for X_n and Y_n maps as is required for the coupled logistic map. It is shown that Genetic Programming based optimization approach has been quite successful in generating map equations for non-linear dynamical systems. The map equations for CLM system have given excellent results (especially for the reduced map equations), whereas for EEG data the model for trend (after fluctuations have been filtered out) are fairly good. Further the synchronized time series pairs show striking similarity in the parameter space when modeled using genetic programming, both for the CLM and the EEG case. As expected, this similarity in parameter space does not hold true for non-synchronized time series. Thus, the structures of map equations generated by GP approach help us understand the effects of parameter mismatches on synchronization.

References

1. Ding, M.L., Ding, E.L., Ditto, W.L., Gluckman, B., In, V., Peng, J.H., Spano, M.L., Yang, W.M.: Control and Synchronization of Chaos in High Dimensional Systems: Review of Some Recent Results. Chaos 7, 644–652 (1997)
2. Strogatz, S.H.: From Kuramoto to Crawford: Exploring the Onset of Synchroniztion in Populations of Coupled Oscillators. Physica D 143, 1–20 (2000)
3. Pikovsky, A., Rosenblum, M., Kurths, J.: Synchronization: A Universal Concept in Nonlinear Sciences. Cambridge University Press, Cambridge (2001)
4. Boccaletti, S., Kurths, J., Osipov, G., Valladares, D.L., Zhou, C.S.: The Synchronization of Chaotic Systems. Physics Reports - Review Section of Physics Letters 366, 1–101 (2002)
5. Andrzejak, R.G., Lehnertz, K., Mormann, F., Rieke, C., David, P., Elger, C.E.: Indications of nonlinear deterministic and finite-dimensional structures in time series of brain electrical activity: Dependence on recording region and brain state. Physical Review E 64, 061907 (2001)
6. Fogel, D.B.: Evolutionary Computation, The Fossil Record. IEEE Press, Los Alamitos (1998)
7. Holland, J.H.: Adaptation in Natural and Artificial Systems, 2nd edn. University of Michigan Press, Ann Arbor (1975)
8. Goldberg, D.E.: Genetic Algorithms in Search, Optimization, and Machine Learning. Addison Wesley Publication, Reading (1989)
9. Mitchell, M.: An Introduction to Genetic Algorithms. MIT Press, Cambridge (1996)
10. Szpiro, G.G.: Forecasting Chaotic Time Series With Genetic Algorithms. Physical Review E 55, 2557 (1997)
11. Abarbanel, H., Brown, R., Sidorovich, J., Tsimring, L.: The Analysis of Observed Chaotic Data in Physical Systems. Rev. of Mod. Phys. 65, 1331 (1993)
12. Ahalpara, D.P., Parikh, J.C.: Genetic Programming Based Approach for Modeling Time Series Data of Real Systems. International Journal of Modern Physics C 19(1), 63–91 (2008)
13. Ahalpara, D.P., Panigrahi, P.K., Parikh, J.C.: Variations in Financial Time Series: Modelling Through Wavelets and Genetic Programming. In: Chatterjee, A., Chakrabarti, B.K. (eds.) Econophysics of Markets and Business Networks. Proceedings of the Econophys-Kolkata III, vol. 35. Springer, Heidelberg (2007)
14. Santhanam, M.S., Patra, P.K.: Statistics of Atmospheric Correlations. Phys. Rev. E64, 016102 (2001)
15. Daubechies, I.: Ten Lectures on Wavelets. SIAM, Philadelphia (1992)
16. Manimaran, P., Panigrahi, P.K., Parikh, J.C.: Multiresolution Analysis of Stock Market Price Fluctuations. Phys. Rev. E72, 046120 (2005)
17. Ahalpara, D.P., Parikh, J.C.: Modeling Time Series Data of Real Systems. International Journal of Modern Physics C 18(2), 235–252 (2007)
18. Press, W.H., Teukolsky, S.A., Vetterling, W.T., Flannery, B.P.: Numerical Recipes in C: The Art of Scientific Computing, 2nd edn., p. 200. Cambridge University Press, Cambridge (1993)
19. Kantz, H., Schreiber, T.: Nonlinear Time Series Analysis. Cambridge University Press, Cambridge (1997)

Memory with Memory in
Tree-Based Genetic Programming

Riccardo Poli[1], Nicholas F. McPhee[2], Luca Citi[1], and Ellery Crane[2]

[1] School of Computer Science and Electronic Engineering, University of Essex, UK
{rpoli,lciti}@essex.ac.uk
[2] Division of Science and Mathematics, University of Minnesota, Morris, USA
mcphee@morris.umn.edu,seidaku@gmail.com

Abstract. In recent work on linear register-based genetic programming (GP) we introduced the notion of Memory-with-Memory (MwM), where the results of operations are stored in registers using a form of soft assignment which blends a result into the current content of a register rather than entirely replace it. The MwM system yielded very promising results on a set of symbolic regression problems. In this paper, we propose a way of introducing MwM style behaviour in tree-based GP systems. The technique requires only very minor modifications to existing code, and, therefore, is easy to apply. Experiments on a variety of synthetic and real-world problems show that MwM is very beneficial in tree-based GP, too.

1 Introduction

In the vast majority of programming models, assignments are entirely destructive in the sense that an instruction that stores a result in a register or in a memory location completely overwrites their previous value. This overwriting model of assignment was carried over to most versions of genetic programming (GP) that had state and assignments (see [8] for a review). This includes linear GP [2], which evolves sequences of (virtual or real) machine code instructions that usually act by destructively writing to registers or memory locations. Other examples include indexed memory [1,3,11], and work on evolving data structures (such as stacks) that have internal state [4,5]. Similarly, systems that use data structures such as stacks in their computational model, such as PushGP [9,10], manipulate internal states with destructive assignments.

This is in contrast to most biological systems, where the state of such a system is rarely if ever completely replaced with a new state with no regard for or "memory" of the previous state. Changes in protein concentrations within a cell, for example, can happen quickly but are typically incremental in nature, each new state being constructed via modification of the previous state rather than the complete replacement of it.

Noting this dichotomy between nature and GP, in previous work [7] we introduced the idea that the results of operations in a linear register-based GP system could be stored in registers using a form of *soft assignment*. Instead of having the new value completely overwrite the old value of the register, these soft assignments combine the old and new values. We called this technique *Memory-with-Memory (MwM)*.

Although there are many approaches that could be taken to combining the old values with the new when performing assignments, in [7] we found that using a simple weighted average of the old value of a register with the new value being assigned

L. Vanneschi et al. (Eds.): EuroGP 2009, LNCS 5481, pp. 25–36, 2009.

worked well. In particular, if v_{old} is the original value of the register, and v_{new} is the new value being assigned to the register, the resulting value v_{result} is given by

$$v_{result} = \gamma v_{new} + (1 - \gamma) v_{old} \qquad (1)$$

where γ is a constant that indicates the *assignment hardness*, allowing users to determine how "soft" the assignment operator is.[1] In [7] we showed that including this new type of assignment in a linear GP system can significantly improve performance on a variety of symbolic regression problems. Also, MwM was found to change the nature of bloat, reducing the amount of ineffective code.

A question that naturally comes to mind is whether an extension of the MwM idea to mainstream tree-based GP is possible. Obviously, MwM could be used in any GP system which includes primitives that read and write into some form of memory: all one needs to do is to make assignments to memory soft. However, memory is rarely used in tree-based GP, where most applications evolve functions with no side effects rather than programs. For these, the standard form of MwM cannot be used.

In this paper we propose an alternative way of introducing MwM type of behaviour in mainstream, tree-based GP. This requires only very minor modifications to existing systems, making it easy to add. We describe the technique and our GP system in Secs. 2 and 3. Sec. 4 shows that MwM is beneficial in a variety of settings, including a real EEG reconstruction problem. We conclude in Sec. 5.

2   Memory with Memory in Tree-Based GP

The MwM technique for register-based GP is essentially a way of ensuring that previously computed results cannot entirely be wiped by a single instruction. In tree-based GP, instructions do not write their results into registers. Instead, they pass them as a return value to other instructions higher up in the tree. Those instructions, in turn, use the results to compute a new return value, and so on. To see how one could apply the MwM idea to this style of GP, we need to dissect MwM into its most elementary components.

With MwM, executing instructions effectively involves two steps: a) a calculation and b) a soft assignment of the result of the calculation into a register or memory location. In tree-based GP, executing instructions also consists of two operations. One is, again, a calculation, while the second consists of passing of the result of the calculation to a parent instruction higher up in the tree. Hence, in a tree-based GP system, because calculation results are not stored in memory, what matters is how the results of calculations are turned into return values. If we allow the result of a calculation to entirely determine the value returned by a node in the tree, we essentially have a *"hard" return operation*. If, instead, we allow the return value of a node to be a blend between the result of a calculation and previously returned results (we will explain what this means in a moment), then we create a *"soft" return operation*. This allows us to achieve the objectives of MwM for tree-based GP systems where instructions have no memory or side effects. For this reason, we will call this technique *MwM for tree-based GP*.[2]

---

[1] If $\gamma = 1$ the assignment is completely "hard" (as in traditional GP).

[2] Although "memory" is not strictly accurate when dealing with return values instead of memory cells, we choose to use this term to preserve the connection to our earlier work.

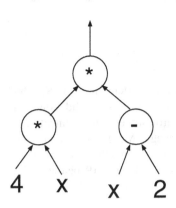

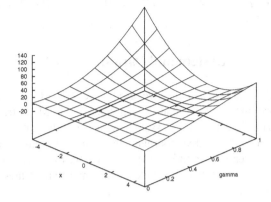

Fig. 1. Example program

Fig. 2. Illustration of how the behaviour of the program in Fig. 1 varies from the constant function 4 to the parabola $4x^2 - 8x$ as γ varies from 0 to 1

An important question that needs answering to turn this idea into practice is: what do we mean by "blending return values with previously returned results"? Many approaches are possible. However, in the light of the positive experience of [7], before trying something sophisticated we thought we should try a simple approach: function nodes in a tree return a weighted average of their first argument with the result of the calculation corresponding to an instruction. In particular, if IN_1, IN_2, etc. represent the inputs to a particular instruction, F, its output, OUT, is given by

$$OUT = \gamma F(IN_1, IN_2, \ldots) + (1 - \gamma)IN_1 \qquad (2)$$

where γ is a constant that indicates the *hardness of the return operations*. In the extreme case where $\gamma = 1$ the return is completely "hard" and so $OUT = F(IN_1, IN_2, \ldots)$ as in traditional GP. Values of $\gamma < 1$ provide a smoother behaviour. Note that there isn't any specific reason for always choosing a weighted sum with the first argument of a function. The choice of argument could be randomised or there could be versions of these softer instructions for every possible argument. In the future we will investigate whether this has an effect on performance. However, in this first exploration we decided to go for the simplest possible strategy.

Fig. 1 shows an example program. We would normally interpret it as a representation of the expression $(4 \times x) \times (x - 2) = 4x^2 - 8x$. However, when MwM is used, the tree in Fig. 1 represents the function $\gamma(a \times b) + (1 - \gamma)a$ where $a = \gamma(4 \times x) + (1 - \gamma)4$ and $b = \gamma(x - 2) + (1 - \gamma)x$. Substitution of a and b into that expression produces $\gamma((\gamma(4 \times x) + (1 - \gamma)4) \times (\gamma(x - 2) + (1 - \gamma)x)) + (1 - \gamma)(\gamma(4 \times x) + (1 - \gamma)4)$ which simplifies to $8\gamma^3 + 4\gamma^2 x^2 - 8x\gamma^3 + 8\gamma x - 8x\gamma^2 - 4\gamma^2 - 8\gamma + 4$. As γ varies from 1 to 0 the expression gradually morphs from the original $4x^2 - 8x$ to the value of the leftmost leaf of the tree, 4.[3] If, for example, we compute this expression for $\gamma = 0.5$, we obtain $x^2 + x$, while

---

[3] This behaviour is consistent with what happens in the original MwM system for linear GP, where for $\gamma = 0$ all programs become identity functions (the input register).

for $\gamma = 0.1$ we obtain $0.04x^2 + 0.712x + 3.168$. This morphing process is illustrated in Fig. 2.

3   GP System and Parameters

For our experiments we used a version of the TinyGP system in Java [8] which we modified so that it can handle MwM instructions and use validation sets.[4] This is a tree-based system with a steady state control strategy, tournament selection and negative tournaments as a replacement strategy. More details and source code can be found in [8]. All that was required to adapt TinyGP to MwM was to change four lines of code of the interpreter (lines 9, 11, 13 and 19 below), which becomes:

```
1     double run() { /* Interpreter */
2       char primitive = program[PC++];
3       double input;
4       if ( primitive < FSET_START )
5         return(x[primitive]);
6       input = run();
7       switch ( primitive ) {
8       case ADD :
9           return( input * ( 1.0 - GAMMA ) + (input + run()) *
                GAMMA );
10      case SUB :
11          return( input * ( 1.0 - GAMMA ) + (input - run()) *
                GAMMA );
12      case MUL :
13          return( input * ( 1.0 - GAMMA ) + (input * run()) *
                GAMMA );
14      case DIV : {
15          double den = run();
16          if ( Math.abs( den ) <= 0.001 )
17              return( input );
18          else
19              return( input * ( 1.0 - GAMMA ) + (input / den)
                  * GAMMA );
20          }
21      }
22      return( 0.0 );
23    }
```

Tab. 1 shows the parameters and primitive sets we used for the experiments described in the following section.

---

[4] The system can also work in a multi-deme configuration where runs execute on different machines in a cluster and pass their best individuals to a central store, which in return passes individuals back to the demes. However, we did not use this feature in the work reported here.

Table 1. Parameter settings used in our experiments

Parameter	Value
Function set	ADD, SUB, MUL, DIV (protected)
Terminal set	100 random constants uniformly distributed in the range $[-5, 5]$ plus 1 to 64 variables (depending on problem), except for the EEG prediction problem where 40 constants in the range $[-1, 1]$ were used
Independent runs	100, 500 or 2000 (depending on problem)
Max initial depth	5
Max size after crossover	10,000
Crossover point selection	uniform
Point mutation rate (per primitive)	0.05
Population size	10,000 or 100,000
Generations	100
Fitness evaluations per run	population size × generations
Crossover rate (per individual)	0.1
Mutation rate (per individual)	0.9
Tournament size	2
Hardness of return operation (γ)	1.0, 0.7, 0.5, 0.3, 0.1
Fitness cases	11 to 5000 (depending on problem)
Fitness	sum of absolute errors
Stopping criterion	fitness $< 0.05 \times$ # fitness cases

4 Problems and Results

To test the behaviour of our tree-based GP version of MwM, we used five classes of test problems: 1) symbolic regression with a sine target function, 2) one symbolic regression and two prediction problems involving the Mackey-Glass time series, 3) a prime prediction problem, 4) symbolic regression with two different polynomial targets, and 5) a real-world problem involving the reconstruction of EEG ear-lobe electrodes from other electrodes. The problems and the results we obtained on them using different values of the parameter γ are described in the following sections. As we will see, except in one case, the use of MwM helps GP to either improve its success rate or the accuracy of its solutions (or both).

4.1 Sine Problem

The sine symbolic regression problem was used for illustration of the TinyGP system in [8]. The problem requires programs to fit the sine function over a full period of oscillation. We used 63 fitness cases obtained by sampling $\sin(x)$ for $x \in \{0.0, 0.1, 0.2, \ldots 6.2\}$. Based on the stopping criterion indicated in Tab. 1, we define a *success* to be a run where the best fitness was less than 3.15, or an average error of less than 0.05 over the 63 test cases.

For each configuration of γ we performed 500 independent runs with populations of size 10,000. Our results are reported in Tab. 2.

Table 2. Success rates and average end-of-run program size vs. hardness of return operation (γ) in the Sine symbolic regression problem

γ	Success rate	Average program size
1.0	0.296	54.88
0.7	0.442	56.24
0.5	**0.618**	67.59
0.3	0.540	88.16
0.1	0.066	140.54

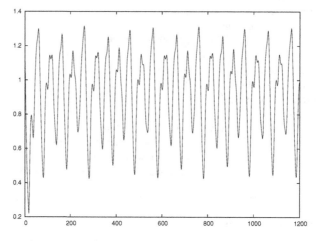

Fig. 3. Mackey-Glass chaotic time series

The results show that most GP configurations that use MwM performed better than the standard GP case ($\gamma = 1$), with the best MwM configuration ($\gamma = 0.5$) effectively doubling the success rate of the system. It is also apparent from the results that as the value of γ decreases, the average size of the programs at the end of the run increases. As we will see, this is a common feature in our experiments with MwM (which we also observed in [7]). This behavior occurs, we theorize, because MwM causes every instruction to contribute to the output of a program, making it possible to continue to incrementally improve a program by appropriately extending it.

4.2 Mackey-Glass Problems

The Mackey-Glass chaotic time series is often used for testing prediction algorithms. We show the first 1,200 samples of this time series in Fig. 3.

In a prediction setting, algorithms are typically required to predict the next sample in the series given any number of previous samples. In principle, then, the estimated sample can be fed back into the algorithm to produce a further sample into the future and so on. However, the time series can also be used as a target for symbolic regression where the independent variable is time, and the dependent variable is the value taken by

Table 3. Success rates and average end-of-run program size vs. hardness of return operation (γ) (left) and box plot of best of run fitnesses (right) in the Mackey-Glass regression problem

γ	Success rate	Average program size
1.0	0.00	62.74
0.7	0.03	73.90
0.5	0.04	75.31
0.3	0.26	87.46
0.1	**0.78**	104.87

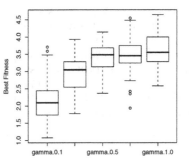

the series at that particular time. In this work we have used the Mackey-Glass chaotic time series for both types of applications.

We tested three configurations: two for prediction and one for regression. For each configuration and value of γ we performed 100 independent runs with populations of 100,000 individuals.

In the symbolic regression problem we gave GP the first 51 samples from the series and asked it to find a function which transformed the sample number (i.e., the numbers 0, 1, etc. up to 50) into the corresponding value in the Mackey-Glass chaotic time series. Tab. 3 shows the results. Again, the addition of MwM helps evolution considerably. In fact, γ = 0.1 makes the problem problem easy, while the results from γ = 1.0 might lead one to deem the problem impossible to solve. As before, we see the relationship between γ and the average end-of-run program size, with smaller γ's leading to bigger programs.

In the first prediction problem we constructed a training set including 1,192 fitness cases. Each fitness case had 8 independent variables representing the values taken by the time series in 8 consecutive samples. The corresponding target was simply the value of the following sample (which we wish to predict). For example, the first fitness case provides the values of the series at times 0 through to 7 to GP and asks GP to find a function that produces as output the value of sample 8. By sliding this 9-sample window over the time series, we obtained the rest of the training set. The results of our experiments are shown in Tab. 4.

All GP configurations were very successful at solving the problem. This is not surprising given the high correlation between samples present in the Mackey-Glass series and our stopping criterion which required absolute average errors per sample of less than 0.05. Looking at the mean best fitnesses, however, we find that, again, MwM helps evolution, as the settings γ = 0.7 and γ = 0.5 provide significant reductions in prediction error over the standard GP case. We also find that excessively soft return operations hinder performance. Size-wise, very small values of γ lead to bigger programs, but size seems to depend less on γ than for other problems (probably because runs were stopped early as a result of being successful).

The second prediction problem was similarly constructed. The difference here was that we used 7 input taps (samples) and taps were not consecutive, but at distances of 1,

Table 4. GP performance for different values of hardness of return operation (γ) in Mackey-Glass prediction problem with consecutive taps

γ	Success rate	Mean best-end-of-run fitness	Average program size
1.0	1.0	2.69012	42.48
0.7	1.0	**1.61918**	35.89
0.5	1.0	1.89431	40.79
0.3	1.0	2.79318	42.26
0.1	0.96	6.44385	98.10

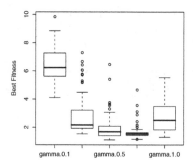

Table 5. GP performance for different values of hardness of return operation (γ) in Mackey-Glass prediction problem with non-consecutive taps

γ	Success rate	Mean best-end-of-run fitness	Average program size
1.0	1.0	3.79617	45.86
0.7	1.0	2.83241	42.25
0.5	1.0	**2.45872**	54.50
0.3	1.0	4.51185	75.97
0.1	1.0	6.0517	105.00

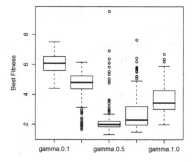

2, 4, 8, 16, 32 and 64 from the sample we wanted to predict. This produced a training set of 1,137 fitness cases. The results of our experiments are shown in Tab. 5.

Again we find that all systems are successful at finding solutions, but that γ = 0.7 and γ = 0.5 help produce better solutions, while excessively small γ's hinder performance. Size-wise, very small values of γ lead to bigger programs, but other values affect size less markedly (again likely because all runs were successful and were stopped early).

4.3 Prime Prediction Problem

The prime prediction problem was originally suggested as a competition for the GECCO 2006 conference. The original version of the problem consisted in evolving a polynomial with integer coefficients such that given an integer value i as input produced the i-th prime number, $p(i)$, for the largest possible value of i. The evolved functions are therefore required to produce consecutive primes for consecutive values of the input i.

Despite its simple statement and the monotonicity of the function to be evolved, it turned out that the problem is extremely difficult and so solutions deviated significantly from the ideal (polynomial with integer coefficients). For example, the winner of the GECCO 2006 competition, David Joslin, proposed the polynomial $1.6272 + 0.7747 \times$

Table 6. Success rates and average end-of-run program size vs. hardness of return operation (γ) in prime prediction problem

γ	Success rate	Average program size
1.0	0.00	62.64
0.7	0.04	86.74
0.5	0.08	87.60
0.3	**0.12**	94.30
0.1	0.02	114.76

$x + 0.05215 \times x^3 - 2.7092 \times 10^{-6} \times x^8 + 1.5748 \times 10^{-14} \times x^{18} - 2.1966 \times 10^{-16} \times x^{20}$
that correctly predicts only the first 8 primes. Walker and Miller, the runner-ups, did much better but with a solution that wasn't even a polynomial.

In our version of the problem we treated the problem as a symbolic regression problem. We provided the first 11 primes as fitness cases. So, the problem requires GP to evolve a program that maps 1 into 2, 2 into 3, 3 into 5, 4 into 7, 5 into 11, and so on. To make the problem easier we did not require programs to exactly match the target output, but to exhibit a total sum of absolute errors of less than $0.05 \times 11 = 0.55$.

We performed 100 runs with populations of size 100,000 for each assignment of γ. Results are shown in Tab. 6. As was the case in the Mackey-Glass problem, MwM increases the success rate for the problem significantly turning a problem which we would have deemed impossible to solve for standard GP into a problem of moderate difficulty. We again see that solutions are bigger for smaller γ's.

4.4 Polynomial Symbolic Regression

We applied the MwM GP system to polynomial symbolic regression problems using two target polynomials: $x^2 + 1.419x + 1.009$ which is of medium difficulty and $8x^5 + 3x^3 + x^2 + 6$ which is much harder.

For each target polynomial the fitness is the sum of the absolute error of the evolved function on 21 evenly spaced test points in the range $[-1,1]$: $\{-1.0, -0.9, -0.8, \ldots, 0.8, 0.9, 1.0\}$. The target then is to minimise this error.

We used population of 10,000 individuals and we performed 500 independent runs for each value of γ. We also did 100 runs (for each γ) with populations of 100,000 for the harder polynomial. The results are shown in Tab. 7.

In this instance, MwM results were mixed. While for the easier polynomial, all GP configurations which used MwM performed better than the standard GP case ($\gamma = 1$), in the harder polynomial MwM did not help. [5]

4.5 EEG Reconstruction Problem

Brain electrical activity is typically recorded from the scalp using Electroencephalography (EEG). This is used in electro-physiology, in psychology, as well as in

---

[5] While it is somehow disappointing to find a problem where MwM does not help given the positive results obtained with it in all other problems, we should not be surprised to find such problems in the light of the no-free lunch theory.

Table 7. Success rates and average end-of-run program size vs. hardness of return operation (γ) in polynomial regression problems

Target $x^2 + 1.419x + 1.009$ (popsize=10,000)		Target $8x^5 + 3x^3 + x^2 + 6$ (popsize=10,000)		Target $8x^5 + 3x^3 + x^2 + 6$ (popsize=100,000)	
γ	Success rate	γ	Success rate	γ	Success rate
1.0	0.888	1.0	**0.116**	1.0	**0.63**
0.7	0.976	0.7	0.030	0.7	0.19
0.5	0.996	0.5	0.010	0.5	0.00
0.3	**1.000**	0.3	0.002	0.3	0.00
0.1	0.982	0.1	0.000	0.1	0.01
γ	Avg Prog Size	γ	Avg Prog Size	γ	Avg Prog Size
1.0	39.81	1.0	47.88	1.0	50.90
0.7	53.70	0.7	54.78	0.7	55.56
0.5	73.90	0.5	63.86	0.5	66.17
0.3	96.16	0.3	74.41	0.3	72.47
0.1	153.62	0.1	110.37	0.1	100.57

brain-computer interfaces research. Voltages at the EEG electrodes are always recorded relative to some reference. Traditionally the reference has been either one ear electrode or the average of the two ears [6]. This is because there is neither muscular activity (which produces very large potentials) nor, obviously, neural activity in the ears. Note that in some EEG equipment, the ear electrodes are actually the reference voltage against which the voltage of other electrodes is measured. In others they are not, but it is still common to refer signals back to the ears to be consistent with the large body of literature on EEG analysis.

As a result of using the ears as references, any changes occurring in the impedance of the ear electrodes can produce huge artifacts in every signal recorded. If, for example, a single electrode detaches slightly, then every other channel can be flooded with noise, leading to huge baseline shifts. It is, therefore, very important to come up with ways of verifying if the ear electrodes are working properly, for example by comparing them against a prediction of what their voltage should be. There are also systems which do not require the use of ear electrodes. In these systems it is then impossible to refer the recorded signals (for example for comparison with other research) back to the ears. In such systems, if one could predict what the ear electrodes would measure, based on the signals recorded from other electrodes, then one could refer signals back to the reconstructed ear electrodes.

In this last set of experiments we compared GP systems with different degrees of MwM to see how well they can construct a soft-sensor for the left ear from real EEG recordings. We constructed a dataset using 64-channel EEG recordings from 5 different subjects acquired over a period of around a month. From each subject's recording we extracted two fragments of approximately 20 seconds each. The fragments were approximately half an hour apart. The original data was sampled at 2KHz. After band-pass filtering we subsampled it at 128 samples per second. We then chose 250 random time steps within each fragment. At each time step we saved the voltages recorded at the 64 channels plus the left ear reference voltage as a fitness case. This gave us a training

Table 8. GP performance for different values of hardness of return operation (γ) in EEG ear-electrode reconstruction problem

γ	Mean Generalisation Fitness	Fitness Std Dev	Std Error of the Mean	Avg Prog Size
1.0	39886.0	885.42	19.80	5.16
0.7	39637.5	1039.81	23.25	5.74
0.5	**39485.6**	1252.81	28.01	6.41
0.3	39831.0	1275.56	28.52	7.47
0.1	40342.9	1101.98	24.64	9.25

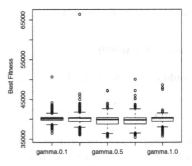

set of 5,000 fitness cases. Using different fragments from the same 5 subjects we also constructed a validation set of 5,000 fitness cases. This set was not used to compute fitness but only to decide when runs should be stopped.

We performed 2,000 runs for each configuration of γ with a population of size 10,000. GP had 64 input variables (the voltage values of the 64 channels) and one output (the left ear voltage). Tab. 8 reports the generalisation results obtained in different configurations of MwM. Again, MwM significantly improves performance (analysis of variance shows that performance differences are statistically highly significant), slightly affecting program size.

5 Conclusions

In this paper we have extended the idea of memory-with-memory, originally applied to linear, register-based GP, to the domain of tree-based GP with instructions without side effects and memory. This setup is very common, and is used, for example, in virtually all symbolic regression applications of tree-based GP. We achieve this by using a "soft" return operation to pass the values computed by instructions up the tree instead of the standard, "hard", return operation which is used in most computer science (including traditional GP). Instead of having the new value completely determine the output of a node, the computed value is first combined (using a weighted average) with the value of the first argument to a function and then returned.

In tests with a variety of symbolic regression and prediction problems, tree-based GP with MwM almost always does as well as traditional GP, while significantly outperforming it in several cases. Particularly striking are the very small values of γ (corresponding to a *very* soft form of value return). In several of the cases the greatest success was obtained with γ values of 0.3, and in one case (the Mackey-Glass regression) γ = 0.1 provided the best performance. These are extremely low values and represent a radically different notion of value return than standard GP. This suggests that these kinds of semantics play a crucial role, and that modifying them can have a powerful impact on system performance.

The data suggest that MwM GP can continue to improve (if slowly) solutions over time where traditional GP may in fact get stuck in local optima. As a result, solutions evolved using MwM tend to be slightly bigger than without it. Unlike with other systems, this cannot be attributed to introns, since with MwM every instruction contributes to a program's output. In future research we will try to understand what makes a problem hard for systems with MwM.

Acknowledgments

We would like to thank EPSRC (grant EP/G000484/1) for financial support.

References

1. Angeline, P.J.: An alternative to indexed memory for evolving programs with explicit state representations. In: Koza, J.R., Deb, K., Dorigo, M., Fogel, D.B., Garzon, M., Iba, H., Riolo, R.L. (eds.) Genetic Programming 1997: Proceedings of the Second Annual Conference, Stanford University, CA, USA, July 13-16, pp. 423–430. Morgan Kaufmann, San Francisco (1997)
2. Banzhaf, W., Nordin, P., Keller, R.E., Francone, F.D.: Genetic Programming – An Introduction; On the Automatic Evolution of Computer Programs and its Applications. Morgan Kaufmann, San Francisco (1998)
3. Brave, S.: Evolving recursive programs for tree search. In: Angeline, P.J., Kinnear Jr., K.E. (eds.) Advances in Genetic Programming 2, ch. 10, pp. 203–220. MIT Press, Cambridge (1996)
4. Bruce, W.S.: Automatic generation of object-oriented programs using genetic programming. In: Koza, J.R., Goldberg, D.E., Fogel, D.B., Riolo, R.L. (eds.) Genetic Programming 1996: Proceedings of the First Annual Conference, Stanford University, CA, USA, July 28–31, pp. 267–272. MIT Press, Cambridge (1996)
5. Langdon, W.B.: Genetic Programming and Data Structures: Genetic Programming + Data Structures = Automatic Programming! Genetic Programming, vol. 1. Kluwer, Boston (1998)
6. Luck, S.J.: An Introduction to the Event-Related Potential Technique. MIT Press, Cambridge (2005)
7. McPhee, N.F., Poli, R.: Memory with memory: Soft assignment in genetic programming. In: Keijzer, M., Antoniol, G., Congdon, C.B., Deb, K., Doerr, B., Hansen, N., Holmes, J.H., Hornby, G.S., Howard, D., Kennedy, J., Kumar, S., Lobo, F.G., Miller, J.F., Moore, J., Neumann, F., Pelikan, M., Pollack, J., Sastry, K., Stanley, K., Stoica, A., Talbi, E.-G., Wegener, I. (eds.) GECCO 2008: Proceedings of the 10th annual conference on Genetic and evolutionary computation, Atlanta, GA, USA, pp. 1235–1242. ACM, New York (2008)
8. Poli, R., Langdon, W.B., McPhee, N.F.: A field guide to genetic programming. Published via (2008) (With contributions by Koza, J.R.), http://lulu.com, http://www.gp-field-guide.org.uk
9. Spector, L., Klein, J., Keijzer, M.: The push3 execution stack and the evolution of control. In: Beyer, H.-G., O'Reilly, U.-M., Arnold, D.V., Banzhaf, W., Blum, C., Bonabeau, E.W., Cantu-Paz, E., Dasgupta, D., Deb, K., Foster, J.A., de Jong, E.D., Lipson, H., Llora, X., Mancoridis, S., Pelikan, M., Raidl, G.R., Soule, T., Tyrrell, A.M., Watson, J.-P., Zitzler, E. (eds.) GECCO 2005: Proceedings of the 2005 conference on Genetic and evolutionary computation, Washington DC, USA, vol. 2, pp. 1689–1696. ACM Press, New York (2005)
10. Spector, L., Robinson, A.: Genetic programming and autoconstructive evolution with the push programming language. Genetic Programming and Evolvable Machines 3(1), 7–40 (2002)
11. Teller, A.: The evolution of mental models. In: Kinnear Jr., K.E. (ed.) Advances in Genetic Programming, ch. 9, pp. 199–219. MIT Press, Cambridge (1994)

On Dynamical Genetic Programming: Random Boolean Networks in Learning Classifier Systems

Larry Bull and Richard Preen

Department of Computer Science
University of the West of England
Bristol BS16 1QY, UK
Larry.bull@uwe.ac.uk, richard.preen@uwe.ac.uk

Abstract. Many representations have been presented to enable the effective evolution of computer programs. Turing was perhaps the first to present a general scheme by which to achieve this end. Significantly, Turing proposed a form of discrete dynamical system and yet dynamical representations remain almost unexplored within genetic programming. This paper presents results from an initial investigation into using a simple dynamical genetic programming representation within a Learning Classifier System. It is shown possible to evolve ensembles of dynamical Boolean function networks to solve versions of the well-known multiplexer problem. Both synchronous and asynchronous systems are considered.

1 Introduction

In 1948 Alan Turing produced a paper entitled "Intelligent Machinery" in which he was first to highlight evolutionary search as a possible means by which to program machines (e.g., see [10] for an overview). In the same paper, Turing also presented a formalism he termed "unorganised machines" by which to represent intelligence within computers. These consisted of two types: A-type unorganised machines, which were composed of two-input NAND gates randomly connected into networks; and, B-type unorganised machines which included an extra triplet of NAND gates on the arcs between the NAND gates of A-type machines by which to affect their behaviour in a supervised learning-like scheme through the constant application of appropriate extra inputs to the network. In both cases, each NAND gate node updates in parallel on a discrete time step with the output from each node arriving at the input of the node(s) on each connection for the next time step. The structure of unorganised machines is very much like an artificial neural network with recurrent connections and hence it is perhaps surprising that Turing made no reference to McCulloch and Pitts' [26] prior seminal paper on networks of binary-thresholded nodes. However, Turing's scheme extended McCulloch and Pitts' work in that he also considered the training of such networks with his B-type architecture. This has led to their also being known as "Turing's connectionism". Moreover, as Teuscher [39] has highlighted, Turing's unorganised machines are (discrete) nonlinear dynamical systems and therefore have the potential to exhibit complex behaviour despite their construction from simple elements. That is, each node in the network is an active, constantly updating entity,

L. Vanneschi et al. (Eds.): EuroGP 2009, LNCS 5481, pp. 37–48, 2009.

which is in contrast to the majority of feedforward and recurrent artificial neural networks. The current work aims to exploit the potential of such systems for general computation through the use of evolutionary search, what is herein termed "dynamical genetic programming" (DGP).

A number of representations have been presented by which to enable the evolution of computer programs, the most common being tree-based LISP S-expressions of course (e.g., [24]). Other forms of Genetic Programming (GP) [24] include the use of machine code instructions (e.g., [3]) and finite state machines (e.g., [13]). Most relevant to the form of GP to be explored in this paper is the small amount of prior work on graph-based representations. Teller and Veloso's [38] "neural programming" uses a directed graph of connected nodes, each with functionality defined in the standard GP way, with recursive connections included. Significantly, each node is executed with synchronous parallelism for some number of cycles before an output node's value is taken. Poli (e.g., [32]) presented a very similar scheme wherein the graph is placed over a two-dimensional grid and executes its nodes synchronously in parallel. Other examples of graph-based GP typically contain sequentially updating nodes (e.g., [28] [30]). Schmidt and Lipson [34] have recently demonstrated a number of benefits from (non-dynamical) graph encodings over traditional trees, such as reduced bloat and increased computational efficiency. The motivating idea behind this work is that Turing's initial scheme can be augmented with elements of more recent discrete dynamical systems research and evolutionary computing to create a flexible and robust approach to the automated design of computer programs for difficult problems. That is, it is proposed that using simulated evolution to shape computer programs capable of rich temporal behaviour in themselves will enable the effective control or prediction of systems which contain complex dynamics. In particular, the use of parameter self-adaptation and asynchronous node/instruction execution is presented for the discrete case.

2 Evolving Discrete Dynamical Systems

The most common form of discrete dynamical system is the Cellular Automaton (CA) [43] which consists of an array of cells (lattice of nodes) where the cells exist in states from a finite set and update their states in parallel in discrete time. Traditionally, each cell calculates its next state depending upon its current state and the states of its closest neighbours. That is, CAs may be seen as a graph with a (typically) restricted topology. Packard [31] was the first to use evolutionary computing techniques to design CAs such that they exhibit a given emergent global behaviour. Following Packard, Mitchell et al. (e.g., [29]) have investigated the use of a Genetic Algorithm (GA) [17] to learn the rules of uniform binary CAs. As in Packard's work, the GA produces the entries in the update table used by each cell, candidate solutions being evaluated with regard to their degree of success for the given task. Andre et al. [2] repeated Mitchell et al.'s work whilst using traditional GP to evolve the update rules. They report similar results. Sipper (e.g., [36]) presented a non-uniform, or heterogeneous, approach to evolving CAs. Each cell of a one- or two-dimensional CA is also viewed as a GA population member, mating only with its lattice neighbours and receiving an individual fitness. He shows an increase in performance over Mitchell et al.'s work by exploiting the potential for spatial heterogeneity in the tasks. Teuscher [39] used a GA to design unorganised machines for simple bit-stream regeneration and pattern classification tasks. In this paper the general approach of evolving a graph of coupled units which update in parallel to exhibit a desired behaviour is cast as performing an arbitrary computation as the emergent phenomenon.

3 Random Boolean Networks

The discrete dynamical systems known as random Boolean networks (RBN) were originally introduced by Kauffman [22] to explore aspects of biological genetic regulatory networks. Since then they have been used as a tool in a wide range of areas such as self-organisation (e.g., [23]), computation (e.g., [27]) and robotics (e.g., [33]). An RBN typically consists of a network of N nodes, each performing a Boolean function with K inputs from other nodes in the network, all updating synchronously. As such, RBN may be viewed as a generalization of both binary CAs and Turing's A-type unorganised machines. Therefore in this paper RBNs are used as the basic representation scheme with which to design dynamical computer programs through evolutionary search. Since they have a finite number of possible states and they are deterministic, the dynamics of RBN eventually fall into a basin of attraction. It is well-established that the value of K affects the emergent behaviour of RBN wherein attractors typically contain an increasing number of states with increasing K. Three phases of behaviour are suggested: ordered when $K=1$, with attractors consisting of one or a few states; chaotic when $K>3$, with a very large numbers of states per attractor; and, a critical regime around $K=2$, where similar states lie on trajectories that tend to neither diverge nor converge and 5-15% of nodes change state per attractor cycle (see [23] for discussions of this critical regime, e.g., with respect to perturbations). Analytical methods have been presented by which to determine the typical time taken to reach a basin of attraction and the number of states within such basins for a given degree of connectivity K (see [23]).

Previously, Van den Broeck and Kawai [42] explored the use of a simulated annealing-type approach to design feedforward RBN for the four-bit parity problem and Lemke et al. [25] evolved RBN of fixed N and K to match an arbitrary attractor. More closely akin to the approach proposed here, Kauffman [23, p.223] describes the use of evolution to design RBN which must play a (mis)matching game where mutation is used to change connectivity, the Boolean functions and N. In this paper, the same degrees of freedom are allowed for the evolutionary search but parameter self-adaptation is included and the Learning Classifier System (LCS) [18] framework is used. In particular, LCS evolve an ensemble of solutions to a given task (see [12] for discussions) wherein divisions of the problem space emerge along with their solution. To date, no temporally dynamic representation scheme has been used within LCS. A number of representations have previously been presented beyond the traditional binary scheme however, including integers [46], real numbers [45], Lisp S-expressions (e.g., [1]), fuzzy logic [41] and neural networks [4]. Thus this paper also represents an initial study into the use of simple forms of dynamical system within LCS. A very small number of studies have considered multiple, coupled RBN (e.g., [9]) but no previous consideration of ensemble scenarios, as here, are known.

4   Dynamical GP in a Simple Learning Classifier System

In this paper a version of the simple accuracy-based LCS termed YCS [5] - which is a derivative of Wilson's XCS [44] - is used. YCS is without internal memory and maintains a rulebase of P initially randomly created rules. Associated with each rule is a predicted payoff value (p), a scalar which indicates the error (ε) in the rule's predicted payoff and an estimate of the average size of the niches (action sets - see below) in which that rule participates (σ). The initial random population has these parameters initialized, somewhat arbitrarily, to 10.

On receipt of an input message, the rulebase is scanned, and any rule whose condition matches the message at each position is tagged as a member of the current match set [M]. An action is then chosen from those proposed by the members of the match set and all rules proposing the selected action form an action set [A]. A version of XCS's explore/exploit action selection scheme will be used here. That is, on one cycle an action is chosen at random and on the following the action with the highest average payoff is chosen deterministically (ties broken randomly). No learning occurs on exploit trials – they are simply used to indicate progress.

The simplest case of immediate payoff reward R is considered here. Reinforcement in YCS consists of updating the error, the niche size estimate and then the payoff estimate of each member of the current [A] using the Widrow-Hoff delta rule with learning rate β:

$$\varepsilon_j \leftarrow \varepsilon_j + \beta(\,|R - p_j| - \varepsilon_j) \tag{1}$$

$$\sigma_j \leftarrow \sigma_j + \beta(\,|[A]| - \sigma_j) \tag{2}$$

$$p_j \leftarrow p_j + \beta(R - p_j) \tag{3}$$

The original YCS employs two discovery mechanisms, a panmictic (standard global) GA and a covering operator. On each time-step there is a probability g of GA invocation. The GA uses roulette wheel selection to determine two parent rules based on the inverse of their error:

$$f_j = (1 / (\varepsilon_j^v + 1)) \tag{4}$$

Here the exponent v enables control of the fitness pressure within the system by facilitating tunable fitness separation under fitness proportionate selection (see [5] for discussions). Offspring are produced via mutation (probability μ) and crossover (single point with probability χ), inheriting the parents' parameter values or their average if crossover is invoked. Replacement of existing members of the rulebase uses roulette wheel selection based on estimated niche size. If no rules match on a given time step, then a covering operator is used which creates a rule with the message as its condition and a random action, which then replaces an existing member of the rulebase in the usual way. Parameter updating and the GA are not used on exploit trials.

In this paper, to aid the generalization process, the panmictic GA is altered to operate within niches (e.g., see [5] for discussions). The mechanism uses XCS's time-based approach under which each rule maintains a time-stamp of the last system cycle upon which it was part of a GA. The GA is applied within the current [A] when the average number of system cycles since the last GA in the set is over a threshold θ_{GA}.

If this condition is met, the GA time-stamp of each rule is set to the current system time, two parents are chosen according to their fitness using standard roulette-wheel selection, and their offspring are potentially crossed and mutated, before being inserted into the rulebase as described above.

To use RBN as the rules within this system the following scheme is adopted. Each of an initial randomly created rule's nodes has K randomly assigned connections, here $1 \le K \le 5$. There are initially as many nodes N_{init} as input fields I for the given task and its outputs O, plus one other, as will be described, i.e., $N_{init}=I+O+1$. The first connection of each input node is set to the corresponding locus of the input message. The other connections are assigned at random within the RBN as usual. In this way the current input state is always considered along with the current state of the RBN itself per network update cycle by such nodes.

Matching consists of executing each rule for T cycles based on the current input. The value of T is typically chosen to be well within the basin of attraction of the RBN. Nodes are initialised randomly. In this study well-known Boolean problems are explored and hence there are only two possible actions, meaning only one output node is required. An extra "matching" node is also required to enable RBNs to (potentially) only match specific sets of inputs. If a given RBN has a logical '1' on the match node, regardless of its output node's state, the rule does not join [M]. This scheme has also been exploited within neural LCS [4]. Thereafter match set and action set processing proceeds as described above. A cover operator has not been found necessary in the tasks explored here.

Due to the need for a possible different number of nodes within the rules for a given task, the representation scheme is of variable length. In this initial study, mutation only is used here and applied to the node's logic and connectivity map at rate μ. Node logic is represented by a binary string and its connectivity by a list of K integers in the range $[1, N]$. Since each rule has a given fixed K value, each node maintains a binary string of length 2^K which forms the entries in the look-up table for each of the possible 2^K input states of that node, i.e., as in Packard's [31] aforementioned work on evolving CAs, for example. These strings are subjected to mutation on reproduction at the self-adapting rate μ for that rule. Hence, within the RBN representation, evolution can define different Boolean functions for each node within a given network rule, along with its connectivity map. Typically, in LCS, as within GAs and GP, the parameters controlling the algorithm are global and remain constant over time. However this is not always the case in evolutionary computing, particularly within Evolution Strategies (ES) [35] where the mutation rate is a locally evolving entity in itself, i.e., it adapts during the search process. Self-adaptive mutation not only reduces the number of hand-tunable parameters of the evolutionary algorithm, it has also been shown to improve performance. The approach has been used to add adaptive mutation to LCS [8], and to control other system parameters, such as the learning rate (e.g., [19]). The results demonstrate that adaptive parameters can provide improved performance, particularly in dynamic environments. Here each rule has its own mutation rate μ, stored as a real number and initially seeded uniform randomly in the range $[0.0,1.0]$. This parameter is passed to its offspring. The offspring then applies its mutation rate to itself using a Gaussian distribution, i.e., $\mu' = \mu \, e^{N(0,1)}$, before mutating the rest of the rule at the resulting rate.

As noted above, this process is also used to enable the number of nodes, i.e., the complexity of the RBN, to vary to match the task. Each rule also contains a second mutation rate γ, adjusted in the same way as μ. Once standard mutation is applied, the probability γ is tested. Upon satisfaction, a new randomly connected node is either added or the last added node is removed. The latter case only occurs if the network currently consists of more than the initial number of nodes. This self-adaptive growth scheme has previously been used within neural LCS (e.g., [20]), within GAs [6], and ES [7]. Evolving variable-length solutions via mutation only has previously been explored a number of times (e.g., [13]). Traditional GP can be seen to primarily rely upon recombination to search the space of possible tree sizes, although the standard mutation operator effectively increases or decreases tree size also.

5 Experimentation

Versions of the well-known multiplexer task are used in this paper. These Boolean functions are defined for binary strings of length $l = x + 2^x$ under which the x bits index into the remaining 2^x bits, returning the value of the indexed bit. The correct classification to a randomly generated input results in a payoff of 1000, otherwise 0.

Figure 1 shows the performance of the RBN-LCS on the 6-bit multiplexer problem with most parameters taken from [5]: $P=2000$, $v=10$, $\theta_{GA}=25$, $\beta=0.2$, $T=100$ and $N_{init} = 8$ (6 inputs, one output, one match node). After [44], performance from exploit trials only is recorded (fraction of correct responses are shown), using a 50-point running average, averaged over ten runs. It can be seen that a near optimal solution is learnt around 30,000 trials with optimality seen around trial 70,000. The average degree of connectivity K converges to around 2, i.e., connectivity evolves to the aforementioned critical regime identified for RBN in general. This behaviour indicates that the evolutionary process is able to identify an appropriate typical topology with which to generate complex behaviour, i.e., in this case a computation. For other tasks, other values of K may prove beneficial; high K may be expected in random number generation, for example. The average error drops significantly, as does the mutation rate, with the growth mutation dropping more slowly and a corresponding slight amount of growth is seen. It can be noted that a growth event under which a new node is added into an RBN is essentially neutral here since the new node receives inputs from the existing nodes (or itself) on addition but only provides inputs to other nodes after subsequent connectivity mutations.

Figure 2 shows the performance of the system with the same parameters on the 11-bit multiplexer problem. It can be seen that an optimal solution is learnt around 200,000 trials in this harder case, with similar behaviour for the various parameters.

In the aforementioned work on evolving CAs, the global behaviour of the lattice is typically used to determine the output of the discrete dynamical system. The same approach has also been explored here such that if a fraction of nodes ϕ within the given rule are in state '0', the rule is said to match the current input and advocate action '0'. Conversely, if the fraction ϕ of nodes are in state '1', the rule is said to match and advocate action '1'. Otherwise the rule is deemed not to match the current input. Under this scheme $N_{init} = 1$. $\phi = 0.75$ is used here. The results in Figure 3 on the 6-bit multiplexer suggest that this is beneficial to the evolutionary process with optimal performance realised around 20,000 learning trials. The average connectivity of rules is approximately 1.75 and the networks typically grow one extra node. Figure 4 shows the approach is again faster on the 11-bit problem. Therefore this "population encoding" scheme appears more robust, as might be expected, although future work will continue to explore this aspect of DGP.

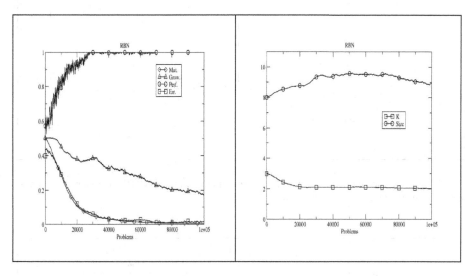

Fig. 1. Showing the behaviour of the RBN representation on the 6-MUX. Left-hand figure shows performance (Perf.), mutation rate (Mut.), network size change probability (Grow.) and the average error of all rules (Err.). The right-hand shows average K and network size.

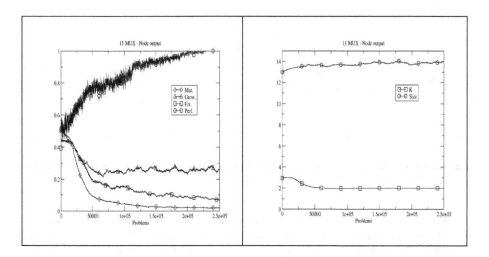

Fig. 2. Showing the behaviour of the RBN representation on the 11-MUX

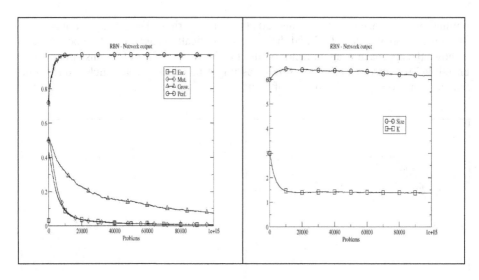

Fig. 3. Showing the behaviour of the RBN representation on the 6-MUX where a percentage of global network state is used to determine the output

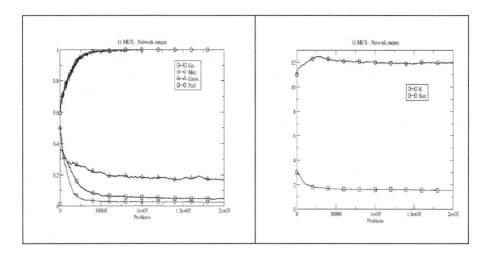

Fig. 4. Showing the behaviour of the RBN representation on the 11-MUX where a percentage of global network state is used to determine the output

As a rough benchmark, whilst direct comparison is perhaps difficult due to the use of a different training scheme, it can be noted that Koza [24] reports ~7.3 x 10^7 evaluations being required with traditional GP to solve the 11-bit problem, i.e., considerably more than the ~1 x 10^5 evaluations seen in Figure 4.

6   Asynchronous DGP in a Learning Classifier System

As noted above, traditional RBN consist of N nodes updating synchronously in discrete time steps but asynchronous versions have also been presented, after [16], leading to a classification of the space of possible forms of RBN [15]. Asynchronous forms of CA have also been explored (e.g., [21]) wherein it is often suggested that asynchrony is a more realistic underlying assumption for many natural and artificial systems. Harvey and Bossomaier [16] showed that asynchronous RBN exhibit either point attractors, as seen in asynchronous CAs, or "loose" attractors where "the network passes indefinitely through a subset of its possible states" [16] (as opposed to distinct cycles in the synchronous case). Thus the use of asynchrony represents another feature of RBN with the potential to significantly alter their underlying dynamics thereby offering another mechanism by which to aid the simulated evolutionary design process for a given task. Di Paolo [11] showed it is possible to evolve asynchronous RBN which exhibit rhythmic behaviour at equilibrium. Asynchronous CAs have also been evolved (e.g., [37]). No prior work on the use of asynchronous RBN for computation is known and neither is any prior work on asynchronous node/instruction updating in any other form of GP.

Asynchrony is here implemented as a randomly chosen node being updated on a given cycle, with as many updates per overall network update cycle as there are nodes in the network before an equivalent cycle to one in the synchronous case is said to have occurred (see [15] for alternative schemes).

Figure 5 shows how, despite their potentially very different underlying dynamics to synchronous RBN, the performance of the two systems on the 11-bit problem is very similar (compare with Figure 4). The only marked difference is in the evolution of smaller networks on average; initial growth is typically lost in the asynchronous case. Similar results are found for the node output scheme (not shown).

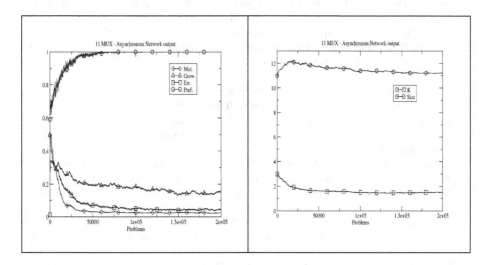

Fig. 5. Showing the behaviour of the asynchronous RBN representation on the 11-MUX where a percentage of global network state is used to determine the output

7 Conclusions

Sixty years after Turing's seminal work and almost forty years after Kauffman's presentation of RBN, this paper explored discrete Boolean forms of dynamical GP. In particular, a form of LCS has been presented with which to design ensembles of dynamical genetic programs. It has here been shown that evolutionary search is able to design ensembles that collectively solve a computational task under the reinforcement learning scheme of LCS. It can be noted that Forrest and Miller [14] used RBN to model the internal rule chaining of traditional LCS.

Current research is exploring many of the myriad of possibilities dynamical GP presents as a form of evolutionary computing by which to solve complex problems. Most importantly, it is being applied to tasks where the evolution of their temporal behaviour can be exploited directly, such as time-series data mining and adaptive control.

References

1. Ahluwalia, M., Bull, L.: A Genetic Programming Classifier System. In: Banzhaf, W., et al. (eds.) Proceedings of the Genetic and Evolutionary Computation Conference – GECCO 1999, pp. 11–18. Morgan Kaufmann, San Francisco (1999)
2. Andre, D., Koza, J.R., Bennett, F.H., Keane, M.: Genetic Programming III. MIT Press, Cambridge (1999)
3. Banzhaf, W.: Genetic Programming for Pedestrians. In: Forrest, S. (ed.) Proceedings of the Fifth International Conference on Genetic Algorithms, p. 628. Morgan Kaufmann, San Francisco (1993)
4. Bull, L.: On Using Constructivism in Neural Classifier Systems. In: Guervós, J.J.M., Adamidis, P.A., Beyer, H.-G., Fernández-Villacañas, J.-L., Schwefel, H.-P., et al. (eds.) PPSN 2002. LNCS, vol. 2439, pp. 558–567. Springer, Heidelberg (2002)
5. Bull, L.: Two Simple Learning Classifier Systems. In: Bull, L., Kovacs, T. (eds.) Foundations of Learning Classifier Systems, pp. 63–90. Springer, Heidelberg (2005)
6. Bull, L.: Coevolutionary Species Adaptation Genetic Algorithms: A Continuing SAGA on Coupled Fitness Landscapes. In: Capcarrère, M.S., Freitas, A.A., Bentley, P.J., Johnson, C.G., Timmis, J., et al. (eds.) ECAL 2005. LNCS, vol. 3630, pp. 322–331. Springer, Heidelberg (2005)
7. Bull, L.: Toward Artificial Creativity with Evolution Strategies. In: Parmee, I. (ed.) Adaptive Computing in Design and Manufacture VIII. IPCC (2008)
8. Bull, L., Hurst, J., Tomlinson, A.: Self-Adaptive Mutation in Classifier System Controllers. In: Meyer, J.-A., et al. (eds.) From Animals to Animats 6, pp. 460–468. MIT Press, Cambridge (2000)
9. Bull, L., Alonso-Sanz, A.: On Coupling Random Boolean Networks. In: Adamatzky, A., Alonso-Sanz, R., Lawniczak, A., Juarez Martinez, G., Morita, K., Worsch, T. (eds.) Automata 2008: Theory and Applications of Cellular Automata, pp. 292–301. Luniver Press (2008)
10. Copeland, J.: The Essential Turing, Oxford (2004)
11. Di Paolo, E.A.: Rhythmic and Non-rhythmic Attractors in Asynchronous Random Boolean Networks. Biosystems 59(3), 185–195 (2001)
12. Drugowitsch, J.: Design and Analysis of Learning Classifier Systems. Springer, Heidelberg (2008)

13. Fogel, L.J., Owens, A.J., Walsh, M.J.: Artificial Intelligence Through A Simulation of Evolution. In: Maxfield, M., Callahan, A., Fogel, L.J. (eds.) Biophysics and Cybernetic Systems: Proceedings of the 2nd Cybernetic Sciences Symposium, pp. 131–155. Spartan Books (1965)

14. Forrest, S., Miller, J.H.: The Dynamical Behaviour of Classifier Systems. In: Schaffer, J.D. (ed.) Proceedings of the Third International Conference on Genetic Algorithms, pp. 304–310. Morgan Kaufmann, San Francisco (1989)

15. Gershenson, C.: Classification of Random Boolean Networks. In: Standish, R.K., Bedau, M., Abbass, H. (eds.) Artificial Life VIII, pp. 1–8. MIT Press, Cambridge (2002)

16. Harvey, I., Bossomaier, T.: Time out of Joint: Attractors in Asynchronous Random Boolean Networks. In: Husbands, P., Harvey, I. (eds.) Proceedings of the Fourth European Artificial Life Conference, pp. 67–75. MIT Press, Cambridge (1997)

17. Holland, J.H.: Adaptation in Natural and Artificial Systems. University of Michigan Press (1975)

18. Holland, J.H.: Adaptation. In: Rosen, R., Snell, F.M. (eds.) Progress in Theoretical Biology, vol. 4, pp. 263–293. Plenum, New York (1976)

19. Hurst, J., Bull, L.: A Self-Adaptive Classifier System. In: Lanzi, P.L., Stolzmann, W., Wilson, S.W. (eds.) IWLCS 2000. LNCS, vol. 1996, pp. 70–79. Springer, Heidelberg (2001)

20. Hurst, J., Bull, L.: A Neural Learning Classifier System with Self-Adaptive Constructivism for Mobile Robot Control. Artificial Life 12(3), 353–380 (2006)

21. Ingerson, T., Buvel, R.: Structure in Asynchronous Cellular Automata. Physica D 10(1-2), 59–68 (1984)

22. Kauffman, S.A.: Metabolic Stability and Epigenesis in Randomly Constructed Genetic Nets. Journal of Theoretical Biology 22, 437–467 (1969)

23. Kauffman, S.A.: The Origins of Order: Self-Organization and Selection in Evolution, Oxford (1993)

24. Koza, J.R.: Genetic Programming. MIT Press, Cambridge (1992)

25. Lemke, N., Mombach, J., Bodmann, B.: A Numerical Investigation of Adaptation in Populations of Random Boolean Networks. Physica A 301, 589–600 (2001)

26. McCulloch, W.S., Pitts, W.: A Logical Calculus of the Ideas Immanent in Nervous Activity. Bulletin of Mathematical Biophysics 5, 115–133 (1943)

27. Mesot, B., Teuscher, C.: Deducing Local Rules for Solving Global Tasks with Random Boolean Networks. Physica D 211(1-2), 88–106 (2005)

28. Miller, J.: An Empirical Study of the Efficiency of Learning Boolean Functions using a Cartesian Genetic Programming Approach. In: Banzhaf, W., Daida, J., Eiben, A.E., Garzon, M.H., Honavar, V., Jakiela, M., Smith, R.E. (eds.) Proceedings of the Genetic and Evolutionary Computation Conference – GECCO 1999, pp. 1135–1142. Morgan Kaufmann, San Francisco (1999)

29. Mitchell, M., Hraber, P., Crutchfield, J.: Revisiting the Edge of Chaos: Evolving Cellular Automata to Perform Computations. Complex Systems 7, 83–130 (1993)

30. Niehaus, J., Banzhaf, W.: Adaption of operator Probabilities in Genetic Programming. In: Miller, J., Tomassini, M., Lanzi, P.L., Ryan, C., Tetamanzi, A.G.B., Langdon, W.B. (eds.) EuroGP 2001. LNCS, vol. 2038, pp. 325–336. Springer, Heidelberg (2001)

31. Packard, N.: Adaptation Toward the Edge of Chaos. In: Kelso, J., Mandell, A., Shlesinger, M. (eds.) Dynamic Patterns in Complex Systems, pp. 293–301. World Scientific, Singapore (1988)

32. Pujol, J., Poli, R.: Efficient Evolution of Asymmetric Recurrent Neural Networks using a PDGP-inspired two-dimensional Representation. In: Banzhaf, W., Poli, R., Schoenauer, M., Fogarty, T.C. (eds.) EuroGP 1998. LNCS, vol. 1391, pp. 130–141. Springer, Heidelberg (1998)

33. Quick, T., Nehaniv, C., Dautenhahn, K., Roberts, G.: Evolving Embedded Genetic Regulatory Network-Driven Control Systems. In: Banzhaf, W., Ziegler, J., Christaller, T., Dittrich, P., Kim, J.T. (eds.) ECAL 2003. LNCS, vol. 2801, pp. 266–277. Springer, Heidelberg (2003)

34. Schmidt, M., Lipson, H.: Comparison of Tree and Graph Encodings as Function of Problem Complexity. In: Proceedings of the Genetic and Evolutionary Computation Conference – GECCO 2007, pp. 1674–1679. ACM Press, New York (2007)

35. Schwefel, H.-P.: Numerical Optimization of Computer Models. Wiley, Chichester (1981)

36. Sipper, M.: Evolution of Parallel Cellular Machines. Springer, Heidelberg (1997)

37. Sipper, M., Tomassini, M., Capcarrere, S.: Evolving Asynchronous and Scalable Non-uniform Cellular Automata. In: Proceedings of the Third International Conference on Artificial Neural Networks and Genetic Algorithms, pp. 66–70. Springer, Heidelberg (1997)

38. Teller, A., Veloso, M.: Neural Programming and an Internal Reinforcement Policy. In: Koza, J.R. (ed.) Late Breaking Papers at the Genetic Programming 1996 Conference, pp. 186–192. Stanford University (1996)

39. Teuscher, C.: Turing's Connectionism. Springer, Heidelberg (2002)

40. Turing, A.: Intelligent Machinery. In: [Copeland, 2004], pp. 395–432 (1948)

41. Valenzuela-Rendon, M.: The Fuzzy Classifier System: a Classifier System for Continuously Varying Variables. In: Booker, L., Belew, R. (eds.) Proceedings of the Fourth International Conference on Genetic Algorithms, pp. 346–353. Morgan Kaufmann, San Francisco (1991)

42. Van den Broeck, C., Kawai, R.: Learning in Feedforward Boolean Networks. Physical Review A 42, 6210–6218 (1990)

43. Von Neumann, J.: The Theory of Self-Reproducing Automata. University of Illinois, US (1966)

44. Wilson, S.W.: Classifier Fitness Based on Accuracy. Evolutionary Comp. 3, 149–175 (1995)

45. Wilson, S.W.: Get Real! XCS with Continuous-Valued Inputs. In: Lanzi, P.-L., Stolzmann, W., Wilson, S.W. (eds.) Learning Classifier Systems: From Foundations to Applications, pp. 209–222. Springer, Heidelberg (2000)

46. Wilson, S.W.: Mining Oblique Data with XCS. In: Lanzi, P.L., Stolzmann, W., Wilson, S.W. (eds.) IWLCS 2000. LNCS, vol. 1996, pp. 158–176. Springer, Heidelberg (2001)

Why Coevolution Doesn't "Work": Superiority and Progress in Coevolution

Thomas Miconi

School of Computer Science
University of Birmingham
Birmingham B152TT, UK
txm@cs.bham.ac.uk

Abstract. Coevolution often gives rise to counter-intuitive dynamics that defy our expectations. Here we suggest that much of the confusion surrounding coevolution results from imprecise notions of superiority and progress. In particular, we note that in the literature, three distinct notions of progress are implicitly lumped together: *local* progress (superior performance against current opponents), *historical* progress (superior performance against previous opponents) and *global* progress (superior performance against the entire opponent space). As a result, valid conditions for one type of progress are unduly assumed to lead to another. In particular, the confusion between historical and global progress is a case of a common error, namely *using the training set as a test set*. This error is prevalent among standard methods for coevolutionary analysis (CIAO, Master Tournament, Dominance Tournament, etc.) By clearly defining and distinguishing between different types of progress, we identify limitations with existing techniques and algorithms, address them, and generally facilitate discussion and understanding of coevolution. We conclude that the concepts proposed in this paper correspond to important aspects of the coevolutionary process.

1 Introduction

The term "coevolution" refers to any situation where various lineages undergo *reciprocal evolutionary change*: lineages evolve in response to each other's evolution. This implies that these lineages evolve on coupled fitness landscape. In biology, coevolution denotes reciprocal evolutionary change only; in artificial coevolution, the term has been slightly extended to cover any situation in which lineages evolve on coupled fitness landscape, rather than the change itself.

Coevolution is of course useful for competitive tasks (such as games), or when no suitable fitness function can be readily devised. Early experiments by Axelrod [1] and especially Hillis [2] have generated much optimism with regard to the potential of coevolution.

The justification for using coevolution in optimisation tasks usually relies on the concept of *arms races*, as introduced in evolutionary biology by Dawkins and Krebs [3]. Rosin and Belew [4] summarise the transposition of this concept to artificial evolution:

L. Vanneschi et al. (Eds.): EuroGP 2009, LNCS 5481, pp. 49–60, 2009.

> The success of a host implies failure for its parasites. When the parasites evolve to overcome this failure, they create new challenges for the hosts; the continuation of this may lead to an evolutionary "arms race" (...) New genotypes arise to defeat old ones. New parasite types should serve as a drive toward further innovation, creating ever-greater levels of complexity and performance by forcing hosts to respond to a wider range of more challenging parasite test cases.

Unfortunately, it quickly became apparent that the actual behaviour of coevolution was more complex than expected. Instead of a reliable march to progress, practitioners of coevolution were faced with counter-intuitive dynamics which frustrated their expectations. The discrepancy between expectations and reality has been aptly summarised by Ficici [5, Chap. 1]:

> [The] gap between the hypothesized potential of coevolutionary algorithms and realized practice remains substantial - the successes of coevolution (of which there are now many) are balanced by frequently encountered and irksome *pathologies* (of which there are also many). These pathologies do not merely deprive us of a satisfying result (as might a local optimum), but more importantly they appear to violate our intuitions of how selection pressure in coevolution is supposed to operate. [...] To grapple with this incongruity between expectation and reality, coevolution researchers have appropriated or invented a myriad of terms, such as *cyclic dynamic, mediocre stable-state, collusion, forgetting, disengagement,* and *focusing.*

Ficici [5,6] has suggested that the critical defect in early (and current) coevolutionary experiments was the lack of an explicit *solution concept.* A solution concept is a way to indicate which members of a given search space (if any) are the solutions that we are looking for. Thus, by defining a solution concept, we explicitly state what exactly we are looking for, before starting to look for it. To quote Ficici, "while this point may seem obvious, years of coevolutionary practice indicate otherwise" [5, Conclusion].

However, most experiments in coevolution did use a certain solution concept, although often in an implicit manner. For example, from reading the discussion in Nolfi and Floreano [7] or Rosin and Belew [4], it is clear that they implicitly target a certain solution concept, namely the "best scoring solution": the expected "solution" is the individual that maximises its expected score against a random opponent. Ficici recognises this, and calls this criterion the "conventional" solution concept. He then shows that different solution concepts (best scoring solution, Pareto dominance, Nash equilibrium, etc.) possess different properties, which were previously not recognised due to a lack of attention to solution concepts.

We believe that the most important oversight in coevolution is not so much the lack of an explicit solution concept (especially considering that most experiments implicitly use a certain solution concept); but rather, the lack of clear definitions for *superiority* and *progress.*

2   The Meaning of "Better"

What does it mean to say that an individual A is "better" (that is, more desirable as an outcome of our search) than an individual B? At first sight, if we consider only competitive tasks on symmetric domains, there seems to be an obvious indicator of superiority, namely the outcome of a head-to-head competition. If A defeats B, then we might say that A is better than B.

However, this simple idea is defeated by the problem of *intransitivity*. If A defeats B and B defeats C, it does not follow that A will also defeat C. Suppose that A is able to defeat B, but that B is able to defeat many more other opponents than A[1]. Then we might not necessarily regard A as superior to B. In addition, in non-symmetric tasks, head-to-head competitions cannot be used directly between members of the same population.

This hints at the fact that superiority between any two individuals A and B cannot be meaningfully assessed by considering A and B alone, in isolation. Rather, the superiority of an individual A over an individual B can only be determined (or even meaningfully discussed) by *comparing* their respective performances against a *common set* of other opponents. The particular set of opponents that we use for this purpose is as important as the particular criterion which is used to calculate superiority. As we will see, using different categories of opponents produces different notions of "superiority", and of "progress", which are often implicitly confused.

To decide superiority between A and B, we therefore need two things:

- A common "reference" set of opponents, against which we collect the respective performance of A and B.
- A comparison criterion, which we use to process the resulting data and decide which of A or B is superior.

Together, these two components define a *superiority relation*, with which we can compare any two individuals A and B.

Notice that the notions of "superiority" and "progress" are immediately connected. Progress occurs when newer individuals are superior to their ancestors, according to whatever procedure we use to determine superiority. Different definitions of superiority (using different criteria or, importantly, different reference sets) are therefore associated with different concepts of progress.

2.1   Comparison Criteria and Solution Concepts

The most obvious comparison criterion is simply the average score, but others can be used. A common example is Pareto dominance: A is superior to B, against the reference set R, if A can defeat all individuals in R that B can - and then some.

---

[1] As Stanley & Miikkulainen point out [8], parasite-host relationship offer frequent examples of large, successful organisms exploited by highly specialised, opportunistic opponents.

While there is a clear similarity between comparison criteria and solution concepts, the two are not equivalent. A solution concept is absolute: within a certain subset of the search space, an individual is either a solution or it is not. By contrast, comparison criteria are supposed to be *ordinal*, since they must be used to compare two individuals. Any comparison criterion is also usable as a solution concept, by simply stating that "solutions" are individuals that are superior or equivalent to any other according to this criterion. But the reverse is not true. For example, the solution concept of Nash equilibrium [6] is not directly usable as a general comparison method between any two individuals: in any given set, only a small number of individuals will be Nash equilibria (or quite possibly none at all if mixed strategies are not considered). If neither A nor B is a Nash equilibrium (which will be the case for the vast majority of individuals) then we cannot distinguish between A and B.

2.2 Opponent Sets

While there has been much discussion of solution concepts, the other component of superiority relations (the set of opponents against which performance is compared) has attracted little attention. We believe that this latter component is the source of much of the confusion that surrounds coevolution. Obviously, the overall result of the comparison depends on the set of opponents that is chosen: comparing performance against different sets of opponents will result in different orderings between individuals. However, in the literature, several different sets of opponents are implicitly considered. This leads to different opponent sets (and thus different orderings) being silently lumped together under the same term "progress", predictably causing much confusion. The next section seeks to tackle this issue in detail.

3   Local, Historical and Global Progress

A close look at the literature reveals that the words "progress" and "superiority" often refer to at least three different concepts. These concepts are distinguished by the three different reference sets of opponents that are (implicitly) considered. These are:

- Local: A is better than B, when comparing performances against the *current* opponents.
- Historical: A is better than B, when comparing performances against *all previously encountered* opponents - that is, current opponents and their ancestors.
- Global: A is better than B when comparing their performances against *all possible opponents* - that is, the entire search space.

The above applies both to superiority and progress. To obtain explicit definitions of local, historical and global progress, we just replace A and B with "newer individuals" and "ancestral individuals", respectively.

Local progress describes a situation where newer individuals are superior to their ancestors, when comparing their respective performances against current opponents. It is the only form of progress that can be automatically and mechanically brought about by natural selection. Because selection occurs precisely by favouring the individuals that are best adapted within the current environment, it is expected that newer individuals will perform at least no worse than their immediate ancestors against current opponents. Importantly, natural selection is a local process, both in space and time. Any longer-term effect must rely on additional causes, which should be made explicit if these effects are expected to occur.

Historical progress occurs when newer individuals are superior to their ancestors, when comparing their performances against all previous opponents - that is, the opponents that have been previously encountered over the history of the population's evolutionary history. It corresponds to the concept expressed by the term "arms race". As is well known, the local progress brought about by natural selection can easily be harnessed to drive historical progress, through the use of an *archive*. An archive is a collection of previously encountered individuals which is maintained and enriched throughout the evolutionary process. The purpose of an archive is simply to keep *past* opponents within the *current* selective environment, thereby making local and historical progress identical.

Global progress occurs when newer individuals are superior to their ancestors, when comparing their performance against the entire search space. This captures the intuitive goal of artificial coevolution. Essentially, global progress is what we are really interested in, and what we would like to observe in our algorithms. Unfortunately, it does not necessarily arise from the former two effects. It is not clear that an algorithm could be devised that would directly impose this kind of progress as mechanically as archives can enforce historical progress: this would involve knowledge of unknown opponents, which is absurd. Therefore, a practical option is to examine the conditions under which local and historical progress can be harnessed to obtain global progress with some degree of reliability.

Using this terminology, we can now uncover important examples of confusion between different types of "progress" in early coevolutionary research.

4   Two Fragile Assumptions in Early Coevolutionary Research

4.1   The "Arms Races" Argument: Extrapolating from Local to Global Progress

As previously discussed, the supposed benefits of coevolution were often justified by invoking the "arms race" hypothesis. One example is Rosin and Belew's description, already quoted above (Section 1). Nolfi and Floreano [7] adopt essentially the same position:

> [The] success of predators implies a failure of the prey and conversely, when prey evolve to overcome the predators they also create a new challenge for them. Similarly, when the predators overcome the new prey

by adapting to them, they create a new challenge for the prey. Clearly the continuation of this process may produce an ever-greater level of complexity...

The difficulty with this argument is a confusion between *local* progress and *global* progress. We intuitively accept that, due to the action of natural selection, newer individuals may outperform their immediate ancestors against current opponents. However, this uncontroversial notion is then silently extended to a rather different idea, namely that they should also outperform their immediate ancestors in terms of global performance (that is, against the entire search space). Through a rhetorical shortcut, the implicit assent that was easily elicited from us for the former proposition, is surreptitiously carried over to the latter.

This fragile argument underlies much of the expectations that surrounded early coevolutionary research - and of the resulting frustration generated by empirical results.

Researchers quickly realised that coevolution does not conform to these expectations (hence the "pathologies" discussed by Ficici). Indeed, in the passage by Nolfi and Floreano quoted above, the authors immediately warn that "this does not necessarily happens, as we will see below". Indeed their paper shows numerous examples of how coevolution fails to behave according to these expectations. By distinguishing between local, historical and global progress, we can pinpoint one source of discrepancy between stated expectations and reality.

4.2   Confusion between Historical Progress and Global Progress

Reliance on the "arms race" concept has had another consequence on coevolutionary research. Arms races are cases of *historical* progress: they imply that newer individuals should outperform their ancestors against ancestral opponents (in the words of Dawkins and Krebs [3]: "that modern predators might massacre Eocene prey"). Due to the lack of distinction between various types of progress, this has had the effect of turning historical progress into the actual goal of coevolutionary research: because the use of coevolution is based on the arms race concept, a successful arms race and a successful coevolutionary run are one single concept. In the weaker form, arms races are seen as mechanically leading to global progress. In the stronger form, arms races are the end in itself of coevolution: historical progress *is* progress, period. Stanley and Miikkulainen [8] make the point bluntly: "In competitive coevolution, the goal is to establish an 'arms race' that will lead to increasingly sophisticated strategies." Ficici and Pollack [9] state that "the key to successful coevolutionary learning is a competitive arms race between opposed participants... in open-ended domains, coevolutionary progress can, theoretically, continue indefinitely".

This is essentially a confusion between *historical* and *global* progress. The historical progress of arms races (better performance against ancestral opponents) is confused with the actual desired outcome, namely global progress (better performance against the entire opponent space).

It turns out that this confusion between historical and global progress is an example of a common error in statistical learning, namely, confusing the *training set* with the *test set*. This is because ancestral opponents constitute the "training set" against which current individuals have been optimised. It is well-known that performance against the training set should not be regarded as a reliable indicator of general performance, especially for comparison purposes. In particular, evaluating the quality of individuals solely from their performance against the training set raises the risk of *overfitting*, in which individuals are strong against the training set, but perform poorly against unseen opponents. Yet this is exactly what is occurring in coevolutionary research, due to the lack of distinction between historical and global progress. As we will see in the next section, this confusion is particularly problematic in the sub-field of analysis and evaluation methods.

This confusion between historical and global progress is deeply entrenched in the literature, despite contrary empirical results. As an example, Nolfi & Floreano [7] have shown empirically that sustained arms races do not necessarily result in optimum performance. In some circumstances, when the authors compared the results of coevolution with and without a "Hall of Fame" (an archive of previous champions), they found out that individuals evolved *with* a Hall of Fame were defeated by individuals evolved *without* it. This occurred despite the fact that the Hall of Fame method had demonstrably enforced a *more* reliable (historical) "progress", in the sense that individuals evolved with a Hall of Fame were significantly better at defeating their own ancestors than those without. Using our terminology, better historical progress had led to poorer global progress. However, in their conclusions (p. 25), the authors still surmise that "co-evolution will lead to a progressive increase in complexity when complete general solutions (i.e. solutions which are successful against all the strategies adopted by *previous opponents*) exist and can be selected by modifying the current solutions" (emphasis added). Even though their own experiments showed that success against previous opponents is not directly related to general performance, the authors still associate the former with the latter. As a result, in the same paper, they propose evaluation and analysis methods for detecting "progress" that target historical progress alone, with predictably damaging consequences. These methods (and others based on the same principle) are discussed in the next section.

5   Application: Analysis and Evaluation Methods

5.1   Assessing Performance and Progress in Artificial Coevolution

The confusion between historical and global progress has had particularly important effects in the sub-field of analysis and evaluation methods for coevolution. Several methods have been proposed to accurately measure the performance of individuals, as well as evaluate the amount of "progress" that has occurred during a coevolutionary run. Two closely related examples are Cliff and Miller's CIAO plots ("Current Individual against Ancestral Opponent" [10]) and Nolfi and Floreano's Master Tournament matrices [7]. The main idea of these methods is to summarise the results of a coevolutionary run as a matrix of coloured dots. The colour of dot (x, y) is dark if the champion of population A at generation x

defeats the champion of population B at generation y; otherwise the dot is left blank. Master Tournament matrices are square matrices that collect the results of all such competitions. CIAO plots are triangular, and only involve (x, y) pairs where $x > y$; they are particularly suited to one-population, symmetric tasks, for which (x, y) and (y, x) represent the same competition. The appearance of the matrices provide information about the "performance" of successive champions, and consequently of how much "progress" has occurred over the run. In particular, Nolfi and Floreano point out that an "ideal situation" would result in a matrix perfectly divided between a white triangle and a black triangle, indicating that every champion of each population is capable of defeating all previous opponents. Of course, this rarely occurs in practice: real coevolutionary runs are likely to result in more mixed results, which may not be so easy to interpret. Cartlidge [11, Chap. 7] uses image processing techniques to improve the readability of real plots. Additionally, we have previously suggested that the simple operation of trimming the matrix by evenly sampling champions (using champions from every n-th generation, rather than champions of all generations) could also result in more readable graphs [12].

Stanley and Miikkulainen [8] take a different approach with their Dominance Tournament technique, which focuses on "dominant" individuals over the course of coevolution. The first dominant individual is the champion of the population at generation 1. After this, dominant individuals are those that can defeat *all* previous dominant individuals. The authors argue that the frequency of appearance of dominant individuals is an indication of how much "progress" occurs during the run.

5.2   Standard Methods Target Historical Progress

These methods share an obvious common point: they all measure the "quality" of an individual according to its performance against *previous opponents* - that is, the sequence of successive opponents against which they have evolved. In this case, the training set is literally being used as a test set.

What's more, coevolutionary runs and algorithms are judged and ranked according to how much historical progress occurs. For example, Master Tournament matrices evenly divided into two perfect triangles of opposite colours are suggested to represent "ideal" situations, because they represent cases of perfect historical progress: each individual is able to defeat all of its ancestral opponents. This implies a 100% performance against the training set, which would be regarded as a worrying sign in traditional machine learning! The same applies to Dominance Tournaments, where greater frequency of appearance of dominant individuals is regarded as desirable.

Of course, in practice, strong historical progress may carry *some* evidence of global progress: if a newer individual defeats all its ancestral opponents, while an older individual defeats none of them, then we may well suspect that the former is actually superior to the latter, not just against these particular opponents, but also against the entire opponent space. However, we must keep in mind that this is merely a reasonable assumption, not a logical necessity; it is easy to devise artificial fitness landscapes for which this would not be true. Furthermore, such perfect historical progress is extremely unlikely in practice, and lesser differences in historical performance may not convey any information about global performance.

5.3 Solution: Co-testing Independent Populations

Fortunately, this problem can be overcome by following a simple rule: when applying these techniques, do not pit individuals against the opponents with which they have coevolved. The opponents used for testing should have evolved independently of the individuals we are trying to test, so as to serve as independent representatives of the opponent space. For this, we may take advantage of the fact that coevolutionary experiments are rarely performed just once; instead, they are usually repeated over several runs, as is necessary to obtain more reliable data in support of whatever hypothesis is being investigated. Thus, when evaluating individuals (or algorithms), we simply need to ensure that we only oppose individuals from different runs - that is, individuals that have not coevolved together.

We have previously suggested several methods based on this principle. The simplest one is simply to cross-validate runs by opposing each successive champion of a given run against all champions of several other runs [12]. If we want to compare the performance of two coevolutionary algorithms, then a more economic method is to perform *equal-effort comparisons*: at any generation (or at every n-th generation), we oppose all the current champions of one algorithm against all the current champions of the other, and we track the respective number of victories for each algorithm over time. An advantage of this method is that it allows for real-time tracking of respective performance. We used this method to investigate the properties of certain algorithms; in particular, we found that increasing the number of opponents for each evaluation did not provide noticeable benefits beyond the unexpectedly low number of 4, and using more than 5 opponents actually decreased performance [13, Chap. 7].

6 "Monotonic Progress"

The distinction between historical and global progress allows us to clarify the statement that certain algorithms enforce continual "progress". As an example, De Jong [14] introduced the MAXSOLVE algorithm to ensure "monotonic progress" under the solution concept of "maximum utility" (that is, maximum expected score against an unseen opponent).MAXSOLVE maintains an archive of learners and an archive of tests. At every cycle, new candidate learners and tests are generated. Candidate learners are only included in the archive if they can defeat more tests (from the test archive) than some of the current members. Candidate tests are only included in the test archive if they are defeated by one or more learners.

Thus, MAXSOLVE involves discarding candidates that might potentially be superior (in a global sense) to those currently known. De Jong justifies this as follows:

We note that a candidate discarded in this manner could potentially solve more unseen tests than the candidates that are maintained. This does not violate the monotonicity of the archive, however, since progress is measured with respect to the set of tests seen so far. If the rejected candidate solution does indeed solve more tests, this will eventually be discovered given sufficient further search, at which point it will be included.

Thus we see that MAXSOLVE does enforce *historical* progress, but does not enforce *global* progress. De Jong points out that any superior solution will eventually be recognised after enough information has been gathered. This is true, but not necessarily usable: it expresses the obvious fact that over time, the set of opponents "seen so far" becomes closer and closer to the entire space of possible opponents, and thus that superiority against previously seen opponent will necessarily become more and more equivalent to superiority against all possible opponents. In this trivial sense, historical progress does converge to global progress - but only after a significant portion (or in fact the totality) of the search space has been explored. In the meantime, as shown by Nolfi & Floreano's experiments (described above) and Ficici's theoretical results (described below), perfect historical progress may well lead the algorithm *away* from global progress, by discarding individuals that are historically inferior, but globally superior.

7 Discussion of Coevolutionary Concepts

A further benefit of the terminology is that it considerably simplifies the discussion of certain important concepts and methods in coevolution. As an example, Ficici [6] has recently introduced the important concept of *monotonicity* in solution concepts. Let us quote Ficici's own definition (edited for clarity):

> We define a monotonic solution concept to be one such that every [individual] C that is a solution to [subsets of the search space] G_A and G_C, where $G_A \supset G_C$, will also be a solution to any [subset] G_B, where $G_A \supset G_B \supset G_C$.

What does this mean, and why does it matter? A monotonic solution concept is one for which, if no information is ever discarded, then solutions that are "demoted" (i.e. no longer considered solutions as exploration progresses and more information is gained) will never be "reinstated" (considered as solutions) again, no matter how much information we gain in the future. The consequence is that, under a monotonic solution concept with no loss of information, newer solution are "really" superior to their predecessors, not just temporarily superior (as was the case with MAXSOLVE). In Ficici's own words [6, Sec. 7] (emphasis original):

> If we do not discard information, then the solutions we obtain at time t are guaranteed to be no worse, in an objective (i.e., global) sense, than

solutions obtained earlier; and, this is true *regardless of the fact that we have only local knowledge of the domain and no knowledge of what strategies we might discover in the future.*

Ficici proceeds to show that certain solution concepts (Nash equilibria) are indeed monotonic, while others (including maximum expected score and Pareto dominance) are not. The main drawback is that Nash equilibria are cumbersome to use in practice, because they require mixed strategies.

It turns out that, using our terminology, monotonic solution concepts can be described in a remarkably succinct manner:

A solution concept is monotonic if, among successive solutions, *historical progress implies global progress.*

Note that this description immediately reveals the important aspect of monotonic solution concepts: if a newer solution is superior to a previous one, based on past information (that is, ancestral opponents), then we know that this superiority is real, as opposed to an illusory result of our limited current knowledge: we will not need to "backtrack" to a previously demoted solution. The fact that our terminology can describe such an important concept with such concision and clarity is a further confirmation that the concepts described in this paper represent deep, important aspects of the coevolutionary process, rather than just arbitrary terminology.

8   Summary and Conclusions

In the literature, at least three different notions of superiority and progress are commonly and implicitly referred to. These are *local* progress (superior performance against current opponents), *historical* progress (superior performance against previously ancestral opponents) and *global* progress (superior performance against the entire opponent space). Early expectations about the behaviour of coevolution were partly based on a rhetorical amalgamation of local and historical progress (the "arms race" argument), though they were quickly reassessed in the light of empirical results. However, the confusion between historical and global progress is still rife across the literature, especially concerning analysis and evaluation methods. As a result, analysis methods are often targeted at historical progress alone (progress against previous opponents), which amounts to using the training set as a test set. Fortunately, this can be easily addressed by carefully separating co-evolved populations during a posteriori analysis and evaluation.

By acknowledging the distinction between local, historical and global progress, we can devise more reliable methods for analysing and evaluating coevolutionary trajectories; we can bring precision to claims that certain algorithms enforce "monotonic" progress; we can describe important concept and techniques in a concise and elegant manner. But more importantly, we realise that much of what we regard as "failures" or "pathologies" are actually natural, predictable outcomes of evolution that only appear counter-intuitive due to our warped understanding and unwarranted expectations.

References

1. Axelrod, R.: The evolution of strategies in the iterated prisoner's dilemma. In: Davis, L. (ed.) Genetic algorithms and simulated annealing, pp. 32–41. Morgan Kaufman, San Francisco (1987)
2. Hillis, W.: Co-evolving parasites improve simulated evolution as an optimization procedure. Physica D 42, 228–234 (1990)
3. Dawkins, R., Krebs, J.R.: Arms races between and within species. Procs of the Royal Society of London, Series B 205, 489–511 (1979)
4. Rosin, C.D., Belew, R.K.: Methods for competitive co-evolution: Finding opponents worth beating. In: Eshelman, L. (ed.) Procs. 6th ICGA. Morgan Kaufmann, San Francisco (1995)
5. Ficici, S.G.: Solution Concepts in Coevolutionary Algorithms. Ph.D thesis, Brandeis University (May 2004)
6. Ficici, S.G.: Monotonic solution concepts in coevolution. In: Procs. GECCO 2005. ACM Press, New York (2005)
7. Nolfi, S., Floreano, D.: Coevolving predator and prey robots: Do "arms races" arise in artificial evolution? Artificial Life 4(4), 311–335 (1998)
8. Stanley, K.O., Miikkulainen, R.: The dominance tournament method of monitoring progress in coevolution. In: Procs GECCO 2002 Workshops. Morgan Kaufmann, San Francisco (2002)
9. Ficici, S.G., Pollack, J.B.: Challenges in coevolutionary learning: arms-race dynamics, open-endedness, and medicocre stable states. In: Procs. ALIFE VI. MIT Press, Cambridge (1998)
10. Cliff, D., Miller, G.F.: Tracking the red queen: Measurements of adaptive progress in co-evolutionary simulations. In: Morán, F., Merelo, J.J., Moreno, A., Chacon, P. (eds.) ECAL 1995. LNCS (LNAI), vol. 929. Springer, Heidelberg (1995)
11. Cartlidge, J.: Rules of Engagement: Competitive Coevolutionary Dynamics in Computational Systems. Ph.D thesis, The University of Leeds (2004)
12. Miconi, T., Channon, A.: Analysing coevolution among artificial creatures. In: Talbi, E.-G., Liardet, P., Collet, P., Lutton, E., Schoenauer, M. (eds.) EA 2005. LNCS, vol. 3871, pp. 167–178. Springer, Heidelberg (2006)
13. Miconi, T.: The Road to Everywhere: Evolution, Complexity and Progress in Natural and Artificial Systems. Ph.D thesis, University of Birmingham (2007)
14. De Jong, E.D.: The MaxSolve algorithm for coevolution. In: Procs. GECCO 2005. ACM Press, New York (2005)

On Improving Generalisation in Genetic Programming

Dan Costelloe and Conor Ryan

BDS Group,
Department of Computer Science and Information Systems,
University of Limerick,
Limerick, Ireland
dan.costelloe@ul.ie, conor.ryan@ul.ie

Abstract. This paper is concerned with the generalisation performance of GP. We examine the generalisation of GP on some well-studied test problems and also critically examine the performance of some well known GP improvements from a generalisation perspective. From this, the need for GP practitioners to provide more accurate reports on the generalisation performance of their systems on problems studied is highlighted. Based on the results achieved, it is shown that improvements in training performance thanks to GP-enhancements represent only half of the battle.

1 Introduction

Improvements to Genetic Programming are discovered and reported at Evolutionary Computation conferences and in articles every year. A typical effort involves the discovery of a new technique followed by its comparison with standard Koza-style GP (often on the same problem). These comparisons tend to show that due to a statistically significant improvement between the proposed method and standard GP on *training* data, the new method is therefore declared to be successful.

Unfortunately, what many such studies fail to address is the performance of the proposed GP-improvements on *unseen* data[1]. The central message of this paper is that approaches such as these must take generalisation performance into account before any declaration of success is made.

It should be stressed that the motivation for writing this paper is not founded in an attempt to belittle the valuable contributions that researchers have made to the GP field. On the contrary: this work aims to increase the robustness of such contributions, making them stronger competitors against other methods within the field of Artificial Intelligence and beyond.

---

[1] Even the seminal, award winning papers by Keijzer [1] and Gustafson *et al* [2] are examples. There a multitude of offenders to be found in the GP literature, however, space requirements do not permit an exhaustive listing. The two examples cited have made significant contributions to the GP field – the fact that these particular studies do not address generalisation performance should not deter from this.

L. Vanneschi et al. (Eds.): EuroGP 2009, LNCS 5481, pp. 61–72, 2009.

The rest of this paper is organised as follows; Section 2 examines a simple GP problem and shows how generalisation performance may be observed using existing, commonly used metrics. In Section 3, a selected GP-improvement technique (Linear Scaling) is compared with standard GP on unseen data. Section 4 describes the combination of Linear Scaling with a further improvement leading to performance gains in terms of generalisation. Finally, some conclusions and further work are discussed in Section 5.

2   On Problem Difficulty

To gain an idea of the difficulty of a GP problem Koza [3] suggests the measurement of the number of individuals that must be processed in order to satisfy the problem's *success predicate* with a chosen probability. A strict success predicate can be defined as the discovery of a 100% correct solution to the problem. A slightly less strict version of the predicate deems a GP-generated program to be successful if it achieves N "hits" (where N is the number of fitness cases and a hit occurs when a predicted outcome and a target outcome fall within a pre-specified error window).

If the amount of processing that must take place to achieve a successful solution to a problem is correlated with problem difficulty, then it follows that the lower the value of this measurement (or the larger the error window is), the easier the problem is for GP to solve. Two obvious factors that influence the amount of processing are the population size M and number of generations G. Koza describes the process of measuring the amount of processing required by estimating the probability $\gamma(M, i)$ that a particular run with population size M first yields a successful individual at generation i.

The *cumulative probability of success*, $P(M, i)$, then refers to the probability of achieving a successful result for all generations between 0 and i.[2] Koza defines

$$R(M, i, z) = \lceil \frac{1 - z}{log(1 - P(M, i))} \rceil$$

as the number of runs required to have a probability $z = 99\%$ of discovering a solution before generation i. Calculating the total number of individuals that must be processed to achieve a successful result is then a matter of straightforward multiplication by the population size and the number of generations (including the initial generation):

$$I(M, i, z) = m(i + 1)R(M, i, z)$$

---

[2] It is calculated experimentally as follows:

$$P(M, i) = \frac{N_s(i)}{T}$$

where T is the total number of runs and $N_s(i)$ is the number of successful runs at generation i.

As noted (and even criticised) elsewhere [4], the above metrics are often used in GP literature to compare new GP-variants. For a more comprehensive study on problem difficulty in GP, the reader is referred to the work of Vanneschi *et al* [5].

2.1 A Simple Problem

If there is one symbolic regression problem that receives the most attention in GP research, it is probably that of (re-) discovering the quartic polynomial. The function is described as follows:

$$f(x) = x^4 + x^3 + x^2 + x \tag{1}$$

Due to the popularity of this problem in GP research, it is used here for the purposes of demonstration.

Making things *easy*. Now that we have a simple problem to study and a metric that can be used to approximate the difficulty of the problem, let us now turn our attention to the task of finding a configuration of the problem settings that achieve a high success rate. This task is achieved experimentally as follows. Using a Koza-style GP system [6], we supply a set of configurations to the system, run each one and then choose the configuration that turns out to be the easiest for GP.

An *easy* configuration is (arbitrarily) defined here as one that produces an average success rate of at least 70% by the end of the run. Let us choose configuration value for population size; using the run parameters shown in Table 1, four sets of 30 independent runs are carried out using population size 50, 100, 200 and 500.

By examining the success rate plots in Figure 1 (left), we can immediately identify two "easy" configurations from the use of populations sizes 200 and 500

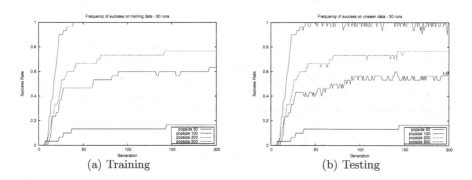

	(a) Training	(b) Testing

Fig. 1. Success rate plots from four sets of runs on the quartic polynomial problem using a training dataset of 21 points in the range [-1:0.1:1] (left) and a test dataset of 11 points in the range [1:0.1:1] (right). Each set of runs uses a different population size.

Table 1. Run settings used to find a population size to use which produces a success rate of at least 70%

Generations	200
Crossover rate	0.95
Mutation rate	0.02
Tournament size	5
Function set	$\{+, -, *, \%, sqr()\}$
Terminal set	$\{x\}$
Raw fitness	MSE(targets, predictions)
Standardised fitness	$\frac{1}{1+MSE}$
Hits criterion	Number of points where the GP program comes within 0.01 of the desired value

since the success rate passes the pre-defined target value of 70% in both cases. Also shown in Figure 1 (right) is the success rate as calculated from the number of hits on the testing dataset. It is encouraging to note that the performance on unseen data holds up almost as well as the training performance, which is a very desirable characteristic for any black box learner/predictor method.

This characteristic is so desirable that it should form the basis of a much more useful comparison between $GP + improvement$ and standard GP for forecasting problems.

3   A Selected Improvement: Linear Scaling

A number of improvements to GP for symbolic regression have been published in recent years [7,8,1,9]. Among these is Keijzer's application of linear scaling [1] to a suite of test problems and the subsequent proof [9] of the superiority of the scaled error measure over standard error measures (such as MSE) on symbolic regression problems. The method works by finding two values, a and b such that

$$a + bg(x) = t + \epsilon$$

It can be proved that the error (ϵ) is reduced via the application of linear scaling for $a \neq 0$ and $b \neq 1$. The process works as follows. If x is a set of independent variables, g is a GP-induced function that produces a set of predictions y and t is a set of target values derived from an unknown function f such that $f(x) = t$, a linear regression on the target values can be performed:

$$b = \frac{\sum_{i=0}^{n} (t - \bar{t})(y - \bar{y})}{\sum_{i=0}^{n} (y - \bar{y})^2} \tag{2}$$

$$a = \bar{t} - b\bar{y} \tag{3}$$

The error measure is then scaled as:

$$MSE(t, a + by) = \frac{1}{N} \sum_{i=0}^{N} ((a + by) - t))^2 \tag{4}$$

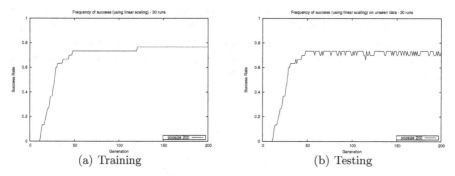

Fig. 2. Success rate plots on the quartic polynomial problem (population size 200) with linear scaling, using a training dataset of 21 points in the range [-1:0.1:1] (left) and a test dataset of 11 points in the range [1:0.1:1] (right)

Due to the relatively low computational cost of calculating a and b, the implementation of this technique has been recommended to GP researchers and some successful results can be found in the literature [10,11]. The technique is not without its apparent limitations, however. In one particular example of its application on a real-world problem [12], the findings suggest that the application of linear scaling leads to over-fitting, resulting in poor forecasting abilities[3].

3.1 Does Linear Scaling Over-Fit?

Using a population size of 200 as chosen from Section 2.1, let us examine the success rate plots when linear scaling is applied to quartic polynomial problem (with all other settings as per Table 1). The results from thirty independent runs are shown in Figure 2. What is clear from this test is that the application of linear scaling to this so-called *easy* problem configuration does not show any evidence of over-fitting. This can be seen in the right plot in the figure: the success rate does not drop off when applied to unseen data, which shows that the GP-derived models can forecast with a comparable amount of accuracy to that observed on the training set. Again, this sort of result is desirable; any experiences with GP-improvement techniques that to *not* follow this pattern should be called into question.

3.2 Comparing Performance

The use of sound statistics is of the utmost importance when comparing approaches such as these. As noted by Luke [4], many examples can be found in GP literature where conclusions regarding successful techniques are founded on questionable statistical practices.

---

[3] It should be pointed out here, however, that the study [12] led to this conclusion based on 20 independent runs. It has been noted elsewhere [4] that at least 30 runs should be carried out before any statistically sound conclusions may be made regarding performance comparisons.

A widely used test to determine whether there is a statistically significant difference between two sets of observations is the paired t-test. This test can be used to compare two methods for example, GP and *improved* GP with all else being equal (same random seeds, run settings and so on). The test is performed as follows; if X and Y are two sets of observations, let

$$\hat{X}_i = (X_i - \bar{X})$$

$$\hat{Y}_i = (Y_i - \bar{Y})$$

$$t = (\hat{X} - \hat{Y})\sqrt{\frac{n(n-1)}{\sum_{i=0}^{n}(\hat{X}_i - \hat{Y}_i)^2}}$$

This statistic has $n - 1$ degrees of freedom and this information is used to estimate the P-value, that is the estimated probability of rejecting the *null hypothesis*. To bring this into context, we could let X be the set of best-of-run values obtained from standard GP applied to a problem and let Y be the set of best-of-run values obtained using *improved* GP. The null hypothesis in this case would be that there is no difference between the two sets of observed values X and Y. The question that the study is asking is: *is there a significant difference in performance between the two methods?* This forms the *alternative hypothesis*, which states that there *is* a significant difference in performance between the methods. To claim statistical significance, the aim is to reject the null hypothesis.

Although the choice of threshold P-value varies across statistical studies, values of 0.05 and 0.01 are typical. In the former case, a P-value obtained that is less than 0.05 can be interpreted (in English) as: "a less than one in twenty chance of being wrong" (about rejecting the null hypothesis).

The only difficulty with the method described above is that the sets of observations X and Y are obtained from a single point in time: the end of the run. It would be statistically much stronger to state that no point over the duration of the run is there evidence that differences in performance are due to chance.

We can use confidence intervals to convey this information – this is statistically equivalent [13] to performing a paired t-test about each set of runs per generation. Using a sample of n (i.e the number of runs) observations of X, the 95% confidence limits are defined as:

$$\bar{X} \pm 1.96\frac{\sigma}{\sqrt{n}}$$

where $\bar{X}$ and σ refer to the mean and standard deviation of the observations. What this measure tells us is that based on the sample provided, we can be 95% certain that the (statistical) population lies inside these limits. To apply this in a GP-context, this information can easily be incorporated into best-of-run plots (using error-bars, where the bars represent the confidence intervals) showing the differences between two competing methods. If there is *no overlap* between the error-bar plots then it is safe to assume that a statistically significant difference in performance is observed.

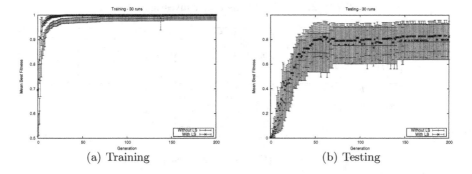

Fig. 3. Mean best fitness plots on the quartic polynomial problem (population size 200). Each plot compares the use of standard GP with and without linear scaling, using a training dataset of 21 points in the range [-1:0.1:1] (left) and a test dataset of 11 points in the range [1:0.1:1] (right). The error-bars shown represent 95% confidence intervals. Note that standardised fitness is used so all y-axis values fall between 0 and 1.

Armed with this technique, we can now assess the performance of linear scaling. A comparison is shown in Figure 3. The figure shows the error bar plots obtained from both training and testing. We can clearly see by inspection (overlapping error-bars) that no significant difference in performance is observed for this particular problem.

This simple demonstration does not provide enough information to conclude that GP with linear scaling does not outperform standard GP. This will be investigated by experimentation in the following section.

3.3 Widening the Scope

Using a set of test problems (most of which can be found in [1]) described in Table 2, let us now examine the performance differences between standard GP and standard GP with linear scaling on a selection of four problems. It has already been shown and also proven that training error is guaranteed to be reduced when linear scaling is applied, so the differences in training performance will not be shown here. Instead, we will examine the generalisation performance on unseen data. Unless otherwise stated, the run settings used for 30 runs of GP both with and without linear scaling are given in Table 3.

Results from the four test problems are summarised[4] in Table 4. In terms of statistical significance, the results are quite disappointing; it is found that in none of four trials examined did the application of linear scaling contribute to a significant performance improvement, which is quite dramatic considering the huge performance gains that would be expected if we were to examine training performance alone.

---

[4] Error-bar plots are not shown here due to space limitations.

Table 2. Test Problems under investigation

#	Description	Training set	Test set
1	$f(x) = x^4 + x^3 + x^2 + x$	401 cases: [0:0.01:4]	81 cases: [1:0.025:3]
2	$f(x) = 0.3x\sin(2\pi x)$	161 cases: [0:0.025:4]	55 cases: [0:0.073:4]
3	$f(x,y) = x^y$	400 random cases: $x,y \in$ [0:1]	100 random cases: $x,y \in$ [0:1]
4	$f(x,y) = 6\sin(x)\cos(y)$	400 random cases: $x,y \in$ [0:1]	100 random cases: $x,y \in$ [0:1]

Table 3. Run settings for the 4 test problems

Population Size	200
Generations	200
Crossover rate	0.7
Mutation rate	0.02
Tournament size	3
Function set	{+, -, *, %, sqr(), sin(), cos(), ln() }
Terminal set	{x}
Raw fitness	MSE(targets, predictions)
Standardised fitness	$\frac{1}{1+MSE}$
Hits criterion	Number of points where the GP program comes within 0.01 of the desired value

Table 4. Summary of results obtained when comparing the generalisation performance of standard GP versus standard GP with Linear Scaling over four test problems. Note that a "No" value means that the improvement to GP has not resulted in performance significantly *better* than standard GP. An example confidence interval illustrating whether or not overlap occurs is also given.

Problem	Significant	95% confidence interval	
		GP	GP with LS
1	No	0.8680 ± 0.1040	0.9587 ± 0.0649
2	No	0.9317 ± 0.0128	0.9505 ± 0.0146
3	No	0.9911 ± 0.0015	0.9957 ± 0.0042
4	No	0.9309 ± 0.0197	0.9644 ± 0.0303

3.4   Discussion

The results obtained from the four test problems would not support any assertion that the use of linear scaling should improve generalisation performance. Variations of the experiments above were also re-run by incorporating the idea of *potency* whereby the level of application of the scaled error measure was both increased and decreased over the duration of the run. In neither case was it possible to improve upon the results from Table 4.

4   Powers Combined: No Same Mates

As indicated previously, the use of Linear Scaling with GP has been shown to offer little performance gains when applied to unseen data. Given that the method has been proven to reduce the error on training data [9] it is unfortunate that the same gains are not evident when it comes to generalisation. However, the results from the experiments above should not necessarily lead us to the conclusion that GP with linear scaling cannot generalise. It could be the case that good generalisation is possible, however not necessarily with the application of Linear Scaling *alone*. It would appear that LS can cause the selection of individuals which match the training data too well – the population quickly becomes saturated with these individuals and the resulting generalisation is poor.

Gustafson *et al* [2] reported a simple GP improvement technique that forced the genetic operators to always use parents with different fitness values. The method was shown to provide a statistically significant performance improvement in the training phase when compared with standard GP. At the core of the work carried out by Gustafson *et al* was the fact that the probability of no change in solution quality increases with the similarity of solutions. By forcing the mating of dissimilar individuals, a significant improvement in solution quality was observed.

Table 5. Summary of results obtained when comparing the generalisation performance of standard GP versus standard GP with Linear Scaling and the No Same Mates technique over four test problems. Note that a "Yes" value means that the improvement to GP resulted in performance significantly *better* than standard GP. An example confidence interval illustrating whether or not overlap occurs is also given.

Problem	Significant	95% confidence interval	
		GP	GP with LS and NSM
1	Yes	0.8680 ± 0.1040	0.9907 ± 0.0047
2	No	0.9317 ± 0.0128	0.9525 ± 0.0112
3	Yes	0.9911 ± 0.0015	0.9975 ± 0.0010
4	Yes	0.9309 ± 0.0197	0.9809 ± 0.0043

In the experiments that follow, we combine this *no same mates* idea with the application of Linear Scaling to see if it results in better generalisation. Using the same experimental settings as given in Table 3, standard GP is compared to GP with Linear Scaling using No Same Mates (NSM). Results from the four test problems are summarised in Table 5.

It can be seen from the table that the incorporation of NSM with Linear Scaling results in a significant improvement over standard GP in three out of four cases. Recall that when standard GP was compared with GP and Linear Scaling, a significant improvement was not observed in any of the four cases. To test that this improvement was due to the *combination* of the two techniques (LS and NSM) and not just NSM alone the experiment was re-run without Linear

Scaling. In this case, no statistically significant improvement in generalisation performance was observed in any case. This is good news, both for Linear Scaling and for the No Same Mates technique. Although the combination of the two techniques has not been an overwhelming success, a step in the right direction is clearly observed. In the case of the second problem, no single configuration of settings was found to significantly outperform (or underperform) another. Furthermore, this problem presented the greatest difficulty for all GP variants tested, resulting in the lowest number of successful individuals discovered.

5   Summary, Conclusions and Future Work

This paper has noted the relative absence of generalisation performance analysis in Genetic Programming research in recent years. This deficit stands in stark contrast to the amount of reported successful GP-improvements stemming from their application to problems consisting solely of training data. The main motive for highlighting this apparent shortcoming has not been to de-value the contributions of GP researchers. Rather, we have aimed to strengthen the foundations by pointing to areas that appear to be lacking with the hope that future research may benefit.

In Section 2, a frequently used metric (Cumulative Frequency of Success) was used to display performance on unseen data for a simple GP problem. We suggest that the success rate plots resulting from unseen data should not differ drastically from those achieved on training data if the method under investigation is to be *relevant* in terms of generalisation. Later, in Section 3, we examined the unseen data success rate on a popular GP improvement technique – Linear Scaling. An initial experiment on the quartic polynomial showed the success rate on unseen data to hold up well compared to the training phase.

This hopeful start lead to a more detailed investigation of differences in performance using commonly used measures of statistical significance in the context of generalisation. The finding (based on a set of test problems) has been quite dramatic: Linear Scaling does not always generalise well – that is to say, not significantly better than standard GP. This is an extremely unfortunate outcome for a method that has been demonstrated to perform so well on training data.

Some valid questions then emerge. Does this mean that Linear Scaling is *bad*? If so, should the practice of using it be discouraged? Not necessarily. It is equally plausible that the technique is simply *too good*. It was hinted in Section 4 that it is causing too much pressure on the selection of individuals that match the training data as closely as possible, with poor generalisation as a consequence.

By combining the application of Linear Scaling with another simple improvement to GP that forces recombination between parents with different fitness values, we have found cases where better generalisation is possible. This second improvement technique appears to be a steadying counter-balance for the aggressive over-fitting characteristics displayed by Linear Scaling. While their symbiotic relationship has not resulted in perfect generalisation scores on all of the test problems studied, we are certainly experiencing a positive outcome which warrants further investigation. It is quite fortunate and somewhat pleasing that such a result has grown from the combination of methods which were intially criticised in Section 1.

Future research directions will therefore include the analysis of generalisation performance of other improvements to GP on a wider suite of problems (including those of prediction on real-world data). We also aim to increase the robustness of Linear Scaling generalisation by examining and refining different types of counter-balancing techniques.

References

1. Keijzer, M.: Improving Symbolic Regression with Interval Arithmetic and Linear Scaling. In: Ryan, C., Soule, T., Keijzer, M., Tsang, E., Poli, R., Costa, E. (eds.) EuroGP 2003. LNCS, vol. 2610, pp. 70–82. Springer, Heidelberg (2003)
2. Gustafson, S., Burke, E.K., Krasnogor, N.: On improving genetic programming for symbolic regression. In: Corne, D., Michalewicz, Z., Dorigo, M., Eiben, G., Fogel, D., Fonseca, C., Greenwood, G., Chen, T.K., Raidl, G., Zalzala, A., Lucas, S., Paechter, B., Willies, J., Guervos, J.J.M., Eberbach, E., McKay, B., Channon, A., Tiwari, A., Volkert, L.G., Ashlock, D., Schoenauer, M. (eds.) Proceedings of the 2005 IEEE Congress on Evolutionary Computation, Edinburgh, UK, vol. 1, pp. 912–919. IEEE Press, Los Alamitos (2005)
3. Koza, J.R.: Genetic Programming: On the Programming of Computers by Means of Natural Selection. MIT Press, Cambridge (1992)
4. Luke, S., Panait, L.: Is the perfect the enemy of the good? In: GECCO 2002: Proceedings of the Genetic and Evolutionary Computation Conference, pp. 820–828. Morgan Kaufmann Publishers Inc., San Francisco (2002)
5. Vanneschi, L., Tomassini, M., Collard, P., Clergue, M.: A survey of problem difficulty in genetic programming. In: Bandini, S., Manzoni, S. (eds.) AI*IA 2005. LNCS, vol. 3673, pp. 66–77. Springer, Heidelberg (2005)
6. Costelloe, D.: gpsr: Genetic Programming for Symbolic Regression, http://gpsr.sourceforge.net
7. Fernandez, T., Evett, M.: Numeric mutation as an improvement to symbolic regression in genetic programming. In: William Porto, V., Saravanan, N., Waagen, D., Eiben, A.E. (eds.) EP 1998. LNCS, vol. 1447, pp. 251–260. Springer, Heidelberg (1998)
8. Topchy, A., Punch, W.F.: Faster genetic programming based on local gradient search of numeric leaf values. In: Spector, L., Goodman, E.D., Wu, A., Langdon, W.B., Voigt, H.-M., Gen, M., Sen, S., Dorigo, M., Pezeshk, S., Garzon, M.H., Burke, E. (eds.) Proceedings of the Genetic and Evolutionary Computation Conference (GECCO 2001), pp. 155–162. Morgan Kaufmann, San Francisco (2001)
9. Keijzer, M.: Scaled symbolic regression. Genetic Programming and Evolvable Machines 5(3), 259–269 (2004)
10. Raja, A., Azad, R.M.A., Flanagan, C., Ryan, C.: Real-time, non-intrusive evaluation of voIP. In: Ebner, M., O'Neill, M., Ekárt, A., Vanneschi, L., Esparcia-Alcázar, A.I. (eds.) EuroGP 2007. LNCS, vol. 4445, pp. 217–228. Springer, Heidelberg (2007)

11. Majeed, H., Ryan, C.: A re-examination of a real world blood flow modeling problem using context-aware crossover. In: Riolo, R.L., Soule, T., Worzel, B. (eds.) Genetic Programming Theory and Practice IV, May 11-13, 2006. Genetic and Evolutionary Computation, ch.14, vol. 5. Springer, Ann Arbor (2006)
12. Valigiani, G., Fonlupt, C., Collet, P.: Analysis of GP improvement techniques over the real-world inverse problem of ocean color. In: Keijzer, M., O'Reilly, U.-M., Lucas, S., Costa, E., Soule, T. (eds.) EuroGP 2004. LNCS, vol. 3003, pp. 174–186. Springer, Heidelberg (2004)
13. Wasserman, L.: All of Statistics: A Concise Course in Statistical Inference (Springer Texts in Statistics). Springer, Heidelberg (2004)

Mining Evolving Learning Algorithms

András Joó

NCRG, Aston University, Birmingham, UK
jooam@aston.ac.uk

Abstract. This paper presents an empirical method to identify salient patterns in tree based Genetic Programming. By using an algorithm derived from tree mining techniques and measuring the destructiveness of replacing patterns, we are able to identify those patterns that are responsible for the increased fitness of good individuals. The method is demonstraded on the evolution of learning rules for binary perceptrons.

Keywords: Genetic programming, perceptron, learning rule, tree mining.

1 Introduction

Although schema theorems for Genetic Programming (GP) have been around for some time (for an overview please see [1]), they describe only the propagation of schemata during the evolutionary run, and give little insight into how to search for patterns that contribute positively to the individuals' fitness. Identifying salient patterns would help in understanding the final solution(s) and would give more insight in the dynamics of the evolutionary process.

The problem we picked – the evolution of learning rules for binary perceptrons – is a moderately hard GP problem. On the one hand it is complex enough to give interesting results, on the other hand teaching perceptrons has a solid theoretical background and the behavior of the optimal learning rule for certain cases is known [2].

The algorithms presented do not use domain specific knowledge and can be used without modification on any tree based GP problem.

2 Related Work

Previous work on the evolution of learning rules for neural networks (NN) [3, 4, 5, 6, 7, 8] focused on obtaining good learning rules for different NN architectures. Relatively little work has been done on understanding and describing the dynamics of the evolution of learning rules. An effort in this direction is the work of Neirotti et al. In [9] they tried to characterize good learning rules by identifying structures that are responsible for the increased fitness. They report on different stages of the evolutionary process that are linked to the spread of certain combinations of variables.

L. Vanneschi et al. (Eds.): EuroGP 2009, LNCS 5481, pp. 73–84, 2009.

Also related to our problem is the work of Tackett [10], who is credited with the first attempt of automatically identifying salient patterns in GP. He presents a half automatic, half manual data mining technique that can be used to identify emerging patterns. The ideas were demonstrated on relatively easy constructional and classification problems.

Other notable attempts in the analysis of patterns in GP include the introduction of the notion of *maximal schema* [11, 12], and following the distribution of building blocks and the tendency of sharing components using simplification and compression [13].

Our method is different in several aspects from the previously mentioned work: (1) it is completely automated; (2) depends on few parameters; (3) uses no domain knowledge; (4) the results are easy to interpret.

3   The Problem

We consider a supervised on-line learning scenario where a binary single layered perceptron tries to learn a linearly separable rule generated by a teacher of similar architecture, through a set of examples. The objective is to minimize the student's *generalization error*, i.e. the probability that an unseen example is misclassified.

3.1   GP Setting

Untyped, tree based, parallel and coarse grained GP [14, 15] is used for the evolution of learning rules. The set of terminals is $\mathbb{T} = \{x_\mu, w_\mu, y_\mu, d_\mu\}$, where x_μ denotes the input vector, w_μ the weight vector, y_μ the perceptron's output, d_μ the desired output. The set of functions is $\mathbb{F} = \{-, +, *, .*, ./, \log, \exp\}$. All these functions are protected, are defined both for vectors and scalars and work like the operators with the same name from MATLAB (e.g. $x.*w = (x_1 w_1, x_2 w_2, \ldots, x_n w_n)$).

The fitness of an individual is given by weighted average of the generalization error taken with negative sign, i.e.

$$F = -\sum_{\mu=1}^{p} \mu e_g(\mu), \tag{1}$$

where the generalization error for a training case μ is calculated as

$$e_g(\mu) = \langle \Theta(-y_\mu d_\mu) \rangle, \tag{2}$$

the average discrepancy between the desired and the actual output after presenting μ training examples to the learning algorithm, and $\Theta(\cdot)$ is the Heaviside function. The fact that the fitness of a learning rule can be measured only indirectly through a set of training cases, results in very noisy fitness values. To avoid the deception of the GP process this might cause, generalization errors are filtered using a Gaussian smoothing filter. The rest of the GP parameters is summarized in Table 1.

Table 1. GP parameter setting

Parameter	Value
Population type	parallel using the island model, elitist, generation-based
Number of islands	16
Migration	frequency 5, rate 5, random migration strategy grid topology
Number of individuals	300 per island
Number of generations	200
Selection	tournament selection, size 3
Tree height	initial 7, maximal 11
Perceptron	single layered, feed-forward, binary, 16 inputs
Training data	generated by a spherical perceptron
Number of training samples	65536

4   The Proposed Method

This section describes the performed steps to identify important patterns in the evolving learning rules.

4.1   Patterns and Related Definitions

We define a *pattern* as a valid GP subtree that might contain *wild characters* in any of its nodes. A wild character matches any single node of the same arity. We consider three types of wild (joker) characters: @ for terminals, # for unary functions and ## for binary functions.

A pattern P is said to be *contained* in a tree T, if T contains at least one subtree T' such that every node of T' matches the corresponding node in P. A pattern can be contained multiple times in a tree. For example the tree $((x + y) * (x - y))$ contains twice the pattern $(x \# \# y)$.

Our pattern definition is similar to the schema definition used by Poli and Langdon in [16] with the exception that our patterns are positionless thus can be present multiple times in a GP individual. We have chosen this kind of definition for two reasons: (1) the Langdon and Poli type schemas are too big (because of the rootedness constraint) and thus too numerous to be analyzed within reasonable time; (2) it is not obvious how to generate and mine for hyperschemas if we allowed them [1].

We define the *order ratio* of a pattern as the proportion of non-wild character nodes in the total number of nodes of a pattern. This measure is more suitable than the *order* (which is equal to the number of non-wild characters in a pattern) because it provides more flexibility when working with patterns of different size.

We define the *frequency* of a pattern in a population of GP trees as the number of occurrences of the pattern in the population divided by the size of the population. Because a pattern can be present multiple times in an individual, the frequency of a pattern can be larger than 1.

The *fitness* of a pattern is the average fitness of all individuals from the population which contain the given pattern at least once.

4.2   The Analysis Process

During the evolutionary run each GP process from the different islands maintains a binary dump file. After each generation the actual populations are appended to these files. The analysis itself is done offline, i.e. after the GP has ended, based on the dump files. The following steps are done for every dump file separately.

Preparation. As a preparatory step we simplify all dumped individuals. To get the very core of each expression we use a modified version of Equivalent Decision Simplification [17]. This ensures that we do not mine in the genotypic space, but in the phenotypic one.

Pattern Growing Algorithm (PGA). As a second step, using an algorithm derived from tree-mining techniques, PGA, (see Section 5), we identify those patterns from each generation whose frequency, height and order ratio are within some user defined ranges. The rationale behind these criteria is the following:

- We know from schema theorems that the number of important patterns increases exponentially (at least at the beginning). Therefore it seems reasonable to hypothesize that there is a certain frequency limit under which patterns are not important, and can be safely filtered out. As one shall see in Section 5, filtering by frequency is a crucial ingredient of the algorithm we use.
- We think that patterns of different height act differently during the GP, thus we separate the analysis of patterns of different height. Moreover interpreting the results is easier if the resulting patterns are divided into different height classes.
- Filtering by order ratio is necessary because low order ratio patterns carry low amounts of usable information. For example the pattern $(@\#\#(@\#\#d))$ is less informative than $(@ * (x * d))$ though both have the same structure and height.

Important patterns occur in fit individuals, thus it is safe to delete all patterns that are not present at least in one of the elite individuals of the evolutionary run.

Post-PGA. As a result of the previous step we have a set of candidate patterns. However, not all of these are equally important and as Tackett [10] notes they might fall in the category of "hitchhikers", i.e. the category of inert patterns that attach themselves to fit individuals, though their contribution to the individuals' fitness is negligible. Tackett used the maximal fitness (and maximal frequency on tie) during the evolutionary run as criterion to sort the patterns. However, the highly oscillatory nature of the pattern fitness makes this idea inadequate to distinguish salient and inert patterns.

Instead we use the following hypothesis. Deleting a salient pattern P from an individual should have destructive effects and should cause significant drop in the individual's fitness. On the other hand, the deletion of an inert pattern should be less destructive, the drop of fitness should not be that dramatic (the fitness might stay the same, or might even increase).

Though a pattern can not be deleted, each subtree it matches can be replaced with a constant expression. An ad-hoc value 1 was used for this purpose. The destructive effect of replacing a pattern P can be quantified by the following formula:

$$\Delta_P \equiv \frac{\langle F'_P \rangle - \langle F_P \rangle}{|\langle F_P \rangle|}, \tag{3}$$

where $\langle F_P \rangle$ is the average fitness of individuals containing P, and $\langle F'_P \rangle$ is the average fitness of individuals containing P after P has been replaced by a constant.

Instead of using the whole population to measure the destructive effects we use a smaller set of individuals $\mathbb{E}$ that has the following properties: (1) it contains each pattern to be tested; (2) each element of it has good fitness so that the potential destructive effects are visible; (3) it has relatively few elements for performance reasons. A handy choice for $\mathbb{E}$ is the set of elites that fulfill all these criteria.

Note that if Δ_P is positive then the presence of P is detrimental, thus P can be discarded; if it is a large negative then the presence of P is beneficial, and P is a salient pattern. However, if it is a small negative value then P is hitchhiker. To separate the salient patterns from the hitchhikers we apply a standard K-means algorithm to the Δ_P profile of the population.

5   Pattern Growing Algorithm

Pattern Growing Algorithm (PGA) is a tree-mining algorithm that incrementally generates all possible patterns until a given height and checks whether these patterns are above a previously given frequency threshold ε_f. Traditional tree mining techniques (for an overview please see [18]) grow patterns from below, i.e. by adding nodes to the bottom of the tree. Because GP trees are constrained by the fact that leaves can not have children, we grow our patterns from above, i.e. by adding roots, and inserting existing patterns underneath.

The algorithm performs the following steps:

1. $k = 1$
2. Gather those unary and binary functions and terminals whose frequency in the population are above the given threshold ε_f. Denote these sets by $\mathbb{U}$, $\mathbb{B}$ and $\mathbb{P}_k$ respectively. Append the unary joker function # to $\mathbb{U}$, the binary joker function ## to $\mathbb{B}$ and the terminal joker character @ to $\mathbb{P}_k$.
3. If $k = \varepsilon_d$ perform filtering by order ratio and height on $\mathbb{P}_k$ and return $\mathbb{P}_k$.
4. $k \leftarrow k + 1$

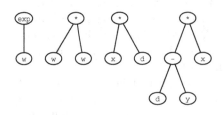

Index	Hash	Elements
0	0	$w{:}3$, $x{:}2$, $d{:}2$, $y{:}1$
1	210	$((d-y)*x){:}1$
2	2	$\exp(w){:}1$
...		
10	10	$(w*w){:}1$, $(x*d){:}1$, $(d-y){:}1$

Fig. 1. Skeleton hashing. The figure presents a population of 4 GP individuals and the corresponding hash table (indices and skeleton hashes are shown for the sake of intelligibility only). Subtrees are stored along with their number of occurrences in the population. If one wants to find the frequency of a pattern, let's say $(@ * @)$, one calculates the skeleton hash of this pattern (10), computes the index (10 mod 11 = 10) and sums the occurrence of those subtrees in row 10 that match the pattern (2). The sum is divided by the number of individuals, which gives 0.5.

5. Grow a set of patterns of height k by placing each element of $\mathbb{P}_{k-1}$ under each element of $\mathbb{U}$. Denote this set of patterns by $\mathbb{P}_k^U$. Filter out those elements from $\mathbb{P}_k^U$ whose frequency in the population is lower than ε_f.
6. Grow a set of patterns of height k by making root nodes each element of $\mathbb{B}$, picking left subtrees from $\mathbb{P}_{k-1}$ and right subtrees from $\bigcup\limits_{i=1}^{k-1} \mathbb{P}_i$. Denote this set of patterns by $\mathbb{P}_k^B$. Take each the element of $\mathbb{P}_k^B$, make a copy of them, swap left and right subtrees of the root, and put back into $\mathbb{P}_k^B$. Filter out those elements from $\mathbb{P}_k^B$ whose frequency in the population is lower than ε_f.
7. Denote $\mathbb{P}_k^B \bigcup \mathbb{P}_k^U$ by $\mathbb{P}_k$.
8. Jump to 3.

As one can observe after each growing step there is a filtering step by the minimum frequency ε_f. This uses the so called *a priori* property of pattern inclusion, namely if a pattern P has frequency f in the population than no pattern containing P can have larger frequency than f. Though, opposed to Tackett's method which has linear complexity, the computational complexity of PGA is exponential (something unavoidable because the inclusion of wildchar characters), the *a priori* property makes PGA usable even for large values of ε_d.

5.1   Skeleton Hashing

Since PGA generates a huge number of patterns whose frequencies have to be calculated, the naive approach of parsing all trees from the population for every single pattern is infeasible. To decrease the number of times when we have to check whether a (sub)tree contains a pattern or not we designed a hashing scheme that helps reducing the computational complexity of PGA.

The hash is used for two purposes. Firstly, to distribute the subtrees of the population into a hashtable H. Each element of H contains a list of pairs (S, n_S), where S is a subtree and n_S is the number of occurrences of S in the population.

The position where a subtree S is chained in H given by σ_S mod $|H|$, where σ_S is the hash value of S and $|H|$ is the size of the hashtable.

Secondly, the hash is used for the fast lookup of those subtrees from H that might match a pattern P, and for the calculation of the frequency of P. These are done performing the following steps:

1. P is hashed and the result is stored in σ_P.
2. σ_P is used to locate the index $i_P = \sigma_P$ mod $|H|$ where subtrees with the same hash value are chained.
3. The list of pairs (S, n_S) from position i_P are parsed, and the occurrence numbers are summed up whenever P matches S. Let us note this number by Σ_P.
4. The frequency of P is given by $\dfrac{\Sigma_P}{|\mathcal{P}|}$, where $|\mathcal{P}|$ is the size of the population.

Because both patterns and subtrees undergo the hashing process at some point, we need a special hashing function that can handle both. The *skeleton hash* function achieves this by encoding only the structure of a subtree or pattern. It is calculated using a preorder traversal without the node content information in the following way:

1. The skeleton hash σ_X of a pattern or subtree X is initialized with an empty string of bits.
2. X is traversed in a preorder way. Whenever the traversal enters a node, σ_X is appended a 1, if exits a node σ_X is appended a 0.
3. If X is of height 1 σ_X is set to 0.

An interesting property of the skeleton hashing is that if a pattern is extended (from above as we do in PGA) then its skeleton hash can be readily extended as well without recalculating the parts we already know. For example if we extended pattern $((x + @) * y)$ with skeleton hash 11010010_2 from above, let us say $\exp(((x + @) * y))$, then the skeleton hash would be $\mathbf{111010010}0_2$.

6   Experiments and Results

PGA depends on relatively few parameters. These are summarized in Table 2. The rationale behind the chosen values is the following. Since we concentrate on patterns of the same height at a time, the minimum and maximum pattern heights are set to the same value from the set $D = \{2, 3, 4, 5\}$. Patterns with height outside of D are either uninformative (i.e. patterns of height 1) or hard to interpret. The minimum value for a pattern's frequency was set to 0.4. This is an empirical value which for the most of the cases provided a good tradeoff between getting too many or too few patterns. Order ratio was required to be in the $[0.8, 1.0]$ interval, to ensure that only very informative patterns are considered for a detailed analysis.

Snippets from the list of patterns of different heights obtained after performing the steps described in Section 4.2 on dump file of a randomly chosen island (no. 5) are presented in Table 3. Patterns are sorted ascendingly according to their Δ_P values (i.e. more significant ones first).

Table 2. PGA parameter setting

Parameter	Value
minimum pattern height	taken from $D = \{2, 3, 4, 5\}$
maximum pattern height	taken from $D = \{2, 3, 4, 5\}$
minimum pattern frequency	0.4
maximum pattern frequency	∞
minimum order ratio	0.8
maximum order ratio	1

Table 3. Snippets of patterns from a randomly chosen island (no. 5) with their Δ_P values

Depth	Pattern	
2	(d-w): -3.68382	x-w): -2.50423
	(y-w): -3.52918	(y*d): -1.19829
	(d.*x): -3.17721	(x-16.000): -0.89499
	exp(w): -2.57351	(d-y): -0.58846
3	(exp(x)*#(w)): -3.69878	(#(x)*exp(w)): -3.68726
	(exp(x)*exp(@)): -3.69592	(exp(y)*#(w)): -3.41594
	(exp(x)*exp(w)): -3.69139	(exp(y)*exp(w)): -3.41097
	(exp(x)##exp(w)): -3.68907	(exp(y)*exp(@)): -3.40978
4	((y-w)*(exp(x)*exp(@))): -4.45944	((y-w)*(exp(x)##exp(w))): -4.4523
	((y##w)*(exp(x)*exp(w))): -4.45882	((y-@)*(exp(x)*exp(w))): -4.45143
	((y-w)*(exp(x)*exp(w))): -4.45621	((y-w)##(exp(x)*exp(w))): -4.45046
	((y-w)*(exp(x)*#(w))): -4.45524	((y-w)*(#(x)*exp(w))): -4.44451

6.1   Classification

To check if a threshold exists that separates salient patterns from inert ones we plot the histogram of the Δ_P values. Note that these values were obtained for a single, randomly chosen island. Because there is a certain genetic flow between the different islands, it is fair to use all the data from the different islands to draw the the cumulative histogram of Δ_P in order to get a more thorough picture (Figures 2, 3, 4 and 5).

Several things can be readily observed from these histograms. The first is that there are no patterns with positive Δ_P values, i.e. no harmful patterns are contained in the list of patterns produced by PGA. The second is that there are two clusters at the two ends of the distributions (more evident on the cumulative histograms). The last one is that the transition between the two phases is more or less continuous.

To separate the cumulative Δ_P data we used the standard K-means algorithm. This resulted in two clusters: one corresponding to the salient patterns (S) and another corresponding to the hitchhiker patterns (H). To get a maximal separation, the decision boundary was chosen to be

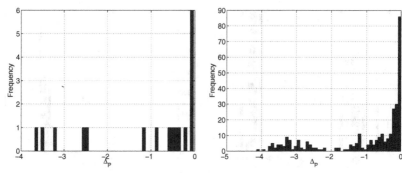

(a) Histogram of Δ_P values for patterns of height 2 from island 5

(b) Cumulative histogram of Δ_P values for patterns of height 2 from 16 islands

Fig. 2. Histograms of patterns of height 2

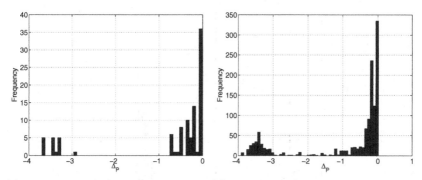

(a) Histogram of Δ_P values for patterns of height 3 from island 5

(b) Cumulative histogram of Δ_P values for patterns of height 3 from 16 islands

Fig. 3. Histograms of patterns of height 3

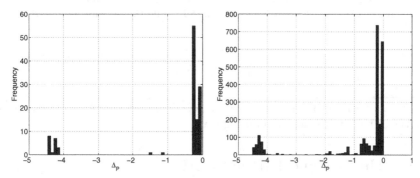

(a) Histogram of Δ_P values for patterns of height 4 from island 5

(b) Cumulative histogram of Δ_P values for patterns of height 4 from 16 islands

Fig. 4. Histograms of patterns of height 4

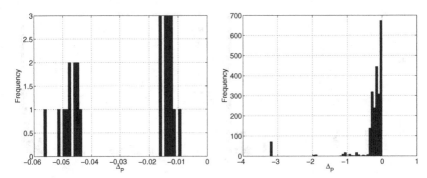

(a) Histogram of Δ_P values for patterns of height 5 from island 5

(b) Cumulative histogram of Δ_P values for patterns of height 5 from 16 islands

Fig. 5. Histograms of patterns of height 5

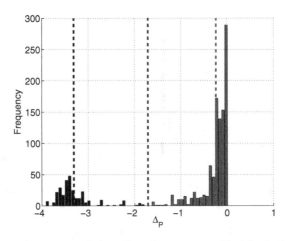

Fig. 6. Cumulative histogram of Δ_P values for patterns of height 3 after the clustering phase. Cluster centers are at -3.3089 and -0.2383, while the decision boundary occurs at -1.7109.

$$\varepsilon_P = \frac{\max(\mathbb{S}) + \min(\mathbb{H})}{2}. \tag{4}$$

Finally this decision boundary is used to separate patterns. An example, using patterns of height 3 is depicted on Figure 6.

7   Conclusions and Further Work

This paper presented a practical method to extract salient patterns from a GP process. This was achieved by generating all patterns that fulfill some frequency, order ratio and height criteria, then applying a series of filters. Though the algorithms were demonstrated on the evolution of learning rules they are general enough to be applied for any tree based GP problem.

The techniques presented here are work in progress, and there is still lot to do in this area. The following two problems are of main importance for us:

1. Checking whether there is a chronological ordering in the appearance of different patterns.
2. Checking the hypothesis whether patterns act differently or not during the different phases of the evolution.

References

[1] Langdon, W., Poli, R.: Foundations of Genetic Programming. Springer, Heidelberg (2002)
[2] Rattray, M., Saad, D.: Globally optimal on-line learning rules for multi-layer neural networks. Journal of Physics A: Mathematical and General 30(22), L771–L776 (1997)
[3] Chalmers, D.J.: The evolution of learning: An experiment in genetic connectionism. In: Touretsky, D.S., Elman, J.L., Sejnowski, T.J., Hinton, G.E. (eds.) Proceedings of the 1990 Connectionist Summer School, pp. 81–90. Morgan Kaufmann, San Francisco (1990)
[4] Baxter, J.: The evolution of learning algorithms for artificial neural networks. Complex Systems, 313–326 (1992)
[5] Bengio, S., Bengio, Y., Cloutier, J.: Use of genetic programming for the search of a new learning rule for neutral networks. In: Proceedings of the 1994 IEEE World Congress on Computational Intelligence, Orlando, Florida, USA, vol. 1, pp. 324–327. IEEE Press, Los Alamitos (1994)
[6] Radi, A., Poli, R.: Genetic programming can discover fast and general learning rules for neural networks. In: Genetic Programming 1998: Proceedings of the Third Annual Conference, University of Wisconsin, Madison, Wisconsin, USA, pp. 314–323. Morgan Kaufmann, San Francisco (1998)
[7] Radi, A., Poli, R.: Discovery of backpropagation learning rules using genetic programming. In: Proceedings of the 1998 IEEE World Congress on Computational Intelligence, Anchorage, vol. 1, pp. 371–375. IEEE Press, Los Alamitos (1998)
[8] Radi, A., Poli, R.: Discovering efficient learning rules for feedforward neural networks using genetic programming. Technical Report CSM-360, Department of Computer Science, University of Essex, Colchester, UK (January 2002)
[9] Neirotti, J.P., Caticha, N.: Dynamics of the evolution of learning algorithms by selection. Physical Review E 67, 041912 (2003)
[10] Tackett, W.A.: Mining the genetic program. IEEE Expert 10(3), 28–38 (1995)
[11] Smart, W., Andreae, P., Zhang, M.: Empirical analysis of GP tree-fragments. In: Ebner, M., O'Neill, M., Ekárt, A., Vanneschi, L., Esparcia-Alcázar, A.I. (eds.) EuroGP 2007. LNCS, vol. 4445, pp. 55–67. Springer, Heidelberg (2007)
[12] Smart, W., Zhang, M.: Empirical analysis of schemata in genetic programming using maximal schemata and MSG. In: 2008 IEEE World Congress on Computational Intelligence, Hong Kong. IEEE Computational Intelligence Society, IEEE Press, Los Alamitos (2008)
[13] McKay, R., Shin, J., Hoang, T.H., Nguyen, X.H., Mori, N.: Using compression to understand the distribution of building blocks in genetic programming populations. IEEE Congress on Evolutionary Computation, 2501–2508 (2007)

[14] Nowostawski, M., Poli, R.: Parallel genetic algorithm taxonomy. In: Nowostawski, M., Poli, R. (eds.) Third International Conference on Knowledge-Based Intelligent Information Engineering Systems, December 1999, pp. 88–92 (1999)

[15] Cantu-Paz, E.: Efficient and Accurate Parallel Genetic Algorithms (Genetic Algorithms and Evolutionary Computation 1), 1st edn. Springer, Heidelberg (2000)

[16] Poli, R., Langdon, W.B.: A new schema theory for genetic programming with one-point crossover and point mutation. In: Genetic Programming 1997: Proceedings of the Second Annual Conference, pp. 278–285. Morgan Kaufmann, San Francisco (1997)

[17] Mori, N., McKay, R., Nguyen, X.H., Essam, D.: Equivalent decision simplification: A new method for simplifying algebraic expressions in genetic programming. In: Proceedings of 11th Asia-Pacific Workshop on Intelligent and Evolutionary Systems (2007)

[18] Chi, Y., Muntz, R.R., Nijssen, S., Kok, J.N.: Frequent subtree mining - an overview. Fundamenta Informaticae 66(1-2), 161–198 (2004)

The Role of Population Size in Rate of Evolution in Genetic Programming

Ting Hu and Wolfgang Banzhaf

Department of Computer Science, Memorial University, St. John's, Canada
{tingh,banzhaf}@cs.mun.ca

Abstract. Population size is a critical parameter that affects the performance of an Evolutionary Computation model. A variable population size scheme is considered potentially beneficial to improve the quality of solutions and to accelerate fitness progression. In this contribution, we discuss the relationship between population size and the rate of evolution in Genetic Programming. We distinguish between the *rate of fitness progression* and the *rate of genetic substitutions*, which capture two different aspects of a GP evolutionary process. We suggest a new indicator for population size adjustment during an evolutionary process by measuring the rate of genetic substitutions. This provides a separate feedback channel for evolutionary process control, derived from concepts of population genetics. We observe that such a strategy can stabilize the rate of genetic substitutions and effectively accelerate fitness progression. A test with the Mackey-Glass time series prediction verifies our observations.

1 Introduction

The search process in *Evolutionary Computation* (EC) systems is a simultaneous process of exploration in parallel and exploitation in depth. Population size is a key factor to maintain population diversity in this process, and thus critical for the performance of an EC method. Recently, population size control has attracted increasing interests in the literature [13]. Population size control is non-trivial and challenging because it is often problem-specific and the interaction among various EC parameters is not completely clear yet. In general, the literature on population size control has two main foci: i) initializing a proper population size a priori and ii) adjusting population size during evolution. In this article, we focus on the latter. Population size adjustment is motivated by the observation that the required population size changes during different stages of evolution [2]. Such an adjustment is usually directed by a feedback loop. This feedback has been implemented through the controlled persistence of individuals or through the measurement of fitness progression, both of which are able to reflect the process of evolution to some extent. In Biology, particularly in the study on *population genetics*, population size has been intensely studied regarding its role in the rate of evolution [15,16]. It is generally accepted that the effect of population size on evolution acceleration is conditioned on the nature of selection at a particular moment rather than on a monotonic relationship. Typically,

L. Vanneschi et al. (Eds.): EuroGP 2009, LNCS 5481, pp. 85–96, 2009.

under *positive selection*, i.e., selection mostly accepting adaptive phenotypes, a large population is favorable for rapid evolution. In contrast, under *negative selection*, i.e., selection mostly eliminating deleterious phenotypes, a small population evolves faster. These two selection conditions can be reflected by the rate of genetic substitutions. Although this perspective is still under debate in the biological community, it is intriguing to study this relationship in EC systems. Thus, we investigate the interplay between population size and rate of evolution in a GP model and see how this originally biological notion translates to artificial systems.

2   Background and Motivation

In this section, we first briefly review studies on population size control in EC. Then, we discuss the relation between population size and the rate of evolution from both an EC and biological point of view.

2.1   Population Size Control

Research on population size control in EC originated from *Genetic Algorithms* (GAs). A number of theoretical contributions on analyzing population size initialization have been published based on Goldberg's seminal "components decomposition approach" and the notion of *building blocks* [8,9]. The essence of these works is that population size should be initialized according to the "complexity" of a specific problem. That is, for a more difficult problem, more diversity of a population is required, and thus a larger population size should be initialized.

Recently, it was realized that even for a given problem instance the required population size can vary during the process of evolution. Therefore, besides a good initial population size, some empirical methods on adjusting population size dynamically have been proposed. Arabas et al. [1] propose the *Genetic Algorithm with Variable Population Size* (GAVaPS) by regulating the *age* and *lifetime* of each individual. Population size fluctuates as a result of removing over-aged individuals and reproducing new ones. Back et al. [2] extend this lifetime notion in their *Adaptive Population size Genetic Algorithm* (APGA) to steady-state GAs. Fernandes and Rosa [5] propose the *Self-Regulated Population size Evolutionary Algorithm* (SRP-EA) to enhance APGA using a diversity-driven reproduction process. Alternatively, Harik and Lobo [10] introduce *parameter-less GA*, where several populations with different sizes evolve in parallel, starting with small population sizes. By inspecting the average fitness of these populations, less fit undersized populations are replaced by larger ones. Eiben et al. [4] suggest to use the pace of fitness improvements as the signal to control population size in *Population Resizing on Fitness Improvement GA* (PRoFIGA).

In GP, determination of an ideal population size is of even greater significance. As with the GA, population size in GP is relevant to its capabilities in finding the target and to its computational efficiency. In particular, it is related to the phenomenon of *bloat*. Poli et al. [17] establish that smaller populations bloat at a slower rate than larger ones. Downing [3] investigates population size in relation to evolvability in GP. Thus, adjusting population size dynamically benefits GP in various ways. A theoretical analysis on population size in GP based on building blocks is conducted by Sastry et al. [18]. The empirical population size adjustment schemes for GAs can also be applied to GP. Moreover, some GP-specific techniques have been employed as well. For example, Wedge and Kell [20] propose the *Genotype-Fitness Correlation* as a landscape metric to predict ideal population sizes in different systems. Tomassini et al. [19] design a dynamic population size GP using fitness progression as signal to delete over-sized and worse-fit individuals or to insert mutated best-fit individuals with certain criteria.

2.2   Population Size and Rate of Evolution

The term *rate of evolution* is understood somewhat differently in EC and in Biology. The goal of evolution is much more explicit in computational systems than in nature. It is to find the fittest solution to a given problem. Therefore, the rate of evolution in EC usually refers to how fast an EC population improves its fitness value, i.e., the *rate of fitness progression*. In this sense, as long as an EC population is able to find solutions, a small size is favored because of a small overhead. Thus, computer scientists have been seeking intelligent population size control schemes to strike a balance between exploration and exploitation during the search process.

In Biology, there is no explicit fitness function to measure, and the rate of evolution comes in a few different flavors, depending on the objects being examined, such as gene sequences, proteins, etc. Because of the infeasibility of defining fitness explicitly in natural systems, fitness is usually reflected by the likelihood that a relevant genetic change is selected. For instance, in molecular biology, the rate of evolution is usually measured by the rate that mutants are accepted and replace former alleles in genetic sequences. Biologists refer to this rate of evolution as *rate of genetic substitutions*. Here, we distinguish the rate of fitness progression and the rate of genetic substitutions to acknowledge the two aspects of rate of evolution. Fitness progression focuses on attaining the goal of the search, while rate of genetic substitutions concentrates on the dynamics of evolution and provides a different tool to study evolutionary processes.

Population geneticists have been identifying the effect of population size on the rate of molecular evolution, i.e., the rate of genetic substitutions. The *Nearly Neutral Theory* of molecular evolution by Ohta [15,16] is regarded as one of the most important principles for modern molecular evolution research. This theory defines both slightly deleterious and slightly advantageous mutations as *nearly neutral mutations*. It extends an earlier insight of Fisher [6] that the probability of a mutant being selected will be low if the outcome of this mutation on phenotypes is far-reaching. The theory predicts that there are a substantial number of nearly neutral mutations in molecular evolution. These nearly neutral mutations would be able to generate adaptation at a later time under certain genetic or

environmental changes. Thus, they play an important role for providing variation potential. In this theory, population size can influence the rate of molecular
evolution by its effects on the chance of accepting a nearly neutral genetic change
through statistical laws. That is, the chance of a random mutant being favored
by selection is slimmer within a larger population. When the majority of mutations are deleterious, a smaller population can evolve faster because more nearly
neutral changes are introduced into the population. In contrast, when mutations are mostly advantageous, evolution is faster in a larger population. When
most mutations are neutral, the rate of evolution is nearly independent of the
population size.

These predictions have been extensively tested and discussed in the biological community. Below are some examples where both increasing and decreasing
population size may accelerate evolution, depending on the link to environments.
Gillespie [7] examines the relation between population size and the rate of genetic
substitutions via computer simulation of several well-known biological models.
While verifying such a relation, he suggests the relation can sometimes be blurred
by the extreme complexity of natural systems. With population size fluctuation
being one of these complicating factors, he further emphasizes the necessity of
studying such fluctuation in population genetics. Woolfit and Bromham [21] compare genetic sequences between island endemic species and closely related main-
land species. This is an example where decreasing population size can accelerate
evolution. In a study on the recent rapid molecular evolution in human genomes,
Hawks et al. [11] hypothesize that the current dramatically growing human population may be the major driving force of new adaptive evolution. They indicate
that a growing population size can provide the potential for rapid adaptive innovations if a population is highly adaptive to the current environment.

3   Adjusting Population Size during Evolution in GP

We propose to apply the ideas from population genetics to a GP system. It
is generally assumed that the fitness of new offspring generated by mutation
or crossover in each generation approximately follows a Gaussian distribution.
According to the *Central Limit Theorem*, the average fitness among a larger
population has a smaller variance [15] (Fig. 1). We use the *selection favoring
degree* S_f, similar to the *selection coefficient* in Biology, to denote the degree of
new offspring being favored by selection. A positive value of S_f of an offspring
implies that it is likely to be accepted, and a negative value of S_f means that
it will most likely be rejected by selection. Further, if the majority of offspring
have positive S_f, selection is referred to as positive (Fig. 1 left). In contrast, the
selection is negative when $S_f < 0$ for most offspring (Fig. 1 right). From this
figure, we observe that, under positive selection, increasing population size can
accelerate the rate of genetic substitutions (left), while decreasing population
size can allow more genetic substitutions under negative selection (right).

Selection acting positively or negatively may vary during different stages of
an evolutionary process in GP, and the rate of genetic substitutions reflects this

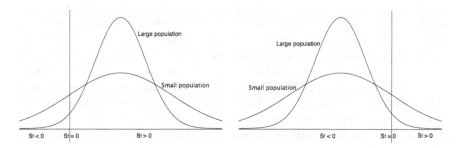

Fig. 1. Selection effects. Positive selection (left) and negative selection (right).

varying selection pressure. Therefore, adjusting population size according to the rate of genetic substitutions is expected to compensate the selection pressure. Thus, evolution can be guided away from stagnation and, further, can achieve better fitness progression. A measurement for the rate of genetic substitutions is reviewed briefly next, followed by our proposed population size adjustment approach.

3.1   Measuring the Rate of Genetic Substitutions

The *nonsynonymous to synonymous substitution ratio* k_a/k_s is a widely accepted measurement of the rate of genetic substitutions in molecular biology [23]. This metric has been applied by us [12] to measure the rate of genetic substitutions in GP. Not all genetic changes are effective in GP. Genetic changes that modify the encoded function can be regarded *nonsynonymous*, while others are regarded as *synonymous* changes. Specifically, in each generation the population is formed by surviving individuals under genetic variations and selection. Then, k_a measures the rate of nonsynonymous genetic substitutions of the process, and k_s measures the rate of neutral changes accepted. The rates k_a and k_s are obtained by count-ing the number of nonsynonymous (synonymous, resp.) genetic substitutions normalized by the total nonsynonymous (synonymous, resp.) sensitivities of in-dividuals to genetic operations in each generation. Therefore, k_a measures the adaptive evolutionary "distance" (adaptive substitution rate) and k_s practically provides the background "clock ticks" (neutral substitution rate). In Biology, the ratio k_a/k_s is regarded as a measure for the rate of genetic substitutions. On one hand, $k_a/k_s > 1$ implies that most genetic changes are adaptive and favored by selection. On the other hand, when $k_a/k_s < 1$, the majority of genetic changes are regarded deleterious and are diminished by selection.

Here, we slightly revise this k_a/k_s ratio to measure the rate of genetic substi-tutions. From one generation to the next, N_a denotes the number of all nonsyn-onymous genetic *changes* (i.e., attempted changes) and M_a counts the number of nonsynonymous genetic *substitutions* (i.e., accepted changes). A *sampled seman-tic test set* different from the training set is fed to an individual before and after a genetic change to test whether this change is nonsynonymous or synonymous. For instance, if a parent and its offspring have the same output for all sampled semantic test cases, the genetic change generating this offspring from the parent is regarded as synonymous. Otherwise, the change is considered nonsynonymous. Thus, $k_a = M_a/N_a$ measures the rate of nonsynonymous genetic substitutions. The synonymous substitution rate k_s can be defined similarly by dividing the number of synonymous genetic substitutions M_s by the number of attempted synonymous genetic changes N_s. The ratio k_a/k_s measures the rate of adaptive genetic substitutions relatively to a background silent genetic substitution rate. The case $k_a/k_s = 1$ corresponds to the situation where nonsynonymous genetic changes are selected at the same rate as neutral changes. When $k_a/k_s > 1$, selection is positive because a larger portion of nonsynonymous changes are favored by selection. In contrast, negative selection is reflected by the case $k_a/k_s < 1$.

3.2   Adaptive Population Size Approach

Next, an adaptive population size scheme is proposed using the k_a/k_s ratio defined above. We adopt truncation selection such that population size adjustment can be achieved easily without duplication or generating random individuals. Typically, at generation t, the current population produces an offspring population of the same size via genetic variations including crossover and mutation. Parents and offspring will compete through tournament selection to yield the next generation, and the population size $P_{size}(t + 1)$ will be adjusted according to the currently observed rate of genetic substitutions $(k_a/k_s)(t)$. Thus, the adaptive population size is regulated in each generation in an attempt to maintain a stable rate of genetic substitutions as follows:

- If $(k_a/k_s)(t) > 1$ (positive selection), we increase the population size proportional to the changes of the rate of genetic substitutions such that,

$$P_{size}(t + 1) = P_{size}(t) \times (1 + |(k_a/k_s)(t) - (k_a/k_s)(t - 1)|).$$

- If $(k_a/k_s)(t) = 1$ (neutral selection), we keep the same population size that,

$$P_{size}(t + 1) = P_{size}(t).$$

- If $(k_a/k_s)(t) < 1$ (negative selection), when $(k_a/k_s)(t)$ is increasing, we increase the population size to suppress further deleterious genetic substitutions, and when $(k_a/k_s)(t)$ is decreasing, we decrease the population size to encourage more genetic substitutions. That is,

$$P_{size}(t + 1) = P_{size}(t) \times (1 + ((k_a/k_s)(t) - (k_a/k_s)(t - 1))).$$

Note that we do not limit the population size by an upper bound. However, a lower bound of population size will be established in applications.

4   Experiments

We expect that dynamic adjustment of population size according to the measured k_a/k_s ratio can maintain a fairly stable rate of genetic substitutions. Since

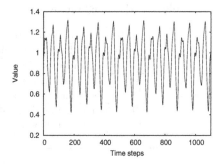

Fig. 2. Mackey-Glass time series

evolution is better guided this way, the performance of a GP system in fitness progression should improve as well. This is verified through simulations and comparisons to fixed-size populations. A tree structure is adopted to encode GP individuals here. The test benchmark will be introduced next, followed by our discussion of experimental results.

4.1 Test Suite

We use the Mackey-Glass chaotic time series prediction as our benchmark problem. Mackey-Glass time series prediction is a difficult modeling problem in machine learning and in GP [14]. It predicts future values of a time series based on history. GP is trained using these historical data. The series is generated using the following recursive function [22],

$$x_{t+1} = x_t - b \times x_t + \frac{a \times x_{t-\tau}}{1 + (x_{t-\tau})^{10}},$$

where $x_0 = 1$, and the parameters are set to

$$a = 0.2, b = 0.1, \tau = 17.$$

Fig. 2 depicts a plot of this function. We use the first 1,001 points as the training set. This problem is considered a difficult one because it does not have a closed-form solution. Thus, it will take GP a long time to converge.

Empirically, a population size between 500 and 1,000 is suitable for this type of problem. Here we conduct experiments in three scenarios. Two of the scenarios have fixed-size populations of 500 and 1,000, the third has an adaptive population size (APS) using our dynamic adjustment approach. It starts with an initial population size of 1,000 and a lower limit 300. The GP configuration is as shown in Table 1. Note that we adopt the number of function evaluations as a control metric although it operates in a generational mode. This allows a fair comparison among different scenarios.

Table 1. GP Configuration

Population size	500/1,000/APS(Adaptive Population Size)
Tree initialization	Ramped-Half-and-Half with limit 6
Function set	$+$, $-$, $\times$, and protective $/$
Terminal variable set	x_1, x_2, ..., and x_{17}, variable x_i denotes the previous point i time steps ago
Terminal constant set	Random ephemeral numbers equally distributed in $[-1, 1]$ with granularity 0.01
Crossover rate	0.9
Mutation rate	0.1
Maximum mutation subtree depth	4
Crossover and mutation method	Subtree replacement
Maximum tree depth	100
Training set	Points from 0 to 1,000 time steps
Fitness function	Root Mean Square (RMS) error
Selection	Tournament with size 4
Sampled sematic test set	20 cases such that $x_i^j = (i + j - 2) \times 0.04$ $(1 \leq i \leq 17, 1 \leq j \leq 20)$, $0 \leq x_i^j \leq 1.4$
Maximum number of evaluations	100,000

4.2 Results and Discussion

We have run GP 200 times for each scenario. Before we present statistical results, we look into the details of a "typical" execution of the APS scenario (Fig. 3). This particular population evolves for 147 generations before it reaches the 100,000 function evaluation number limit. In the figure, we plot (a) best fitness, (b) k_a/k_s ratio, (c) average tree size, and (d) population size over generations. We observe that the k_a/k_s ratio stays well under 1, which implies that selection is negative over time. This concurs with the general understanding that random genetic changes are mostly deleterious and with the property of the k_a/k_s ratio in Biology [23]. The population size drops from an initial 1,000 to approximately 700 after 20 generations, and stabilizes at 650∼700 afterwards. Also notice that, as evolution progresses, the best fitness improves but at a slower rate and average tree size increases, which is expected for tree GP. Normally, bloat would slow down the rate of genetic substitutions due to the introduction of redundant substructures. However, this is successfully alleviated by adjusting the population size to stimulate evolution so that there is a steady k_a/k_s ratio. This is verified by our next close study of the interaction between the k_a/k_s ratio and population size.

In Fig. 4, we depict the response of the k_a/k_s ratio change to population size adjustment, derived from the data recorded from 200 runs of the APS scheme. Using the recorded population size and the k_a/k_s ratio of each run, we quantify the correlation between the way they change over generations using a sequence of 1's and -1's. For a generation compared to the previous, if both population size and the k_a/k_s ratio increase, or if both population size and the k_a/k_s ratio

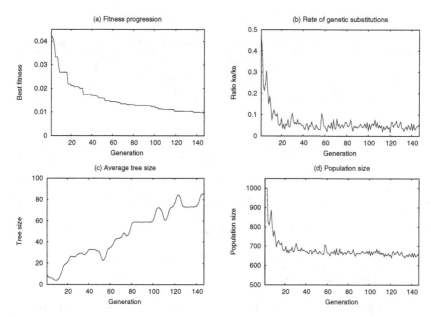

Fig. 3. An example run with adaptive population size according to the rate of genetic substitutions k_a/k_s ratio

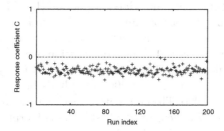

Fig. 4. Correlations between population size and k_a/k_s ratio

decrease, we have a 1. Otherwise, we have a -1. Therefore, the number of 1's in the produced sequence records the number of occurrences where the change of the k_a/k_s ratio positively correlates to that of the population size; while -1's indicate negative correlation. We define the *response coefficient C* as $C = (2n-l)/l$, where n is the number of 1s in the sequence and l is the sequence length. Thus, if a run has $C = 0$, its k_a/k_s ratio is independent of the change of population size. Alternatively, a positive value of C indicates a positive correlation between the changes of the k_a/k_s ratio and that of the population size. On the other hand, a negative value of C suggests a negative correlation. The figure presents the coefficients for all the 200 simulation runs. Clearly, they are all well below the level of 0. This is indeed our intention of dynamically adjusting population size to stabilize the rate of genetic substitution as stated in Section 3.2.

Table 2. Mean Best Fitness ($\times 10^{-3}$) Comparison

	Mean Best Fitness	Standard Deviation	Median	95% Confidence Interval
Psize 500	13.980	6.941	12.36	[13.153, 14.806]
Psize 1,000	12.902	5.406	12.98	[12.195, 13.610]
APS	12.053	4.750	11.56	[11.226, 12.888]

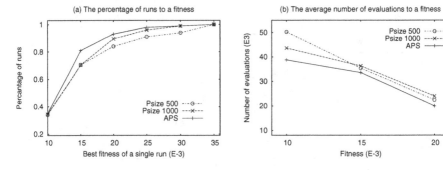

Fig. 5. Fitness progression with fixed population sizes (Psize 500, Psize 1000) and adaptive population size (APS)

Our next observation is that fitness progression can also be accelerated by our population size adjustment scheme. Here, we adopt the three most commonly used metrics to measure the performance of an EC model. They are mean best fitness, success rate, and average number of evaluations to a solution. Table 2 presents the mean best fitness and standard deviation over 200 runs for the three scenarios of fixed population size 500, 1,000, and APS. Clearly, APS achieves an even better fitness than maintaining 1,000 individuals but its population size is mostly between 500 and 1,000. The cumulative success rates of these three groups are depicted in Fig. 5(a). We focus on the best fitness of a run once it terminates. The figure plots the percentage of the total 200 runs of each scenario that yield a better fitness than a given threshold between 0.01 and 0.035. Apparently, APS has the highest percentage for all of the cases, which indicates its superiority over fixed population size strategies. Fig. 5(b) reveals a dual measure for the three scenarios. In this chart, we compare the average number of evaluations needed for a simulation run to achieve fitness levels of 0.01, 0.15, and 0.02. For a given fitness level, APS always incurs less computation overhead, and this difference becomes greater as the fitness requirement gets higher.

5   Conclusion and Future Work

In this article, we investigated the role of population size for rate of evolution in a GP system. We distinguished between rate of genetic substitutions and rate of fitness progression, which describe two aspects of an evolutionary process. The measurement of rate of genetic substitutions was revised and adopted. We transferred an idea from population genetics that population size has varying effects on rate of genetic substitutions under differing selection regimes. We proposed and tested that dynamically adjusting population size can effectively stabilize the rate of genetic substitutions even during late stages of an evolutionary process. It has been further verified that this strategy can also successfully accelerate the rate of fitness progression.

Our observations suggest some future research. First, we would like to implement this idea in GP with different representation structures, e.g., Linear GP and Graph GP, to see if similar conclusions hold in these GP systems. Applications to other branches of Evolutionary Computation, e.g., GAs, are also expected. Second, we will investigate this adaptive population size scheme in changing environments, since dynamic environments may affect the regime of selection more dramatically than static environments. Third, we propose to apply our method to other benchmarks in order to test how it fares compared to other well-known adaptive population size strategies.

Acknowledgements

We thank NSERC for support under Discovery Grant RGPIN 283304-07.

References

1. Arabas, J., Michalewicz, Z., Mulawka, J.: GAVaPS – a genetic algorithm with varying population size. In: Proceedings of IEEE Congress on Evolutionary Computation (CEC 1994), pp. 73–78. IEEE Press, Los Alamitos (1994)
2. Back, T., Eiben, A.E., van der Vaart, N.A.L.: An empirical study on GAs "without parameters". In: Deb, K., Rudolph, G., Lutton, E., Merelo, J.J., Schoenauer, M., Schwefel, H.-P., Yao, X. (eds.) PPSN 2000. LNCS, vol. 1917, pp. 315–324. Springer, Heidelberg (2000)
3. Downing, R.M.: On population size and neutrality: Facilitating the evolution of evolvability. In: Ebner, M., O'Neill, M., Ekárt, A., Vanneschi, L., Esparcia-Alcázar, A.I. (eds.) EuroGP 2007. LNCS, vol. 4445, pp. 181–192. Springer, Heidelberg (2007)
4. Eiben, A.E., Marchiori, E., Valkó, V.A.: Evolutionary algorithms with on-the-fly population size adjustment. In: Yao, X., Burke, E.K., Lozano, J.A., Smith, J., Merelo-Guervós, J.J., Bullinaria, J.A., Rowe, J.E., Tiño, P., Kabán, A., Schwefel, H.-P. (eds.) PPSN 2004. LNCS, vol. 3242, pp. 41–50. Springer, Heidelberg (2004)
5. Fernandes, C., Rosa, A.: Self-regulated population size in evolutionary algorithms. In: Runarsson, T.P., Beyer, H.-G., Burke, E.K., Merelo-Guervós, J.J., Whitley, L.D., Yao, X. (eds.) PPSN 2006. LNCS, vol. 4193, pp. 920–929. Springer, Heidelberg (2006)
6. Fisher, R.A.: Genetical Theory of Natural Selection. Clarendon, Oxford (1930)
7. Gillespie, J.H.: The role of population size in molecular evolution. Theoretical Population Biology 55(2), 145–156 (1999)
8. Goldberg, D.E.: Sizing populations for serial and parallel genetic algorithms. In: Proceedings of the 3rd International Conference on Genetic Algorithms, pp. 70–79. Morgan Kaufmann, San Francisco (1989)

9. Goldberg, D.E., Deb, K., Clark, J.H.: Genetic algorithms, noise, and the sizing of populations. Complex Systems 6(4), 333–362 (1992)
10. Harik, G.R., Lobo, F.G.: A parameter-less genetic algorithm. In: Proceedings of the Genetic and Evolutionary Computation Conference (GECCO 1999), pp. 258–267. Morgan Kaufmann, San Francisco (1999)
11. Hawks, J., Wang, E.T., Cochran, G.M., Harpending, H.C., Moyzis, R.K.: Recent acceleration of human adaptive evolution. Proceedings of the National Academy of Sciences 104(52), 20753–20758 (2007)
12. Hu, T., Banzhaf, W.: Nonsynonymous to synonymous substitution ratio k_a/k_s: Measurement for rate of evolution in evolutionary computation. In: Rudolph, G., Jansen, T., Lucas, S., Poloni, C., Beume, N. (eds.) PPSN 2008. LNCS, vol. 5199, pp. 448–457. Springer, Heidelberg (2008)
13. Lobo, F.G., Lima, C.F.: A review of adaptive population sizing schemes in genetic algorithms. In: Proceedings of the 2005 workshops on Genetic and Evolutionary Computation (GECCO 2005), pp. 228–234. ACM, New York (2005)
14. Oakley, H.: Two scientific applications of genetic programming: stack filters and non-linear equation fitting to chaotic data. In: Kinnear, K.L., Kinnear Jr., K.E. (eds.) Advances in Genetic Programming, pp. 369–389. MIT Press, Cambridge (1994)
15. Ohta, T.: Population size and rate of evolution. Journal of Molecular Evolution 1(4), 305–314 (1972)
16. Ohta, T.: The nearly neutral theory of molecular evolution. Annual Reviews in Ecology and Systematics 23(1), 263–286 (1992)
17. Poli, R., McPhee, N.F., Vanneschi, L.: The impact of population size on code growth in GP: analysis and empirical validation. In: Proceedings of the 10th Annual Conference on Genetic and Evolutionary Computation (GECCO 2008), pp. 1275–1282. ACM, New York (2008)
18. Sastry, K., O'Reilly, U.-M., Goldberg, D.E.: Population sizing for Genetic Programming based upon decision making. In: O'Reilly, U.-M., Yu, T., Riolo, R., Worzel, B. (eds.) Genetic Programming Theory and Practice II, pp. 49–66. Kluwer Academic Publishers, Dordrecht (2004)
19. Tomassini, M., Vanneschi, L., Cuendet, J.: A new technique for dynamic size populations in genetic programming. In: Proceedings of IEEE Congress on Evolutionary Computation (CEC 2004), pp. 486–493. IEEE Press, Los Alamitos (2004)
20. Wedge, D.C., Kell, D.B.: Rapid prediction of optimum population size in genetic programming using a novel genotype - fitness correlation. In: Proceedings of the 10th Annual Conference on Genetic and Evolutionary Computation (GECCO 2008), pp. 1315–1322. ACM, New York (2008)
21. Woolfit, M., Bromham, L.: Population size and molecular evolution on islands. Proceedings of The Royal Society B 272(1578), 2277–2282 (2005)
22. Working Group on Data Modeling Benchmarks. Created on July 20, 2002, http://neural.cs.nthu.edu.tw/jang/benchmark/
23. Yang, Z., Bielawski, J.P.: Statistical methods for detecting molecular adaptation. Trends in Ecology and Evolution 15(12), 496–503 (2000)

Genetic Programming Crossover: Does It Cross over?

Colin G. Johnson

Computing Laboratory
University of Kent
Canterbury, Kent, CT2 7NF
England
C.G.Johnson@kent.ac.uk

Abstract. One justification for the use of crossover operators in Genetic Programming is that the crossover of program syntax gives rise to the crossover of information at the semantic level. In particular, a fitness-increasing crossover is presumed to act by combining fitness-contributing components of both parents. In this paper we investigate a particular interpretation of this hypothesis via an experimental study of 70 GP runs, in which we categorise each crossover event by its fitness properties and the information that contributes most strongly to those fitness properties. Some tentative evidence in support of the above hypothesis is extracted from this categorisation.

1 Introduction

Descriptions of genetic programming (GP) typically include a discussion of why the crossover and mutation operators are used (see e.g. [1,12]). Such discussion goes beyond saying that such operators are used because they are analogies with biological processes: it gives a justification for why such operators are considered to be effective in carrying out search.

Typically the explanation given for the crossover operator is something akin to this. A crossover operator takes components of two successful parent solutions and produces a new child solution which combines features of the two parents. If "bad" features are combined, then the child will be selected out; however if "good" features are combined, then the child will have higher fitness than either of its parents and will have a high chance of being chosen by a selection algorithm

There is lots of experimental work which shows that GP is effective on particular problems (see e.g. the overview in [1]), and that various versions of these operators work better or worse in ways which are broadly consistent with the above hypothesis. There is also theoretical work which provides some arguments that GP works as hypothesised (see [6] for an overview). However there seems to be little work which actually looks in detail at the results of operators in GP runs.

The aim of this paper is therefore to do this. We will define more carefully what is meant by "combining features", and then check experimentally whether

L. Vanneschi et al. (Eds.): EuroGP 2009, LNCS 5481, pp. 97–108, 2009.
© Springer-Verlag Berlin Heidelberg 2009

successful crossovers do result from the combination of features. In particular, the paper focuses on the results of crossover on the results of executing the program, rather than just on the program code.

The remainder of the paper is structured as follows. The next section reviews existing work in this area. A formalization of the ideas above is given, and an experiment proposed. The results of runs of this experiment on seven different test problems is given, and the results discussed. Finally, some weaknesses in the definitions are explored and suggestions for future work given.

2   Background

This section reviews some relevant background on crossover. At the core of the issues of interest is the distinction between the effect of crossover on syntax and semantics. GP defines a biologically-inspired syntax for crossover, and we expect that some notion of crossover in the semantics of the solution. It can be noted that biology does the same; operators, after all, act directly on genotypes not on phenotypes.

When the representation is simple, there is no conflict between the syntax and the semantics. Consider, for example, a GA for timetables which uses a grid of times and days as its representation for a particular room. Define (uniform) crossover by taking each grid square in turn and taking at random the value occupied by the first or the second parent; and define mutation by choosing a grid square and choosing another value. Clearly crossover and mutation are doing basically what is described in the above loose descriptions. There is no complex syntax→semantics map in which this could be "lost".

By contrast, in GP (and other more complicated representations, e.g. those with epistatic interactions), this syntactic crossover gives no guarantee of consequent semantic crossover. Two parents can both have regions of fitness, and there can be no way to carry out a crossover between them so that the child contains both regions. By contrast, in a simple GA representation, the fitness of a particular part of the chromosome makes a particular contribution to aggregate fitness, regardless of other parts. A sub-program that creates high fitness for certain fitness cases in one program can fail to do so when placed in a different context.

A number of authors have explored this connection, theoretically or experimentally. Koza [5] discusses informal analogies with the *schema theory* for genetic algorithms. This has been formalised further by Poli and Langdon [11,6]. In particular, in [11], they carry out experiments to show the effect of different crossover operators in terms of the amount of (genotypic) material exchanged.

A number of papers have explored variants on crossover. These are interesting from the point of view of this paper not in terms of the results as such, but from the evaluation methodology. For example, O'Reilly and Oppacher [9] explore the idea of using hill-climbing to choose between a number of crossover-based moves; Ryan [13] explores the idea of *disassortative mating*, where the two parents are chosen using different criteria; and, Beadle and Johnson [2] explore the idea of preventing crossovers where the semantics of the child are the same as one of the parents. In all of these the influence of the variant of crossover is evaluated indirectly, via its effect on fitness performance. Whilst this is valuable in terms of evaluating the *quality* of the operator, and therefore whether it is sensible to use it, it helps less with *understanding* whether the operator works as designed.

Our experiments below are one version of an attempt to do this at the phenotypic level. Our methods are similar to those carried out by Nordin, Banzhaf and Francone [7,8], in that they are statistical analyses of all the crossovers in GP runs. In those papers, the aim is to study the fitness change in crossover; their conclusions are that most crossovers have one of two outcomes: a large reduction in fitness, or zero fitness change. In our work we similarly analyse a large collection of crossovers, but the aim of the analysis is different.

3 Does Crossover Cross over (in Practice)?

The experimental results in the work discussed above are typically characterised by a population-level approach to analysis: make a modification to the operator, and see what the end result is. In the work below we take a different approach, by looking at each time an operator is used and comparing the parents with the child. Some informal analysis of this kind has been carried out by e.g. Koza [5]. However this work simply takes a small number of particular examples of parent/child triples and analyses them by hand.

The approach taken below is to gather a dataset consisting of certain measurements made concerning every crossover that happens during a GP run. This dataset is then analysed to examine hypotheses about why crossover works.

Informally stated, the hypothesis that we are examining is as follows:

Crossover hypothesis. Crossover works by combining features of the two parents in the child. More precisely, in a successful crossover features in the input-output behaviour of the program that contribute most strongly to the fitness in each parent will contribute to regions of high fitness in the child

Exactly what is meant by "features" is left vague, and a number of possible investigations are possible depending on the definition chosed. In the examples below we will formalise this in one particular away (defining features as subsets of the fitness cases), and then explore it in the context of symbolic regression.

4 Experimental Details

This experiment looks at the crossovers in a GP run, and studies whether crossover leads to the combination of the strongest features in the two parent programs. More formally, we look at whether the fitness cases that have the strongest fitness in the child program overlap with those in the parent programs.

4.1   Methods

This experiment consists of the analysis of crossovers in a number of GP runs. The GP system used was the TinyGP system ([12,10]). No modification has been made to the GP code as such. However, a number of sections have been added to the program to extract information about crossovers, and to analyse that information. The modified version of the code can be downloaded from http://www.cs.kent.ac.uk/people/staff/cgj/software.html.

For a particular crossover, let $p1$ and $p2$ represent the parents and c the child. Define f_{p1} and f_{p2} respectively to be the fitnesses of the parents, and f_c to be the fitness of the child. A crossover will be termed *positive* if $(f_c > f_{p1})\&(f_c > f_{p2})$, *negative* if $(f_c < f_{p1})\&(f_c < f_{p2})$, and *neutral* otherwise. A crossover is said to be *syntactically-identical* if the program text of $p1$ and $p2$ are the same, and *fitness-case-identical* if $p1$ and $p2$ both produce the same values on all fitness cases. Clearly all syntactically-identical crossovers are also fitness-case-identical.

The aim of the analysis is to investigate whether the strongest fitness contributors in each parent both contribute to fitness in the child. This is measured by considering the contribution to fitness of the fitness-cases in the parent and the child; a crossover will be regarded as crossing over fitness if the strongest contributors to fitness in both parents contribute towards the fittest parts of the child. An idea of this is given in figure 1. This is formalised as follows. Let the number of fitness cases be n. Take a proportion b (this will be 20% in this experiment) and form a set of those fitness cases that rank in the top $b \times n$, for each of the parents and the child: call these sets s_{p1}, s_{p2} and s_c.

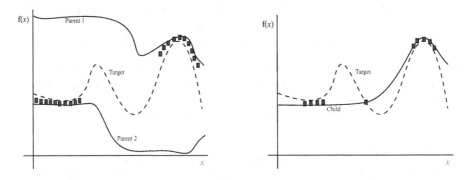

Fig. 1. An example of a good crossover: the best parts of the parents are also found in the best parts of the child

Finally, calculate the amount of intersection between the sets representing the fittest regions for the two parents and children. Let $n_1 = |s_{p1} \cap s_c|$ and $n_2 = |s_{p2} \cap s_c|$. Define a lower threshold ℓ and an upper threshold u; in this experiment $\ell = b \times n \times 0.1$ and $u = b \times n \times 0.4$ (these were chosen following some initial informal experimentation; an future investigation will examine changes to these parameter values).

Table 1. The functions used for symbolic regression in the experiments

Ref. Number	Function	Range	Number of Steps
1.	$f(x) = 0.3x \sin 2\pi x$	$[-2, 2]$	101
2.	$f(x) = x^3 \exp -x \cos x \sin x$	$[0, 10]$	101
	$(\sin x^2 \cos x - 1)$		
3.	$f(x) = \sum_{i=1}^{x} 1/i$	$[1, 50]$	50
4.	$f(x) = \log x$	$[1, 100]$	100
5.	$f(x) = \sqrt{x}$	$[0, 100]$	101
6.	$f(x) = \text{arcsinh}(x)$	$[0, 100]$	101
7.	$f(x) = \sin x$	$[0, 6.2]$	63

Table 2. Parameters used for the GP algorithm in the experiments

Parameter	Value
Population Size	100000
Crossover Probability	0.9
Probability of Mutation per Node	0.05
Number of Generations	100× population size
Selection type	Size-2 tournament selection
Replacement Strategy	steady-state
Problem Type	1-variable symbolic regression
Function Set	$+, -, \times, \div$ (protected)
Terminal Set	x, constants (generated randomly in $[-5, 5]$)

Now define a 0-*crosser* to be a crossover where $(n_1 \leq \ell)\&(n_2 \leq \ell)$, i.e. where regions that contributed most strongly to the fitness of both parents do not contribute strongly to the fitness of the child. Similarly, a 1-crosser is defined to be a crossover where $((n_1 \leq \ell)\&(n_2 \geq u))|((n_1 \geq u)\&(n_2 \leq \ell))$, i.e. one of the fitness-contributing regions (i.e. subsets of fitness cases) from the parent is important in creating the fitness of the child, and one is not. Finally, a 2-crosser is a crossover where $(n_1 > u)\&(n_2 > u)$, i.e. both of the regions that caused the parents to be fit also cause the child to be fit. Note that this is not an exhaustive partition of the space of all possible crossovers; we ignore the middle-ground where there is a small amount of intersection between both. The results below will distinguish between those 2-crossers where $s_{p1} = s_{p2} = s_c$ and those where there are differences. This is to check that genuine information exchange is going on, and that we are not just counting as 2-crossers only those crossovers which are effectively identical at the relevant sample points. These are referred to as *2-crossers(S)* in the case where the values at all the sample points are identical, and *2-crossers(D)* in the case where at least one is different.

We carry out these experiments on a set of one-dimensional symbolic regression problems. These are mostly the one-dimensional problems used by Keijzer [3], which in turn are taken from [4,14,15]. These functions are given in Table 1.

For each function, 10 runs of a GP system with parameters set as in Table 2 were carried out. These are the default values in the TinyGP system. Despite this being

Table 3. Various measures from the crossovers (functions 1 and 2)

Function 1 $f(x) = 0.3x \sin 2\pi x$

Total # of Crossovers:	989711
# of Syntactically-Identical Crossovers:	97111
# of Fitness-Case-Identical Crossovers:	173643

Total # of positive crossovers	22532
Total # of neutral crossovers	257685
Total # of negative crossovers	709502

Counts:

	Negative	Neutral	Positive
0-Crossers	37615	1466	684
1-Crossers	5696	6637	1786
2-Crossers(S)	122797	26824	1398
2-Crossers(D)	426317	202962	16845

Proportions:

	Negative	Neutral	Positive
0-Crossers	0.053	0.006	0.033
1-Crossers	0.008	0.026	0.085
2-Crossers(S)	0.171	0.103	0.066
2-Crossers(D)	0.602	0.789	0.733

Function 2 $f(x) = x^3 \exp{-x} \cos x \sin x(\sin x^2 \cos x - 1)$

Total # of Crossovers:	989963
# of Syntactically-Identical Crossovers:	103871
# of Fitness-Case-Identical Crossovers:	129642

Total # of positive crossovers	23820
Total # of neutral crossovers	221973
Total # of negative crossovers	744170

Counts:

	Negative	Neutral	Positive
0-Crossers	42913	1598	697
1-Crossers	7046	6975	1940
2-Crossers(S)	101285	15768	1630
2-Crossers(D)	433855	170407	17022

Proportions:

	Negative	Neutral	Positive
0-Crossers	0.058	0.007	0.032
1-Crossers	0.009	0.031	0.087
2-Crossers(S)	0.135	0.072	0.072
2-Crossers(D)	0.583	0.768	0.700

Table 4. Various measures from the crossovers (functions 3 and 4)

Function 3 $f(x) = \sum_{i=1}^{x} 1/i$

Total # of Crossovers:	990308
# of Syntactically-Identical Crossovers:	132284
# of Fitness-Case-Identical Crossovers:	148843
Total # of positive crossovers	24543
Total # of neutral crossovers	226276
Total # of negative crossovers	739489

Counts:

	Negative	Neutral	Positive
0-Crossers	297101	14592	2171
1-Crossers	109637	80765	7365
2-Crossers(S)	30797	20490	2373
2-Crossers(D)	80522	63577	8385

Proportions:

	Negative	Neutral	Positive
0-Crossers	0.400	0.066	0.094
1-Crossers	0.149	0.360	0.299
2-Crossers(S)	0.042	0.092	0.100
2-Crossers(D)	0.110	0.275	0.337

Function 4 $f(x) = \log x$

Total # of Crossovers:	989741
# of Syntactically-Identical Crossovers:	118612
# of Fitness-Case-Identical Crossovers:	139199
Total # of positive crossovers	24339
Total # of neutral crossovers	210944
Total # of negative crossovers	754457

Counts:

	Negative	Neutral	Positive
0-Crossers	302442	15531	2594
1-Crossers	125091	86635	6962
2-Crossers(S)	23559	16535	2766
2-Crossers(D)	54276	43396	6987

Proportions:

	Negative	Neutral	Positive
0-Crossers	0.400	0.075	0.115
1-Crossers	0.166	0.411	0.285
2-Crossers(S)	0.031	0.079	0.110
2-Crossers(D)	0.072	0.204	0.282

Table 5. Various measures from the crossovers (functions 5 and 6)

Function 5 $f(x) = \sqrt{x}$

Total # of Crossovers:	989948
# of Syntactically-Identical Crossovers:	113295
# of Fitness-Case-Identical Crossovers:	127623
Total # of positive crossovers	27818
Total # of neutral crossovers	225301
Total # of negative crossovers	736827

Counts:

	Negative	Neutral	Positive
0-Crossers	277874	11488	2079
1-Crossers	110455	79601	7312
2-Crossers(S)	27205	17420	2444
2-Crossers(D)	47924	50063	8710

Proportions:

	Negative	Neutral	Positive
0-Crossers	0.377	0.051	0.081
1-Crossers	0.150	0.356	0.262
2-Crossers(S)	0.365	0.078	0.093
2-Crossers(D)	0.065	0.220	0.306

Function 6 $f(x) = \operatorname{arcsinh}(x)$

Total # of Crossovers:	990366
# of Syntactically-Identical Crossovers:	105236
# of Fitness-Case-Identical Crossovers:	116170
Total # of positive crossovers	28481
Total # of neutral crossovers	226120
Total # of negative crossovers	735765

Counts:

	Negative	Neutral	Positive
0-Crossers	220624	14599	2486
1-Crossers	101593	77143	7988
2-Crossers(S)	21078	15640	2353
2-Crossers(D)	59345	52766	9090

Proportions:

	Negative	Neutral	Positive
0-Crossers	0.300	0.065	0.094
1-Crossers	0.139	0.339	0.270
2-Crossers(S)	0.029	0.071	0.089
2-Crossers(D)	0.081	0.232	0.317

Table 6. Various measures from the crossovers (function 7)

Function 7 $f(x) = \sin x$

Total # of Crossovers:	990024
# of Syntactically-Identical Crossovers:	188128
# of Fitness-Case-Identical Crossovers:	201428
Total # of positive crossovers	15187
Total # of neutral crossovers	201163
Total # of negative crossovers	773674

Counts:

	Negative	Neutral	Positive
0-Crossers	125859	4270	742
1-Crossers	54842	29412	2306
2-Crossers(S)	49847	23109	871
2-Crossers(D)	88051	56990	5562

Proportions:

	Negative	Neutral	Positive
0-Crossers	0.161	0.022	0.051
1-Crossers	0.070	0.150	0.155
2-Crossers(S)	0.065	0.114	0.058
2-Crossers(D)	0.115	0.278	0.361

an experiment purely concerned with crossover, we have retained mutation in the system because the aim of the experiment is to study crossover in the context of a "normal" GP run. Data from each crossover was stored and subsequently analysed in order to obtain the results presented in the following section.

4.2   Results and Discussion

Three sets of results are given. The first of these are in tables 3, 4, 5 and 6, which gives the results for a number of measures from across the run. The numbers presented are means from the ten runs. The numbers in the *proportions* tables represent the proportion of negative crossings which are 0-crossers, 1-crossers, et cetera, then the proportion of neutral crossings which are 0-crossers, 1-crossers, et cetera. Note that these columns don't add up to 1.0, as there are some crossovers that don't fit any of the 4 definitions.

Figure 2 shows the number of 0-, 1- and 2-crossers (of both types) over time. These two sets of plots are generated by grouping the crossovers into blocks of 100, and counting the number of crossovers at each point in each block. These numbers are then averaged across the ten runs at each point on the x-axis. Typically an early spike of 1-crossers is followed by a spike of 0-crossers, followed in some cases by a large number of 2-crossers(D)s. It is difficult to draw any clear conclusions from the patterns of behaviour in these graphs.

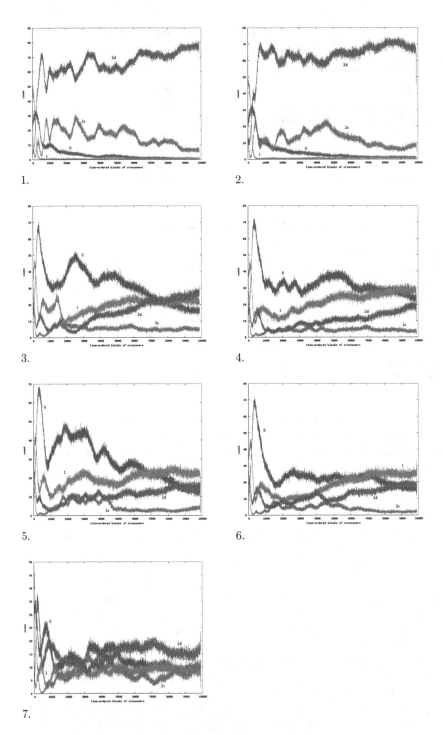

Fig. 2. The change over time in the evolutionary run of the proportion of each type of crossover for each of the 7 functions. Mean over 10 runs.

The most interesting results are in the tables of proportions. Our initial hypothesis was that crossover works by bringing together the fitness-contributing regions of each of the parents. If this were to be the case (as opposed to crossover simply being a mutation operator), we would expect the proportion of crossovers in the fitness-positive 2-crosser(D) category to be larger than those in the corresponding negative section. This is confirmed in each experiment, as is the other piece of supporting evidence, i.e. that the proportion of 0-crossers in the negative column is larger than in the proportion in the positive column.

However, it is noticeable that in each experiment, there are many crossovers where the best fitness cases are brought together in the child (2-crossers) and yet there is an overall decline in fitness.

5 Conclusions and Future Work

We have formalised the idea that a fitness-increasing crossover will combine the best features of the two parents, and shown some tentative support that this is what occurs when fitness-increasing crossovers happen. However, these experiments are rather limited in scope, and there are a number of further experiments and refinements which will be carried out in future work.

One important limitation of the definition that we have given is that a crossover of the type depicted in figure 1 can occur, but it not recognised because one of the regions in the child is not in its best fitness-producing region. Two approaches to these would be (1) to look at improvements in the strongest fitness-contributing regions in the parent or (2) replace the notion of a fixed proportion of strongest fitness cases with a (perhaps adaptive) notion of a fitness-improvement threshold. Another aspect to consider is the intrinsic easyness of some fitness cases in certain problems.

The experiments above have been carried out on a limited domain (they are all symbolic regression problems) and therefore it would be interesting to extend this analysis to other problem domains. Also, it would be interesting to analyse the different types of crossover at different times during the run (in more detail than the time-series plots given in this paper).

A larger-scale criticism of this approach is that it is focused on the fitness cases as individuals, and does not attempt to extract higher-scale features of the problem. An alternative approach, such as studying a formal semantics of the program [2], might be of use in approaching that question.

This paper has focused on this as an analysis approach. However, this could be turned around and used as a way of creating more crossovers of the type which appear to give positive results more frequently.

Finally, there are many other aspects of GP that that could be studied via this general idea of *instrumenting* the activities that occur during a GP run. The most obvious topic to tackle using this approach would be mutation.

References

1. Banzhaf, W., Nordin, P., Keller, R.E., Francone, F.D.: Genetic Programming: An Introduction. Morgan Kaufmann, San Francisco (1998)
2. Beadle, L., Johnson, C.G.: Sematically driven crossover in genetic programming. In: Proceedings of the 2008 IEEE World Congress on Computational Intelligence. IEEE Press, Los Alamitos (2008)
3. Keijzer, M.: Improving symbolic regression with interval arithmetic. In: Ryan, C., Soule, T., Keijzer, M., Tsang, E.P.K., Poli, R., Costa, E. (eds.) EuroGP 2003. LNCS, vol. 2610, pp. 70–82. Springer, Heidelberg (2003)
4. Keijzer, M., Babovic, V.: Genetic programming, ensemble methods and the bias/variance tradeoff - introductory investigations. In: Poli, R., Banzhaf, W., Langdon, W.B., Miller, J., Nordin, P., Fogarty, T.C. (eds.) EuroGP 2000. LNCS, vol. 1802, pp. 76–90. Springer, Heidelberg (2000)
5. Koza, J.R.: Genetic Programming: On the Programming of Computers by means of Natural Selection. Series in Complex Adaptive Systems. MIT Press, Cambridge (1992)
6. Langdon, W.B., Poli, R.: Foundations of Genetic Programming. Springer, Heidelberg (2001)
7. Nordin, P., Banzhaf, W.: Complexity compression and evolution. In: Eshelman, L. (ed.) Genetic Algorithms: Proceedings of the Sixth International Conference (ICGA 1995), Pittsburgh, PA, USA, pp. 310–317. Morgan Kaufmann, San Francisco (1995)
8. Nordin, P., Francone, F., Banzhaf, W.: Explicitly defined introns and destructive crossover in genetic programming. In: Rosca, J.P. (ed.) Proceedings of the Workshop on Genetic Programming: From Theory to Real-World Applications, Tahoe City, California, USA, July 9, pp. 6–22 (1995)
9. O'Reilly, U.-M., Oppacher, F.: Hybridized crossover-based search techniques for program discovery. In: Proceedings of the 1995 World Conference on Evolutionary Computation, Perth, Australia, November 29- December 11995, vol. 2, pp. 573–578. IEEE Press, Los Alamitos (1995)
10. Poli, R.: TinyGP software (visited November 2008), http://cswww.essex.ac.uk/staff/rpoli/TinyGP/
11. Poli, R., Langdon, W.B.: On the search properties of different crossover operators in genetic programming. In: Koza, J.R., Banzhaf, W., Chellapilla, K., Deb, K., Dorigo, M., Fogel, D.B., Garzon, M.H., Goldberg, D.E., Iba, H., Riolo, R. (eds.) Genetic Programming 1998: Proceedings of the Third Annual Conference, University of Wisconsin, Madison, Wisconsin, USA, July 22-25, 1998, pp. 293–301. Morgan Kaufmann, San Francisco (1998)
12. Poli, R., Langdon, W.B., McPhee, N.F.: A field guide to genetic programming (2008), http://lulu.com http://www.gp-field-guide.org.uk (With contributions by Koza, J.R.)
13. Ryan, C.: Pygmies and civil servants. In: Kinnear Jr., K.E. (ed.) Advances in Genetic Programming, pp. 243–263. MIT Press, Cambridge (1994)
14. Salustowicz, R., Schmidhuber, J.: Probabilistic incremental program evolution. Evolutionary Computation 5(2), 123–141 (1997)
15. Sanchez, L.: Interval-valued GA-P algorithms. IEEE Transactions on Evolutionary Computation 4(1), 64–72 (2000)

Evolution of Search Algorithms
Using Graph Structured Program Evolution

Shinichi Shirakawa and Tomoharu Nagao

Graduate School of Environment and Information Sciences,
Yokohama National University, 79-7, Tokiwadai, Hodogaya-ku,
Yokohama, Kanagawa, 240-8501, Japan
shirakawa@nlab.sogo1.ynu.ac.jp, nagao@ynu.ac.jp

Abstract. Numerous evolutionary computation (EC) techniques and
related improvements showing effectiveness in various problem domains
have been proposed in recent studies. However, it is difficult to design
effective search algorithms for given target problems. It is therefore essential to construct effective search algorithms automatically. In this paper,
we propose a method for evolving search algorithms using Graph Structured Program Evolution (GRAPE), which has a graph structure and is
one of the automatic programming techniques developed recently. We apply the proposed method to construct search algorithms for benchmark
function optimization and template matching problems. Numerical experiments show that the constructed search algorithms are effective for
utilized search spaces and also for several other search spaces.

1 Introduction

Automatic programming, which generates computer programs automatically, has
been actively investigated in recent studies. Genetic programming (GP) [1] is a
typical example of automatic programming, which was originally introduced by
Koza. GP evolves computer programs, which usually have a tree structure, and
searches for a desired program using a genetic algorithm (GA). Various representations for GP have been proposed thus far, including graph representation. The
advantages of a graph structure are that it can represent complex structures compared with tree structures, and that it can contain time-series information in its
graph structure. Graph Structured Program Evolution (GRAPE) [2,3] is one of a
number of automatic programming techniques developed recently. The representation of GRAPE is a graph structure, therefore, it can represent complex programs
(e.g., branches and loops) using this structure. The genotype of GRAPE is in the
form of a linear string of integers. GRAPE succeeds in generating complex programs automatically (e.g., factorial, exponentiation, list sorting, etc.). The study
described in this paper is based on this GRAPE technique.

Numerous evolutionary computation (EC) techniques and related improvements, showing their effectiveness in various problem domains such as function
optimization and combination problems, have been proposed. In addition, no
free lunch (NFL) theorems [4] show that all black-box search algorithms have the

L. Vanneschi et al. (Eds.): EuroGP 2009, LNCS 5481, pp. 109–120, 2009.

same average performance over the entire set of optimization problems. Therefore, it is essential to construct a search algorithm that is efficient in its approach to certain problems. In general, it is difficult to instantly construct efficient search algorithms for the given problems. Moreover, the automatic construction of search algorithms requires the use of key techniques. In this context, Meta GA, which optimizes the parameters of GA using GA, was proposed. Parameters such as population size, crossover rate, and mutation rate can be optimized in Meta GA. In addition, several techniques for the evolution of evolutionary algorithms have been investigated [5,6,7,8]. A method for evolving evolutionary algorithms, which evolves generation alternation procedures, was proposed by Oltean [6]. Dioşan and Oltean proposed a method for evolving crossover operators for evolutionary function optimization using GP [7] and a method for evolving the structure of a particle swarm optimization (PSO) algorithm [8].

Recently, the researches on hyper-heuristics, which are the methods to combine heuristics to generate new ones, have attracted increasing attentions. Several literatures on hyper-heuristics using GP are existed. Burke et al. generated a heuristic for the bin-packing problem using GP [9]. Kumar et al. showed that GP can evolve a set of heuristics for biobjective 0/1 knapsack problem [10]. Pillay and Banzhaf generated a heuristic for the timetabling problem [11]. The motivation of this work is the same as hyper-heuristics.

Our aim is to construct more efficient search strategies automatically so that they resemble human-designed search algorithms. In this study, we generate search algorithms by evolving the programs of moving agents in the search spaces using GRAPE. In particular, we focus on function optimization problems. A search algorithm is represented by a graph structure. The parameters are updated at each node. We apply the proposed method to benchmark function optimization problems and template matching problems. After evolving the search algorithms, we apply them to test problems that were not used during their evolution in order to verify their efficiency and versatility.

An overview of GRAPE is presented in the next section. In Section 3, we describe our proposed method, the evolution of search algorithms using GRAPE. Next, in Section 4, we apply the proposed method to the problem of the automatic construction of search algorithms and show several experimental results. Finally, in Section 5, we present our conclusions and possible future work.

2   Graph Structured Program Evolution (GRAPE)

GRAPE [2,3] constructs graph-structured programs automatically. The features of GRAPE are summarized as follows:

- Arbitrary directed graph structures.
- Ability to handle multiple data types using a data set.
- Genotype of an integer string.

A graph-structured program is composed of an arbitrary directed graph of nodes and a data set. The data set through flows the directed graph and is processed at each node. In GRAPE, each node has two parts, one for processing and the other for branching. The processing component executes several types of processing using the data set, e.g., arithmetic calculation and Boolean calculation. After processing is complete, the next node is selected. The branching component decides the next node according to the data set. When the output node is reached, the GRAPE program outputs data and the program terminates. The representation of GRAPE is a graph structure; therefore, it can represent complex programs (e.g., branches and loops) using this structure. There are several data types in the GRAPE program: integer data type, Boolean data type, list data type, and so on. The GRAPE program handles multiple data types using the data set for each type. To adopt an evolutionary method, genotype-phenotype mapping is used in GRAPE. This genotype-phenotype mapping method is similar to Cartesian GP (CGP) [12]. The GRAPE program is encoded in the form of a linear string of integers. The genotype is an integer string, which denotes a list of node types, connections, and arguments. Although the genotype in GRAPE is a fixed-length representation, the number of nodes in the phenotype can vary but is bounded (not all of the nodes encoded in the genotype have to be connected). This allows the existence of inactive nodes. GRAPE has been shown to generate a variety of programs successfully, such as factorial, Fibonacci sequence, exponentiation, list reversing, and list sorting.

3   Evolution of Search Algorithms Using GRAPE

In this study, an action program of searching agents is evolved by GRAPE. The agents search for optimal parameters in the search space. The search algorithm is achieved using the agents control program represented by GRAPE. The agents decide the next search point by moving through the search space in this model. An example of the GRAPE program is shown in Figure 1. The data set (parameters) of GRAPE contains the parameters of the target problem. The data set flows the

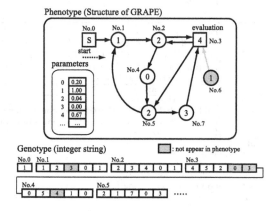

Fig. 1. Structure of GRAPE (phenotype) and the genotype, which denotes a list of node types, connections, and arguments

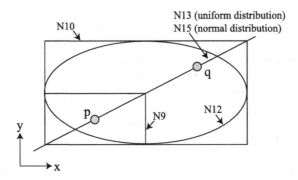

Fig. 2. Distributions of generated parameters using the prepared node functions (two dimensions)

Table 1. Node functions used in the experiments

Id	# Connection	Argument	Definition
			For a random selected parameter p_r
N1	1	–	Initialize the parameter p_r with uniform random number.
N2	1	q	$p_r = q_r$
N3	1	q	$p_r = p_r + u(p_r - \alpha\|p_r - q_r\|, p_r + \alpha\|p_r - q_r\|)$
N4	1	q	$p_r = u(min(p_r, q_r) - \alpha\|p_r - q_r\|, max(p_r, q_r) + \alpha\|p_r - q_r\|)$
N5	1	q	$p_r = N(p_r, (\alpha\|p_r - q_r\|)^2)$
N6	1	q	$p_r = N(\frac{(p_r+q_r)}{2}, (\alpha\|p_r - q_r\|)^2)$
			For the all parameters p_i ($i = 1, 2, 3, ...$)
N7	1	–	Initialize the all parameters p_i with random number.
N8	1	q	$p_i = q_i$
N9	1	q	$p_i = p_i + u(p_i - \alpha\|p_i - q_i\|, p_i + \alpha\|p_i - q_i\|)$
N10	1	q	$p_i = u(min(p_i, q_i) - \alpha\|p_i - q_i\|, max(p_i, q_i) + \alpha\|p_i - q_i\|)$
N11	1	q	$p_i = N(p_i, (\alpha\|p_i - q_i\|)^2)$
N12	1	q	$p_i = N(\frac{(p_i+q_i)}{2}, (\alpha\|p_i - q_i\|)^2)$
			Other types
N13	1	q	$p_i = (p_i - \alpha de_i) + R \cdot ((q_i + \alpha de_i) - (p_i - \alpha de_i))$ d: the distance between p_i and q_i, $e_i = \frac{(q_i - p_i)}{\|q_i - p_i\|}, R = u(0.0, 1.0)$
N14	1	q	$p_i = p_i + ze_i$ $z = N(0, (\alpha d)^2)$, d: the distance between p_i and q_i, $e_i = \frac{(q_i - p_i)}{\|q_i - p_i\|}$
N15	1	q	$p_i = \frac{(p_i+q_i)}{2} + ze_i$ $z = N(0, (\alpha d)^2)$, d: the distance between p_i and q_i, $e_i = \frac{(q_i - p_i)}{\|q_i - p_i\|}$
N16	2	–	Decide the next node with rondom number. If $u(0.0, 1.0) < 0.5$, then connection 1 is chosen. Else connection 2 is chosen.
			Evaluation node
EVAL	3	–	Evaluate the current parameters p. And if global best is updated, then connection 1 is chosen. Else if local best is updated, then connection 2 is chosen. Else connection 3 is chosen.

p_i: ith parameter of agent p.
q_i: ith parameter of partner agent q.
(q is best global parameters ($q0$) or best local parameters ($q1$)
or parameters of other randomly selected agents ($q2$).)
$u(x, y)$: uniformly distributed random number among $[x, y]$.
$N(\mu, \sigma^2)$: normally distributed random number.
Parameter $\alpha = 0.5$.

directed graph and is updated at each node. Following this, the parameters are evaluated when the evaluation node is reached, and then it moves to the next node. After the evaluations of the predefined numbers, the search finishes.

Table 1 shows the node functions used in the experiments. Examples of node functions used in the experiments are the initialization of parameters with uniformly distributed random numbers, the substitution of other parameters, and the update of the parameters using two parameter sets with either uniformly distributed random numbers or normally distributed random numbers. When the parameters are updated, the agent (p) selects the best global parameters $(q0)$, the best local parameters $(q1)$, or the parameters of the random selected agent $(q2)$, shown as parameters "q" in Table 1. The best global parameters are the best parameters of all agents; The best local parameters are the best parameters of the agent. These nodes are based on operators that are usually used in real-coded genetic algorithms [13,14]. For instance, N16 is the $BLX - \alpha$ [13]. Examples of generated parameters using these node functions are shown in Figure 2. This figure shows examples in the case of two-dimensional parameters. Moreover, the evaluation node (EVAL) has three branching components, which correspond to an updating global best, an updating local best, and another case scenario.

A better search algorithm is evolved based on the results using the search algorithm represented by GRAPE. While the operator at each node is relatively simple, it can construct complex search algorithms using a combination of these operators.

The fitness assignment of each individual is based on the results of the search for target search spaces, e.g., the summed best fitness for several search simulations. In this manner, a search algorithm, which is efficient for specific problems, is obtained.

4   Experiments and Results

In this section, we apply the proposed method to the evolution of search algorithms. The target problems are the benchmark function optimization and the template matching problems. Template matching is an important real-world problem. After evolving the search algorithm, we apply it to test problems that are not used during evolution in order to verify its efficiency and versatility.

4.1   Experimental Settings

The parameters used in the experiments are listed in Table 2. In order to avoid the problem caused by the non-reaching structures of the evaluation node, we limited the number of execution steps to 50. When the program reaches the execution limit, the individual is assigned the worst fitness value. We used uniform crossover and mutation as the genetic operators. The Minimal Generation Gap (MGG) [15,16] is used as the generation alternation model. The MGG model is a steady-state model proposed by Satoh et al. The MGG model has a desirable convergence property maintaining the diversity of the population and shows higher performance than other conventional models in a wide range of

Table 2. Parameters used in the experiments

Parameters for GRAPE	
Number of generations	10000
Population size	100
Children size (for MGG)	20
Crossover rate	0.6
Uniform crossover rate (P_c)	0.1
Mutation rate (P_m)	0.02
Maximum number of nodes	50
Execution step limits	50
Parameters for search simulation	
Number of agents	1, 10, 50
Number of function evaluations	10000 (benchmark functions)
	5000 (template matching)
Number of trials (N_t)	5

applications. GRAPE used with the MGG model shows higher performance than GRAPE used with Simple GA in a variety of program evolutions [2].

4.2   Benchmark Function

The Rastrigin-d function is used in order to evaluate the search algorithm represented by GRAPE. The Rastrigin-d function is defined in equation (1) and is a multimodal function.

$$f(x_1, ..., x_n) = 10n + \sum_{i=1}^{n} \{(x_i - d)^2 - 10cos(2\pi(x_i - d))\}$$

$$(-5.12 \leq x_i \leq 5.12)$$

(1)

where n is the number of dimensions. This function has the global minimum value 0 at $(d, ..., d)$. In this experiment, we use $n = 20$ and $d = 1.0$.

The parameters are randomized with the domain of definition. Then, search is performed using the algorithm represented by GRAPE for a predefined number of evaluations. We use 1, 10, and 50 as the number of agents in order to verify the influence of this parameter. If the number of agents is greater than 1, all agents refer to the same search algorithm represented by GRAPE (this model is homogeneous). Equation (2) is used as the fitness function of each individual in GRAPE.

$$fitness = -\sum_{i=1}^{N_t} f_i$$

(2)

where f_i is the best evaluation value of the ith search simulation, and N_t is the number of trials. The higher the numerical value, the better is the performance. We tested the evolution of the search algorithms using these experimental settings and obtained a search algorithm represented by GRAPE, which is efficient with respect to the Rastrigin-d function.

Figure 3 (a) shows an example of the obtained structure (graph-structured search algorithm) constructed by the proposed method when the number of agents is 10. Unnecessary nodes are deleted in this structure. The processing branches to two through the difference in the evaluation of the structure. This

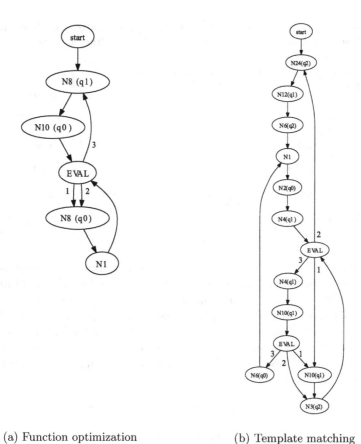

(a) Function optimization   (b) Template matching

Fig. 3. Obtained search algorithm structures for (a) a benchmark function optimization problem and (b) a template matching problem using GRAPE (number of agents is 10)

structure is relatively simple. However, we can see a more complex structure that is difficult to understand at sight in other experiments.

Next, we apply the obtained search algorithms to other test functions. We use the famous seven test functions in Table 3. The experimental condition is as follows:

 − Number of dimensions n: $10, 20, 30$
 − Number of trials: 50
 − Number of function evaluations: 5.0×10^5

We use a sufficient value as the number of function evaluations.

For comparison with conventional search algorithms, we compare the performance of our method with that of particle swarm optimization (PSO) [17,18]. PSO is one of a number of search algorithms introduced by Kennedy and Eberhart. It is categorized as an evolutionary algorithm and is based on an analogy of the movement of the flight of a flock of birds. Each particle represents a searching point

Table 3. Test functions used in the experiments. Sphere, Ridge, and Rosenbrock are unimodal functions. Rastrigin, Ackley, and Schwefel are multimodal functions. n is the number of dimensions.

Name	Functions	Range of x_i	f_{min}
Sphere	$f = \sum_{i=1}^{n} x_i^2$	$[-100, 100]$	0
Ridge	$f = \sum_{i=1}^{n} (\sum_{j=1}^{i} x_j)^2$	$[-100, 100]$	0
Rosenbrock	$f = \sum_{i=1}^{n-1} \{100(x_{i+1} - x_i^2)^2 + (x_i - 1)^2\}$	$[-30, 30]$	0
Rastrigin	$f = 10n + \sum_{i=1}^{n} \{x_i^2 - 10cos(2\pi x_i)\}$	$[-5.12, 5.12]$	0
Ackley	$f = 20 - 20exp(-0.2\sqrt{\frac{1}{n}\sum_{i=1}^{n} x_i^2}) + e - exp(\frac{1}{n}\sum_{i=1}^{n} cos(2\pi x_i))$	$[-32, 32]$	0
Schwefel	$f = \sum_{i=1}^{n} -x_i sin(\sqrt{|x_i|})$	$[-500, 500]$	$-418.98 * n$

in search space. The particles start at a random initial position and search for the minimum or maximum of a given objective function by moving through the search space. Each particle includes information relating to "a velocity vector," "a current position vector," and "own best position vector." Moreover, all particles have the same information, i.e., "a position vector of the best particle of all (global best)." At each iteration, a new velocity vector and a new position vector for each particle update provides a better solution according to these four factors. The updating equations are

$$v_i = \omega v_i + c_1 r_1(\hat{x} - x_i) + c_2 r_2(\hat{x}_g - x_i) \tag{3}$$

$$x_i = x_i + v_i \tag{4}$$

where x_i is the ith particle's position vector, V_i is the ith particle's velocity vector, $\hat{x}$ is the ith particle's previous best position vector, and $\hat{x}_g$ is the global best vector. The parameters ω, c_1, and c_2 are the coefficients of weight. We set $\omega = 0.72, c_1, c_2 = 1.49$. These values are recommended in [18]. The variables r_1 and r_2 are random numbers in the range $[0.0, 1.0]$. In the experiments, the number of particles is 50.

Table 4 shows the search results for 50 different runs with the same parameter set. According to these results, the evolved search algorithms perform significantly better than the conventional PSO algorithm. The evolved search algorithms are effective not only in relation to the Rastrigin function for the training search space, but also in terms of other test functions. It also shows better performance in a variety of problem dimensions. Although the search algorithms are generated using only the Rastrigin-d function (20 dimensions), they can be useful for other types of test functions.

4.3   Template Matching

Template matching is an important technique for investigating similarities between two patterns, predefined patterns (template) and target patterns. Template

Table 4. The average performances for benchmark functions determined by applying the evolved search algorithm and PSO (average over 50 trials with the same parameter set). n is the number of dimensions. The values shown in the parentheses are calculated using standard deviation.

Function	n	# agents = 1	# agents = 10	# agents = 50	PSO
Sphere	10	2.36×10^{-32} (6.58×10^{-32})	1.72×10^{-116} (8.25×10^{-116})	4.10×10^{-119} (1.38×10^{-118})	0.0 (0.0)
	20	7.05×10^{-28} (1.51×10^{-27})	2.54×10^{-81} (7.96×10^{-81})	2.41×10^{-84} (7.24×10^{-84})	4.94×10^{-324} (0.0)
	30	3.91×10^{-25} (7.38×10^{-25})	2.92×10^{-65} (1.29×10^{-64})	1.77×10^{-67} (9.35×10^{-67})	6.32×10^{-179} (0.0)
Ridge	10	2.10×10^{-30} (1.06×10^{-29})	3.25×10^{-114} (2.17×10^{-113})	8.62×10^{-117} (3.84×10^{-116})	1.22×10^{-276} (0.0)
	20	4.00×10^{-25} (2.04×10^{-24})	3.27×10^{-79} (1.11×10^{-78})	1.28×10^{-82} (3.27×10^{-82})	5.33×10^{2} (1.63×10^{3})
	30	1.03×10^{-22} (1.66×10^{-22})	5.64×10^{-62} (3.75×10^{-61})	2.53×10^{-65} (9.50×10^{-65})	4.07×10^{3} (5.05×10^{3})
Rosenbrock	10	8.22 (7.48)	4.36 (5.95)	4.74 (6.30)	3.74×10^{3} (1.78×10^{4})
	20	9.75 (8.26)	4.90 (6.04)	7.14 (7.88)	9.07×10^{3} (2.73×10^{4})
	30	1.20×10^{1} (1.49×10^{1})	8.65 (7.92)	7.49 (1.15×10^{1})	1.11×10^{4} (2.94×10^{4})
Rastrigin	10	7.39×10^{-15} (1.07×10^{-14})	4.52×10^{-14} (6.14×10^{-14})	1.11×10^{-14} (1.34×10^{-14})	6.27 (3.35)
	20	4.04×10^{-14} (4.14×10^{-14})	3.16×10^{-13} (2.76×10^{-13})	7.22×10^{-14} (5.25×10^{-14})	3.79×10^{1} (1.18×10^{1})
	30	1.51×10^{-13} (9.69×10^{-14})	2.80×10^{-12} (9.27×10^{-12})	2.58×10^{-13} (1.37×10^{-13})	9.38×10^{1} (2.91×10^{1})
Ackley	10	1.68×10^{-14} (5.36×10^{-15})	3.42×10^{-14} (2.76×10^{-14})	1.44×10^{-14} (6.27×10^{-15})	4.00×10^{-15} (3.61×10^{-2})
	20	4.80×10^{-14} (1.58×10^{-14})	1.57×10^{-13} (2.88×10^{-13})	3.42×10^{-14} (1.11×10^{-14})	3.29×10^{-1} (2.23×10^{-2})
	30	2.29×10^{-13} (1.11×10^{-13})	4.23×10^{-13} (1.10×10^{-12})	5.69×10^{-14} (1.20×10^{-14})	1.79 (8.86)
Schwefel	10	-4.19×10^{3} (9.09×10^{-13})	-4.19×10^{3} (1.66×10^{1})	-4.19×10^{3} (9.09×10^{-13})	-3.52×10^{3} (2.40×10^{2})
	20	-8.38×10^{3} (1.82×10^{-12})	-8.38×10^{3} (1.82×10^{-12})	-8.38×10^{3} (1.82×10^{-12})	-6.29×10^{3} (4.76×10^{2})
	30	-1.26×10^{4} (7.28×10^{-12})	-1.26×10^{4} (7.28×10^{-12})	-1.26×10^{4} (7.28×10^{-12})	-8.90×10^{3} (5.73×10^{2})

matching is also useful for the detection of an object by determining its position, angle, and magnification ratio in an image. For the detection of a two-dimensional object, four parameters (x-coordinate, y-coordinate, angle, and magnification ratio) must be optimized in template matching. We use a simple objective function: the normalized sum of absolute difference between the template image and the target image. The equation for the objective function in the experiments is as follows:

$$f(x, y, rate, angle) = \frac{1}{W \cdot H \cdot V_{max}} \sum_{i=1}^{W} \sum_{j=1}^{H} |f_{i'j'} - t_{ij}|$$
$$i' = rate * ((i - \tfrac{W}{2}) * cos(angle) + (j - \tfrac{H}{2}) * sin(angle)) + x$$
$$j' = rate * ((i - \tfrac{W}{2}) * sin(angle) + (j - \tfrac{H}{2}) * cos(angle)) + y$$

$$(5)$$

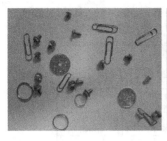

Template image

Target image 1   Target image 2

Fig. 4. Target images and the template image used in the experiments. Target image 1 is used for constructing the search algorithms. Target image 2 is applied to the evolved search algorithm.

Table 5. The average performances for template matching problems found by applying the evolved search algorithm and PSO (average over 50 trials with the same parameter set). Best/Worst refers to the fitness of the best individual in the best/worst run, respectively. The number of evaluations is 10000.

	# agents=1	# agents=10	# agents=50	PSO
Target image 1				
Average	0.0890	0.0889	0.0890	0.105
Stddev	0.00501	0.00578	0.00424	0.0137
Best	0.0846	0.0845	0.0845	0.0858
Worst	0.107	0.107	0.103	0.134
Target image 2				
Average	0.0735	0.0726	0.0714	0.0793
Stddev	0.00684	0.00328	0.00281	0.0113
Best	0.0698	0.0698	0.0698	0.070
Worst	0.116	0.0776	0.0790	0.132

where t_{ij} is the brightness of the template image, f_{ij} is the brightness of the target image, W and H denote the image size of the template image, and V_{max} is the maximum brightness. This objective function should be minimized.

The template images and the target image used in the experiment are shown in Figure 4. These are grayscale images, with the size of the template image being 64×64 pixels, and the size of the target images 640×480 pixels. The ranges of each parameter are as follows:

- Positions of template image: $0 \leq x \leq X, 0 \leq y \leq Y$ (X and Y denote the image size of the target image)
- Magnification ratio: $1.0 \leq rate \leq 2.0$
- Angle: $-180° \leq angle \leq 180°$

The parameters used in the experiment are shown in Table 2. Target image 1 and the template image in Figure 4 are used to evolve the search algorithm represented by GRAPE. Equation (2) is also used as the fitness function of each individual in GRAPE. By these settings, we obtain a search algorithm, which is effective to template matching. Figure 3 (b) shows an example of the obtained structure (graph-structured search algorithm) constructed by the proposed method when the number of agents is 10.

The performances of the evolved search algorithms and PSO are shown in Table 5. The number of evaluations is 10000. We use the two target images shown in Figure 4. The evolved search algorithms perform better than the conventional PSO algorithm for both target images. Therefore, we can conclude that general search algorithms for template matching can be constructed automatically.

5   Conclusions and Future Work

In this paper, we proposed a new method for the automatic construction of search algorithms based on GRAPE, which evolves graph-structured programs. We applied the proposed method to evolve search algorithms for function optimization and template matching problems and confirmed that efficient search algorithms for each problem were obtained by the proposed method. From the experimental results, the performances of evolved search algorithms were found to outperform a conventional PSO algorithm. In future work, we will apply the proposed method to other types of problems. Further, we will propose a method for evolving search algorithms with a heterogeneous model.

References

1. Koza, J.R.: Genetic Programming: On the Programming of Computers by Means of Natural Selection. MIT Press, Cambridge (1992)
2. Shirakawa, S., Ogino, S., Nagao, T.: Graph Structured Program Evolution. In: Proceedings of the Genetic and Evolutionary Computation Conference (GECCO 2007), London, UK, vol. 2, pp. 1686–1693. ACM Press, New York (2007)
3. Shirakawa, S., Nagao, T.: Evolution of sorting algorithm using graph structured program evolution. In: Proceedings of the 2007 IEEE International Conference on Systems, Man and Cybernetics (SMC 2007), Montreal, Canada, pp. 1256–1261 (2007)
4. Wolpert, D., Macready, W.: No free lunch theorems for optimization. IEEE Transactions on Evolutionary Computation 1(1), 67–82 (1997)
5. Tavares, J., Machado, P., Cardoso, A., Pereira, F.B., Costa, E.: On the evolution of evolutionary algorithms. In: Keijzer, M., O'Reilly, U.-M., Lucas, S., Costa, E., Soule, T. (eds.) EuroGP 2004. LNCS, vol. 3003, pp. 389–398. Springer, Heidelberg (2004)
6. Oltean, M.: Evolving evolutionay algorithm using linear genetic programming. Evolutionary Computation 13(3), 387–410 (2005)
7. Dioşan, L., Oltean, M.: Evolving crossover operators for function optimization. In: Collet, P., Tomassini, M., Ebner, M., Gustafson, S., Ekárt, A. (eds.) EuroGP 2006. LNCS, vol. 3905, pp. 97–108. Springer, Heidelberg (2006)
8. Dioşan, L., Oltean, M.: Evolving the structure of the particle swarm optimization algorithms. In: Gottlieb, J., Raidl, G.R. (eds.) EvoCOP 2006. LNCS, vol. 3906, pp. 25–36. Springer, Heidelberg (2006)

9. Burke, E.K., Hyde, M.R., Kendall, G.: Evolving bin packing heuristics with genetic programming. In: Runarsson, T.P., Beyer, H.-G., Burke, E.K., Merelo-Guervós, J.J., Whitley, L.D., Yao, X. (eds.) PPSN 2006. LNCS, vol. 4193, pp. 860–869. Springer, Heidelberg (2006)

10. Kumar, R., Joshi, A.H., Banka, K.K., Rockett, P.I.: Evolution of hyperheuristics for the biobjective 0/1 knapsack problem by multiobjective genetic programming. In: Proceedings of the Genetic and Evolutionary Computation Conference (GECCO 2008), Atlanta, GA, USA, pp. 1227–1234. ACM Press, New York (2008)

11. Pillay, N., Banzhaf, W.: A genetic programming approach to the generation of hyper-heuristics for the uncapacitated examination timetabling problem. In: Neves, J., Santos, M.F., Machado, J.M. (eds.) EPIA 2007. LNCS, vol. 4874, pp. 223–234. Springer, Heidelberg (2007)

12. Miller, J.F., Thomson, P.: Cartesian genetic programming. In: Poli, R., Banzhaf, W., Langdon, W.B., Miller, J., Nordin, P., Fogarty, T.C. (eds.) EuroGP 2000. LNCS, vol. 1802, pp. 121–132. Springer, Heidelberg (2000)

13. Eshelman, L.J., Schaffer, J.D.: Real coded genetic algorithms and interval-schemata. Foundations of Genetic Algorithms 2, 187–202 (1993)

14. Ono, I., Kobayashi, S.: A real-coded genetic algorithm for function optimization using unimodal normal distribution crossover. In: Proceedings of the 7th International Conference on Genetic Algorithms, East Lansing, MI, USA, pp. 246–253. Morgan Kaufmann, San Francisco (1997)

15. Satoh, H., Yamamura, M., Kobayashi, S.: Minimal generation gap model for considering both exploration and exploitations. In: Proceedings of the IIZUKA 1996, pp. 494–497 (1996)

16. Deb, K., Anand, A., Joshi, D.: A computationally efficient evolutionary algorithm for real-parameter optimization. Evolutionary Computation 10(4), 371–395 (2002)

17. Kennedy, J., Eberhart, R.C.: Particle swarm optimization. In: Proceedings of the IEEE International Conference on Neural Networks, Perth, Australia, vol. IV, pp. 1942–1948 (1995)

18. Eberhart, R.C., Shi, Y.: Comparing inertia weights and constriction factors in particle swarm optimization. In: Proceedings of the 2000 Congress on Evolutionary Computation (CEC 2000), vol. 1, pp. 84–88 (2000)

Genetic Programming for Feature Subset Ranking in Binary Classification Problems

Kourosh Neshatian and Mengjie Zhang

School of Engineering and Computer Science
Victoria University of Wellington, P.O. Box 600, Wellington, New Zealand
{kourosh.neshatian,mengjie.zhang}@ecs.vuw.ac.nz

Abstract. We propose a genetic programming (GP) system for measuring the relevance of subsets of features in binary classification tasks. A virtual program structure and an evaluation function are defined in a way that constructed GP programs can measure the goodness of subsets of features. The proposed system can detect relevant subsets of features in different situations including multimodal class distributions and mutually correlated features where other ranking methods have difficulties. Our empirical results indicate that the proposed system is good at ranking subsets and giving insight into the actual classification performance. The proposed ranking system is also efficient in terms of feature selection.

1 Introduction

Feature ranking is a common approach to feature selection in which features are ranked based on their usefulness and relevance to a problem [1]. An advantage of feature ranking over other feature selection methods is that it gives more information about the underlying nature of the problem by measuring the relative importance of features. This capability in turn increases the interpretability of the solutions.

Most feature ranking methods fall into the filter approach category [2,3,4], and can only measure the goodness of one single feature. This includes all feature ranking measures from the information theoretic domain such as information gain (IG), gain ratio, mutual information and the likes. These measures cannot provide any explicit way of measuring the goodness of a group (subset) of features. They are usually combined with different search techniques by which good features are added to a candidate subset gradually (one by one) and irrelevant ones are eliminated.

Another limitation of many of these methods is that they can only consider simple types of relation between a feature and the target class. For example in the logistic regression model [5], the relationship is assumed to be linear; in most of the information theoretic measures, it is assumed that instances can be classified by setting a split point along the feature axis.

Genetic programming (GP) is good at exploring the search space of features (variable terminals) and also constructing compound features. GP has already

L. Vanneschi et al. (Eds.): EuroGP 2009, LNCS 5481, pp. 121–132, 2009.
© Springer-Verlag Berlin Heidelberg 2009

been used for individual feature ranking/selection by evolving a set of discriminative decision stumps with certain assumptions [6,7,8,9]. One assumption which is common among many other ranking methods like principle component analysis [10] is that the best subset of features of size n is obtained by adding the feature ranked n to the best subset of size $n - 1$. However, this assumption is not necessarily true.

1.1   Goals

In this research we address, using GP, the problem of feature subset ranking explicitly. We consider binary classification tasks with numerical features. In particular, the goal of this research is to use GP for non-wrapper feature ranking in a way that

1. the goodness of a subset of features can be evaluated (as opposed to most of the existing feature ranking methods which can only evaluate single features),
2. the proposed system can detect those good features that are not normally detected by existing methods, and
3. the provided ranking scheme for the subsets of features gives a good insight into the actual classification performance.

The rest of the paper is organised as follows. In section 2 we propose a GP relevance measure (GPRM) for evaluating and ranking subsets of features. In section 3, we evaluate the proposed method by applying it to three benchmark dataset and see how the provided ranking is correlated to the actual classification performance. We also measure the performance of the proposed system in a feature selection scenario. Section 4 concludes the paper and gives some future work directions.

2   GP for Feature Subset Ranking

We propose GPRM by introducing a virtual structure for program trees and an evaluation (relevance) function which is used for evaluating subsets of features. Through some case studies, we describe how the proposed system can handle multiple features and how it can find those good features that are usually missed by other relevance measures. We then propose an architecture for using GPRM for feature ranking.

2.1   Program Trees: A Virtual Structure

We extend the concept of feature relevance measure functions by proposing a virtual structure for GP program trees. Figure 1 shows such a structure for a GP program; it measures the relevance of a subset of features to a classification task. At the top (root) of the tree, there is a relevance measure function denoted by RM. This function measures the relevance of its right subtree to the class label

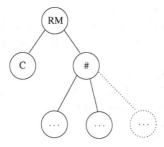

Fig. 1. A virtual structure for a GP program for measuring the usefulness of a subset of features

variable denoted by C. The node RM can be a very simple function in terms of the types of relationships it can detect. However, by providing a good subprogram as its right subtree, we can discover more complex relationships between the features used in the subtree and the class variable. For example, RM could be a linear correlation function for one single feature, but with a sophisticated subtree underneath, we would be able to detect nonlinear relationships between a subset of features and the class variable. We use the power of GP to evolve a rich subtree which leads to a high relevance at the root node. We then regard the contributing variable terminals in that subprogram as a subset of features and the output of the program as the goodness of that subset. This structure is virtual in the sense that the top node of the tree does not take part in any genetic operations and so, in practice, it can be implemented as part of the GP tree evaluation process rather than the GP tree representation.

2.2   Relevance Measure

In the filter approach to feature selection for classification tasks, as a widely-used hypothesis, a good feature is considered to be highly related to the class variable [11]. That is, knowing the value of a related feature should change the probability distribution of the class variable. We are looking for a relevance measure (function) that can be used in the node RM. Since the class label is a nominal (categorical) variable, the function used to measure this relevance should be capable of handling this type of data. Information theoretic measures like information gain (IG) and gain ratio can be used only if the feature being measured is nominal itself or has already been discretised. These methods are also limited to measuring the correlation between the class labels and a single feature rather than a group (subset) of features.

To measure the correlation between a continuous feature and a binary class variable, one could use the logistic regression (LR) model

$$\log\left(\frac{\pi(x)}{1 - \pi(x)}\right) = \alpha + \beta x \tag{1}$$

where $\pi(x)$ is the probability of an example belonging to a particular class given the value of feature x, and the right hand side of the equation is a linear approximation of the logit function. The magnitude of the coefficient β is used as an indicator of linear correlation, where a value of zero shows no linear correlation between the continuous feature x and the class label [12]. The parameter β and constant coefficient α can be estimated by maximizing the likelihood function through the Newton-Raphson method, but it is too expensive a procedure to be considered as a candidate function for the node RM.

Here, we define a binary relevance function (BR) that measures the linear relationship between a nominal and a numeric variable but is computationally cheaper. We define BR to be

$$BR(x,c) = \left(\frac{Cov(x, B(c))}{\sigma(x)\sigma(B(c))} \right)^2 \tag{2}$$

which is actually the square of Pearson's correlation between a numeric random variable x and a function B, with numerical range, of a nominal random variable c. $Cov(\cdot, \cdot)$ and $\sigma(\cdot)$ denote, the covariance function and the standard deviation, respectively. The squared form is used because we only consider the magnitude of relevance and not the direction (sign). For a binary classification task with $c \in \{C_A, C_B\}$, we define $B(c)$ to be

$$B(c) = \begin{cases} +\sqrt{\frac{n_B}{n_A}}, & c = C_A \\ -\sqrt{\frac{n_A}{n_B}}, & c = C_B \end{cases} \tag{3}$$

where n_A and n_B are the numbers of examples belonging to class C_A and class C_B, respectively, and $n_A + n_B = n$ which is the total number of examples in the training set. The function $B(c)$ is a standardized variable with the following expected value and variance:

$$E(B(c)) = p(C_A)\sqrt{\frac{n_B}{n_A}} - p(C_B)\sqrt{\frac{n_A}{n_B}} = \frac{1}{n}(n_A\sqrt{\frac{n_B}{n_A}} - n_B\sqrt{\frac{n_A}{n_B}}) = 0 \tag{4}$$

where $p(.)$ denotes the class distribution. Consequently,

$$\sigma^2(B(c)) = E(B^2(c)) = \frac{n_A}{n}\frac{n_B}{n_A} + \frac{n_B}{n}\frac{n_A}{n_B} = 1 \tag{5}$$

Therefore, the empirical BR can, given that the total number of instances n is large enough, be simplified to

$$BR(x,c) = \frac{(\sum_{i=1}^{n}(x_i - \bar{x})B(c_i))^2}{n\sum_{i=1}^{n}(x_i - \bar{x})^2} \tag{6}$$

where x_i is the i-th observation (the output of the subprogram in Fig 1 for the i-th instance), c_i is the class label of the i-th observation, and $\bar{x}$ is the sample mean. The complexity of computing BR is $O(n)$, which makes it a good candidate for the node RM.

To show why BR is a good alternative to LR for feature ranking, we generate a set of artificial data. We consider one feature, x, and a binary class variable

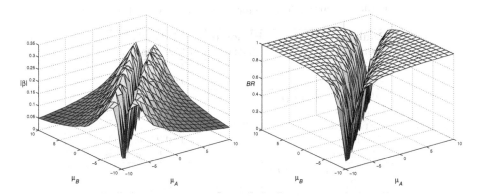

Fig. 2. Magnitude of parameter β in LR (left) and BR function (right) with respect to the mean of two classes

$c \in \{C_A, C_B\}$. Instances of classes C_A and C_B are distributed based on the same distribution but different parameters. We chose normal distributions with means μ_A and μ_B each having 21 values, namely, $\{0, \pm 1, \ldots, \pm 10\}$, and variances $\sigma_A = \sigma_B = 1$. In total we have 441 (21×21) artificial datasets, each containing one feature with 200 instances (100 for each class).

Figure 2 shows the BR function and the magnitude of the parameter β from the logistic regression with respect to μ_A and μ_B. In both graphs, there is a valley-like area along the diagonal where the means of the two classes are close to each other. In this area, the instances of the two classes are almost mixed together and the knowledge of the value of x would not be helpful in discriminating the instances. As we go towards the areas off the diagonal, where the distance between the means becomes larger, the magnitude starts increasing in both figures. On the other hand, there is a difference between these two functions. When the means of classes get far away from each other, $|\beta|$ in LR starts decreasing, indicating a lower relevance. However, from a classification point of view, as the margin between the instances of two classes provided by a feature increases, the feature is considered to be better. In contrast to LR, the BR function returns larger values as the distance between the two classes increases. So BR is a better measure than LR for feature ranking.

2.3   Case Studies

We set up a number of experiments on some artificial data to investigate what type of relevance the GPRM is able to capture that the other methods are not. In particular, we study a bimodal class distribution scenario in which GPRM is able to capture non-linear changes in the class probability, and a binary classification problem with two correlated features in which GPRM can measure the goodness of a subset of features rather than individuals. Single GP runs are conducted in each case and the results are compared to those of other methods. Detailed values regarding this comparison are presented in table 1.

Table 1. Three relevance measures on two case studies

Ranking Method	Bimodal Distribution	Mutually Correlated
Logistic Regression		
$\|\beta\|$ coefficient	x: 0.03	x: 0.15, y: 0.13
Information Gain (IG)		
Entropy	0.61	0.69
Split point	x: 2.8	x: 7.4, y: 4.6
Gain	x: 0.23	x: 0.29, y: 0.00
Gain rate	x: 38%	x: 42%, y: 0.0%
GP Relevance Measure		
GPRM	x: 82%	$\{x, y\}$: 87%

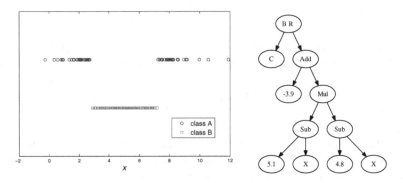

Fig. 3. A bimodal class distribution along feature x in a binary classification problem (Left) The evolved GP program tree for measuring the goodness of x in this scenario (Right)

Bimodal class distribution. Figure 3 shows a binary classification problem with a single feature x, where the distribution of one of the classes, A, is bimodal. This feature is obviously good because by setting up an interval around instances of class B, the classification problem can be solved. However, since the class probability does not change linearly with respect to x, the LR method does not consider x to be a good feature, returning a β coefficient close to zero (0.03, table 1). The IG method tries to find the best split point (in this case 2.8, table 1) using which, instances from different classes can be separated. However, as all the instances cannot be separated around one split point, IG does not report this feature to be a very good one. In contrast, the GPRM evolves a program tree like the one in Fig 3 which transforms the relationship to a linear form that can be detected by the BR function. These results show how this feature is dismissed as irrelevant by methods like LR and IG, but it is successfully detected by the GPRM measure. Notice that the gain rate and GPRM are calculated differently, but they can be regarded as an indicator of how good a feature is for the problem.

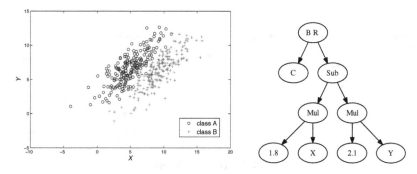

Fig. 4. A binary classification task presented with respect to two of its features, x and y, which are mutually correlated (Left) The evolved GP program tree for measuring the goodness of the subset $\{x, y\}$ (Right)

Mutually-correlated features. Figure 4 shows another binary classification problem with two features, x and y, that are correlated. These two features can be very useful for this classification task as a (straight) line passing through the boundary of the two classes can identify the class of instances. However, to those relevance measures which consider each feature individually, neither feature is necessarily good. LR returns a $|\beta|$ of less than 0.2 for each of these features and the gain rate of IG for x is pretty low and for y is zero. On the other hand, the GPRM maximises the BR function by finding an appropriate program tree. This program tree is shown in Fig 4. The subprogram, as the right child of the root node, actually constructs (by combining x and y) a meta-feature along which the instances are easily separable. The relevance of $\{x, y\}$ is measured to be 87% by GPRM.

2.4   System Architecture

Given a dataset, including a set of features, for a classification task, we use GP to find the GPRM value of different subsets of features. We use the BR function in equation (6) as the RM node and the output of each program is regarded as its fitness. We let GP find individual programs which can maximize this function. Figure 5 depicts the overall architecture of the system. Each program in the population uses a subset of features from the training data and generates a relevance value (GPRM). This value is regarded as the relevance of the subset (to the task) and is stored in a *rank table*. The rank table keeps track of the subsets of features and the relevance of each subset. The table is persistent among different GP runs. Once GP finds a better GPRM for a given subset, the corresponding row in the table will be updated. Over several runs, the proposed GP method evolves a set of good programs, each of which contains a subset of features, and evaluates the relevance of these subsets of features to the classification task. For evaluation of the method, we use each subset in the rank table to create a new dataset (by projection) and then use it for classification. We then, based on the

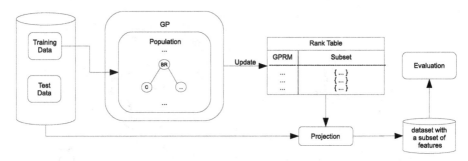

Fig. 5. System architecture

actual classification performance (accuracy) of the subsets of features on the test set, analyse how effective the GPRM method is in determining the goodness of the subsets of features.

3   Empirical Results

3.1   Datasets

We used three datasets with a relatively large number of features from the UCI machine learning repository [13] in the experiments. Table 2 summarizes the main characteristics of these datasets.

Table 2. Datasets

Problem	# Features	# Nominal	# Instances	# Classes
Johns Hopkins Ionosphere	34	0	351	2
Sonar	60	0	208	2
W. Breast Cancer (Diagnostic)	30	0	569	2

3.2   Settings and Implementation Details

Because of our design, we adopt the standard tree-based genetic programs [14], each of which produces a single floating point between zero and one (after applying the BR function) as the relevance measure of the subset of features being used in the program. There is one variable terminal for each feature in the problem. A number of randomly generated constants are also used as terminals. The function set consists of four elementary binary arithmetic operators: $+, -, \times, \div$. The $\div$ operator is "protected" which means it returns zero for division by zero. The ramped half-and-half method [15] is used for generating programs in the initial population and for the mutation operator. An elitist policy is taken by keeping the best individual of the generation. The population size is 1024. The maximum program tree depth is set to 6. The evolution is terminated, at the latest, after the 50th generation or when a GPRM value of 1 is found. We conduct 50 GP runs with different random seeds for each training set. This number of runs allows us to explore the feature space properly. The platform is implemented in Java and we use grid computing to conduct parallel GP runs. Two types of classifiers are used in our experiments, namely the J48 implementation of C4.5 decision tree [16,17] and the SMO version of the SVM classifier [18]. We use the Weka [17] library for the classification and evaluation processes.

3.3   Evaluation

As there is no explicit test data for the above-mentioned datasets, we adopt a 10-fold cross-validation approach. For each given training fold we conduct 50 GP runs and keep the subsets with a GPRM of higher than 0.25. For each feature subset, we train an SVM classifier and a decision tree and measure their performance on the test data. Therefore, for each subset of features, we have one GPRM value and two classification performance measures (one with SVM and the other with DT). The values for the GPRM are obtained using the training data while performance (accuracy) values are obtained using the test data.

Figure 6 shows the GPRM and classification values obtained for all feature subset processed during 500 GP run (10-fold cross validation × 50 runs each) over three datasets. There are more than 20,000 feature subsets processed for each dataset. Each processed subset is represented by two points in each plot, one dark for SVM classifier and one light for the decision tree, forming two clouds of data. The plots indicate that there is roughly a linear relationship between GPRM and the classification performance, where the higher the GPRM value, the better the classification performance. This linear relation is particularly obvious in the ionosphere and sonar datasets. In the breast cancer dataset, SVM exhibits a strong linear relationship while DT exhibits a weaker linear relationship which is due to the fact that DT does not generally perform as good as SVM on this dataset [6].

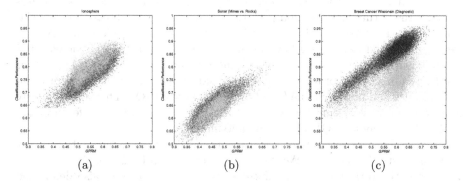

Fig. 6. GPRM vs. classification performance of SVM (the dark cloud) and decision tree (light cloud on top of the dark one) for three datasets: (a) ionosphere, (b) sonar, and (c) breast cancer

Table 3. Correlation coefficient between GPRM and classification performance

Dataset	s	$\rho_{\text{GPRM,SVM}}$	$\rho_{\text{GPRM,DT}}$
Ionosphere	22592	0.86	0.78
Sonar	20972	0.84	0.81
W. Breast cancer	29266	0.93	0.64

Table 3 illustrates quantitative measurements of the linear relationship between the GPRM and the classification performances (classification accuracy rate on the test set) of SVM and DT. Column s represents the total number of subsets of features by the end of the GP runs. The next two columns show the coefficient of correlation between the GPRM and the classification performance yielded by the classifiers SVM and DT, respectively. Almost all the cases show a strong correlation between the GPRM and the classification performance. We conduct a test to determine the statistical significance of our results. The test statistic for testing the significance of the correlation coefficient ρ is $T = \frac{\rho\sqrt{s-2}}{\sqrt{1-\rho^2}}$, where T has a t-distribution with $s - 2$ degrees of freedom [19]. For the given values of s and ρ in Table 3, p-values corresponding to the above test statistic are all smaller than 0.01, which implies that all the estimated correlation coefficient values in the table are statistically significant at a 99% confidence level.

3.4   Selection Performance

In many classification applications, one is interested in finding the subset of features which maximizes the classification performance. This is actually the main goal of feature selection. An ideal feature ranking mechanism would be the one that would rank such a subset as the highest (rank 1). However, in practice, it is hard to provide such a ranking mechanism, since the definition of the best subset of features and its properties are not unique among different classification algorithms.

The correlation between GPRM and the actual classification performance gives us an image of how the calculated relevance of the subsets of features matches the actual classification performance using those subsets. The correlation is quantified over all the explored subsets. For feature selection, however, we are interested in knowing how quickly (with how much effort) one can find the best subset of features given a ranking scheme based on GPRM. In other words, we are looking for a measure that considers only the highest classification performance and how it can be found by the highest GPRM. This is particularly important because during the course of a GP run, the subsets of features are explored in a way that GPRM can be maximised.

Suppose that the best subset of features is ranked at the r_0-th position rather than the first (the highest). We define an error rate to be $Error\ Rate = \frac{r_0-1}{s}$ where s is the total number of ranked subsets of features. We also define the selection performance to be $1 - Error\ Rate$. Table 4 shows the selection performance of the proposed GP system on the test datasets. It shows that in almost all cases, GPRM can be quite effective in finding the best subset of features.

Table 4. GPRM Selection Performance

Dataset	for SVM	for DT
Ionosphere	99%	82%
Sonar	99%	100%
W. Breast cancer	99%	99%

The selection performance gives a perspective quite different from the correlation measures. It shows that even for datasets and classifiers that GPRM and the classification performance are not highly correlated (like the Breast Cancer dataset and the DT classifier), the highest classification performance is obtained for a subset of features which is ranked very close to the first. That is, even an exhaustive wrapper search starting from the highest rank (from right to left on the GPRM axis) could result in finding the best subset very quickly.

4   Conclusions

GP has the capability of finding hidden relationships between a subset of features and the target class variable. We proposed a system to quantify the strength of this relationship. A new relevance measure is used with GP programs to calculate the goodness of the subset of features being used in that program. We found that GP can recognise relevant features in situations where many other measures cannot. We particularly realised that GP is good at handling situations where the class variable has a multimodal distribution or the features are mutually correlated.

Our empirical results show that, although the proposed system does not wrap any classifier for measuring the relevance of subsets of features, the resulting ranking is highly correlated to the actual classification performance. In addition, the best subset of features can be found in the first percentile of high-ranked features, in most cases.

As future work, we will test this method on more and larger data sets with more examples (e.g. with thousands of examples). This work will be extended to multiple-class classification tasks by finding an appropriate function that can replace the BR function. We will also compare this approach with the existing GP approaches for feature ranking such as [6] for the actual classification performance. Another extension is to investigate whether GP can find a good definition for a relevance measure on its own via decomposing an RM function into its elementary components and using them in the primitive function set of GP.

References

1. Jong, K., Mary, J., Cornuéjols, A., Marchiori, E., Sebag, M.: Ensemble feature ranking. In: Boulicaut, J.-F., Esposito, F., Giannotti, F., Pedreschi, D. (eds.) PKDD 2004. LNCS, vol. 3202, pp. 267–278. Springer, Heidelberg (2004)
2. Ruiz, R., Riquelme, J.C., Aguilar-Ruiz, J.S.: Fast feature ranking algorithm. In: Knowledge-Based Intelligent Information and Engineering Systems, pp. 325–331 (2003)

3. Biesiada, J., Duch, W., Kachel, A., Maczka, K., Palucha, S.: Feature ranking methods based on information entropy with parzen windows. In: International Conference on Research in Electrotechnology and Applied Informatics (REI 2005), pp. 109–119 (2005)
4. Lin, T.H., Chiu, S.H., Tsai, K.C.: Supervised feature ranking using a genetic algorithm optimized artificial neural network. Journal of Chemical Information and Modeling 46, 1604–1614 (2006)
5. Cheng, Q., Varshney, P., Arora, M.: Logistic regression for feature selection and soft classification of remote sensing data. Geoscience and Remote Sensing Letters 3, 491–494 (2006)
6. Neshatian, K., Zhang, M.: Genetic programming for feature ranking in classification problems. In: Li, X., et al. (eds.) SEAL 2008. LNCS, vol. 5361, pp. 544–554. Springer, Heidelberg (2008)
7. Davis, R.A., Charlton, A.J., Oehlschlager, S., Wilson, J.C.: Novel feature selection method for genetic programming using metabolomic 1h NMR data. Chemometrics and Intelligent Laboratory Systems 81, 50–59 (2006)
8. Liu, H., Yu, L.: Toward integrating feature selection algorithms for classification and clustering. IEEE Trans. on Knowl. and Data Eng. 17(4), 491–502 (2005)
9. Lin, J.Y., Ke, H.R., Chien, B.C., Yang, W.P.: Classifier design with feature selection and feature extraction using layered genetic programming. Expert Syst. Appl. 34(2), 1384–1393 (2008)
10. Jolliffe, I.T.: Principal Component Analysis (2002)
11. Hall, M.A.: Correlation-based feature selection for discrete and numeric class machine learning. In: Proceedings of the Seventeenth International Conference on Machine Learning table of contents, pp. 359–366. Morgan Kaufmann Publishers Inc., San Francisco (2000)
12. Agresti, A., Agresti, A.: Categorical Data Analysis. Wiley, Chichester (2003)
13. Asuncion, A., Newman, D.: UCI machine learning repository (2007), http://www.ics.uci.edu/~mlearn/MLRepository.html
14. Koza, J.R.: Genetic Programming II: Automatic Discovery of Reusable Programs. MIT Press, Cambridge (1994)
15. Koza, J.R.: Genetic Programming: On the Programming of Computers by Means of Natural Selection. MIT Press, Cambridge (1992)
16. Quinlan, J.R.: C4.5: Programs for Machine Learning. Morgan Kaufmann, San Francisco (1993)
17. Witten, I.H., Frank, E.: Data Mining: Practical Machine Learning Tools and Techniques, 2nd edn. Morgan Kaufmann, San Francisco (2005)
18. Keerthi, S.S., Shevade, S.K., Bhattacharyya, C., Murthy, K.R.K.: Improvements to platt's SMO algorithm for SVM classifier design. Neural Comp. 13, 637–649 (2001)
19. Lowry, R.: Concepts and Applications of Inferential Statistics. VassarStat (2008)

Self Modifying Cartesian Genetic Programming: Fibonacci, Squares, Regression and Summing

Simon Harding[1], Julian F. Miller[2], and Wolfgang Banzhaf[1]

[1] Department of Computer Science, Memorial University, Canada
{simonh,banzhaf}@cs.mun.ca
http://www.cs.mun.ca
[2] Department Of Electronics, University of York, UK
jfm7@ohm.york.ac.uk
http://www.elec.york.ac.uk

Abstract. Self Modifying CGP (SMCGP) is a developmental form of Cartesian Genetic Programming(CGP). It is able to modify its own phenotype during execution of the evolved program. This is done by the inclusion of modification operators in the function set. Here we present the use of the technique on several different sequence generation and regression problems.

Keywords: Genetic programming, developmental systems.

1   Introduction

In biology, the process whereby genotypes gives rise to a phenotype can be regarded as a form of self-modification. At each decoding stage, the expressed genes, the local environment and the cellular machinery influence the subsequent genetic expression [1,2]. The concept of self-modification can be a unifying way of looking at development that allows the inclusion of genetic regulatory processes inside single cells, graph re-writing and multi-cellular systems.

In evolutionary computation self-modification has not been widely considered, but Spector and Stoffel examined its potential in their ontogenetic programming paper [3]. It also has been studied in the the graph re-writing system of Gruau [4] and was implicitly considered in Miller [5]. However, recently, much work in computational development has concentrated at a multi-cellular level and the aim has been to show that evolution could produce developmental cellular programs that could construct various cellular pattern [6]. A common motivation for evolving developmental programs is that they may allow the evolution of arbitrarily large systems which are infeasible to evolve using a direct genetic representation. However many of the proposed developmental approaches are not *explicitly* computational in that often one must apply some other mapping process from the developed cellular structure into a computation. Further discussion of our motivation, and how it relates to previous work, can be found in [7].

In our previous work, we showed that by utilizing self-modification operations within an existing computational method (a form of genetic programming,

L. Vanneschi et al. (Eds.): EuroGP 2009, LNCS 5481, pp. 133–144, 2009.

called Cartesian Genetic Programming, CGP) we could obtain a system that
(a) could develop over time in interaction with environmental inputs and (b)
would at every stage provide a computational function [7]. It could stop its own
development, if required, without external input. Another interesting feature of
the approach is that, in principle, programs could be evolved which allow the
replication of the original code.

Here we demonstrate SMCGP on a number of different tasks. The problems
illustrate different aspects and capabilities of SMCGP, and are intended to be
representative of the types of problems we will investigate in future.

2   Self Modifying CGP

2.1   Cartesian Genetic Programming (CGP)

Cartesian Genetic Programming represents programs as directed graphs [8].
Originally CGP used a program topology defined by a rectangular grid of nodes
with a user defined number of rows and columns. However, later work on CGP
always chose the number of rows to be one, thus giving a one-dimensional topology, as used in this paper. In CGP, the genotype is a fixed-length representation
and consists of a list of integers which encode the function and connections of
each node in the directed graph.

CGP uses a genotype-phenotype mapping that does not require all of the
nodes to be connected to each other, resulting in a bounded variable length
phenotype. This allows areas of the genotype to be inactive and have no influence
on the phenotype, leading to a neutral effect on genotype fitness called neutrality.
This type of neutrality has been investigated in detail [8,9,10] and found to be
beneficial to the evolutionary process on the problems studied.

2.2   SMCGP

In this paper, we use a slightly different genotype representation to previously
published work using CGP.

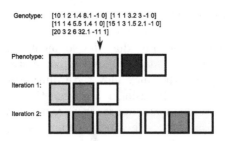

Fig. 1. The genotype maps directly to the initial graph of the phenotype. The genes
control the number, type and connectivity of each of the nodes. The phenotype graph
is then iterated to perform computation and produce subsequent graphs.

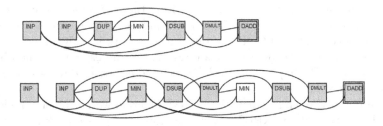

Fig. 2. Example program execution. Showing the DUP(licate) operator being activated, and inserting a copy of a section of the graph (itself and a neighboring functions on either side) elsewhere in the graph in next iteration. Each node is labeled with a function, the relative address of the nodes to connect to and the parameters for the function (see Section 2.4).

As in CGP an integer gene (in our case the first) encodes the function the node represents. This is followed by a number of integer connection genes that indicate the location in the graph where the node takes its inputs from. However in SMCGP there are also three real-valued genes that encode parameters required for the node function. Also there is a binary gene that indicates if the node should be used as an program output. In this paper all nodes take two inputs, hence each node is specified by 7 genes. An example genotype is shown in Figure 1.

Like CGP, nodes take their inputs in a feed-forward manner from either the output of a previous node or from a program input (terminal). The actual number of inputs to a node is dictated by the arity of its function. However, unlike previous implementations of CGP, nodes are addressed *relatively* and specify how many nodes back in the graph they are connected to. Hence, if the connection gene is 1 it means that the node will connect to the previous node in the list, if the gene has value 2 then the node connects 2 nodes back and so on. All such genes are constrained to be greater than 0, to avoid nodes referring directly or indirectly to themselves.

If a gene specifies a connection pointing outside of the graph, i.e. with a larger relative address than there are nodes to connect to, then this is treated as connecting to zero value. Unlike classic CGP, inputs arise in the graph through special functions. This is described in section 2.3. The relative addressing of connection genes allows sub-graphs to be placed or duplicated elsewhere in the graph whilst retaining their semantic validity.

This encoding is demonstrated visually in Figure 2.

The three node function parameter genes are primarily used in performing modification to the phenotype. In the genotype they are represented as real numbers but be cast (truncated) to integers if certain functions require it.

Section 4 details the available functions and any associated parameters.

2.3   Inputs and Outputs

The way we handled inputs in our original paper on SMCGP was flawed. We found, it did not scale well as sub-graphs became disconnected from inputs, as self-modifying functions moved them away from the beginning of the graph and they lost their semantic validity. The new input strategy required two simple changes from conventional CGP and our previous work in SMCGP.

The first, was to make all negative addressing return false (or 0 for non-binary versions of SMCGP). In previous work [7], we used negative addresses to connect nodes to input values.

The second was to change how the INPUT function works. When a node is of type INP (shorthand for INPUT), each successive call gets the next input from the available set of inputs. If the INP node is called more times than there are inputs, the counting starts from the beginning again, and the first node is used.

Outputs are handled slightly differently to inputs. We added another gene to the SMCGP node that decides whether the phenotype could use that node as an output. In previous work, we used the last n-nodes in the graph to represent the n-outputs. However, as with the inputs, we felt this approach did not scale as the graph changes size. When an individual is evaluated, the first stage is to identify the nodes in the graph that have their output gene set to 1. Once these are found, the graph can be evaluated from each of these nodes in a recursive manner.

If no nodes are flagged as outputs, the last n nodes in the graph are used as the n-outputs. Essentially, this reverts the system back to the previous approach. If there are more nodes flagged as outputs than are required, then the leftmost nodes that have flagged outputs are used until the required number of outputs is reached. If there are fewer nodes in the graph than required outputs, the individual is deemed to be corrupt and it is not evaluated (it is given a bad fitness score to ensure that it is not selected for).

2.4   Evaluation of the SMCGP Graph

From a high level perspective, when a genotype is evaluated the process is as follows. The initial phenotype graph is a copy of the genotype graph. This graph is then executed, and if there are any modifications to be made, they alter the phenotype graph.

The genotype is invariant during the entire evaluation of the individual. All modifications are made to the phenotype which starts out as a copy of the genotype. In subsequent iterations, the phenotype will usually gradually diverge from the genotype.

The encoded graph is executed in the same manner as standard CGP, but with changes to allow for self-modification. The graph is executed by recursion, starting from the output nodes down through the functions, to the input nodes. In this way, nodes that are unconnected ('junk') are not processed and do not affect the behavior of the graph at that stage.

For non-self modification function nodes the output value, as in GP in general, is the result of the mathematical operation on input values.

On executing a self-modification node, a comparison is made of the two input values. If the second value is less than the first, the second value is passed through. Otherwise, the node is activated and the first value passed through and the self-modification function in that node is added to a "To Do" list of pending modifications. This makes the execution of the self modifying function dependent on the data passing through the program.

After each iteration, the "To Do" list is parsed, and all manipulations are performed (provided they do not exceed the number of operations specified in the user defined "To Do" list length). The parsing is done in order of the instructions being appended to the list, i.e. first in is first to be executed.

The length of the list can be limited as manipulations are relatively computationally expensive to perform. Here, we limit the length to just 2 instructions. All graph manipulation functions require a number of parameters, as described in section 4.

3 Evolutionary Algorithm and Parameters

We use an (1+4) evolutionary strategy for the experiments in this paper. We bootstrap the process by testing a population of 50 random individuals. We then select the best individual and generate four offspring. We test these new individuals, and use the best of these to generate the next population. If there is more than one best in the population and one of these is the parent, we always choose the *offspring* to encourage neutral drift (see section 2.1).

We have used a relatively high (for CGP) mutation rate of 0.1. This means that each gene has a probability of 0.1 of being mutated. Mutations for the function type and relative addresses themselves are unbiased; a gene can be mutated to any other valid value.

For the real-valued genes, the mutation operator can choose to randomize the value (with probability 0.1) or add noise (normally distributed, sigma 20).

Evolution is limited to 10,000,000 evaluations. Trials that fail to find a solution in this time are considered to have failed.

The evolutionary parameter values have not been optimized, and we would expect performance increases if more suitable values were used.

4 Function Set

The function set is defined in two parts. The first is a set of modification operators, as shown in table 1. These are common to all data types used in SMCGP. The functions chosen are intended to give coverage to as many modification operations as possible. The remainder of the set are the computational operations. The data type these functions manipulate is determined by the problem definition. Table 2 contains definitions for all the numerical (i.e. non-modifying) operators that are available. Depending on the experiment, different sub-sets of this set are used.

The way self modifying functions act is defined by 4 variables. The three genes that are double precision numbers, here we call them "parameters" and denote them P_0, P_1, P_2. The other variable is the integer position of the node in the phenotype graph that contained the self modifying function (i.e. the leftmost

Table 1. Self modification functions

Function name	Description
Duplicate and scale addresses (DU4)	Starting from position $(P_0 + x)$ copy (P_1) nodes and insert after the node at position $(P_0 + x + P_1)$. During the copy, c_{ij} of copied nodes are multiplied by P_2.
Shift Connections (SHIFTCONNECTION)	Starting at node index $(P_0 + x)$, add P_2 to the values of the c_{ij} of next P_1 nodes.
Shift By Mult Connections (MULTCONNECTION)	Starting at node index $(P_0 + x)$, multiply the c_{ij} of the next P_1 nodes by P_2 nodes.
Move (MOV)	Move the nodes between $(P_0 + x)$ and $(P_0 + x + P_1)$ and insert after $(P_0 + x + P_2)$.
Duplication (DUP)	Copy the nodes between $(P_0 + x)$ and $(P_0 + x + P_1)$ and insert after $(P_0 + x + P_2)$.
Duplicate, Preserving Connections (DU3)	Copy the nodes between $(P_0 + x)$ and $(P_0 + x + P_1)$ and insert after $(P_0 + x + P_2)$. When copying, this function modifies the c_{ij} of the copied nodes so that they continue to point to the original nodes.
Delete (DEL)	Delete the nodes between $(P_0 + x)$ and $(P_0 + x + P_1)$.
Add (ADD)	Add P_1 new random nodes after $(P_0 + x)$.
Change Function (CHF)	Change the function of node P_0 to the function associated with P_1.
Change Connection (CHC)	Change the $(P_1 mod 3)$th connection of node P_0 to P_2.
Change Parameter (CHP)	Change the $(P_1 mod 3)$th parameter of node P_0 to P_2.
Overwrite (OVR)	Copy the nodes between $(P_0 + x)$ and $(P_0 + x + P_1)$ to $(P_0 + x + P_2)$, replacing existing nodes in the target position.
Copy To Stop (COPYTOSTOP)	Copy from x to the next "COPYTOSTOP" function node, "STOP" node or the end of the graph. Nodes are inserted at the position the operator stops at.

node is position 0), we denote this x. In the definitions of the SM functions we often need to refer to the values taken by node connection genes (which are all relative addresses). We denote the jth connection gene on node at position i, by c_{ij}.

There are several rules that decide how addresses and parameters are treated:

- When P_i are added to the x, the result is treated as an integer.
- Address indexes are corrected if they are not within bounds. Addresses below 0 are treated as 0. Addresses that reach beyond the end of the graph are truncated to the graph length.
- Start and end indexes are sorted into ascending order (if appropriate).
- Operations that are redundant (e.g. copying 0 nodes) are ignored, however they are taken into account in the ToDo list length.

Table 2. Numeric and other functions

Function name	Description
No operation (NOP)	Passes through the first input.
Add, Subtract, Multiply, Divide (DADD, DSUB, DMULT, DDIV)	Performs the relevant mathematical operation on the two inputs.
Const (CONST)	Returns a numeric constant as defined in parameter P_0.
$\sqrt{x}, \frac{1}{\sqrt{x}}$ Cos, Sin, TanH, Absolute (SQRT, DRCP, COS, SIN, TANH, DABS)	Performs the relevant operation on the first input (ignoring the second input).
Average (AVG)	Returns the average of the two inputs.
Node index (INDX)	Returns the index of the current node. 0 being the first node.
Input count (INCOUNT)	Returns the number of program inputs.
Min, Max (MIN, MAX)	Returns the minimum/maximum of the two inputs.

5   Experiments

5.1   Mathematical Functions and Sequences

We can use SMCGP to produce numeric sequences, where the program provides the first number on the first iteration, the 2nd on the next and continues to generate the next number in a sequence as we iterate the program. For these sequences, the input value to the program is fixed to 1. This forces the program to modify itself to produce a new program that produces the next digit. We demonstrate this ability on two sequences of integers; Fibonacci and a list of square numbers.

Squares. In this task, a program is evolved that generates a sequence of squares 0,1,2,4,9,16,25,... without using multiplication or division operations. As Spector (who first devised this problem) points out, this task can only be successfully performed if the program can modify itself - as it needs to add new functionality in the form of additions to produce the next integer in the sequence [3]. Hence, conventional genetic programming, including CGP, will be unable to find a general solution to this problem.

The function set for this experiment includes all the self modification functions and DADD, CONST, INP, MIN, MAX, INDX, SORT and INCOUNT.

Table 3 shows a summary of results for the squares problem (based on 110 runs). Programs were evolved to produce the first 10 terms in the sequence, with the fitness score being the number of correctly produced numbers. After successfully evolving a program that generates this sequence, the programs were tested on their ability to generalize to produce the first 100 numbers in the sequence. It was found that 84.3% of solutions generalised.

We found that as the squaring program iterates, the length of the phenotype graph increased linearly. However, the number of active nodes inside the graph fluctuated a lot on early iterations but stabilized after about 15 iterations.

Table 3. Evaluations required to evolve a program that can generate the squares sequence

Avg. Evaluations	Std. dev.	Min. evaluations	Max. evaluation	% generalize
141,846	513,008	392	3,358,477	84.3

Table 4. Success and evaluations required to evolve programs that generate the Fibonacci sequence. The starting condition and the length appear to have little influence on the the success rate of time taken. Percentage of solutions that generalize to solve up to 74 terms.

Start	Max. Iterations	% Success	Avg Evals	% Generalise
0 1	12	89.1	1,019,981	88.6
0 1	50	87.4	774,808	94.5
1 2	12	86.9	965,005	90.4
1 2	50	90.8	983,972	94.4

Fibonacci. Koza demonstrated that recursive tree structures could be evolved that generate the Fibonacci sequence [11]. Huelsbergen evolved machine language programs in approximately 1 million evaluations, and found that all his evolved programs were able to generalise [12]. Nishiguchi [13] successfully evolved recursive solutions with a success rate of 70%. The algorithm quickly evolves solutions in approximately 200,000 evaluations. However, the authors do not appear to test for generalisation. More recently, Agapitos and Lucas used a form of object oriented GP to solve this problem [14]. They tested their programs on the first 12 numbers in the sequence, and tested for generality. Generalization was achieved with 25% success and on average required 20 million evaluations. In addition, they note that their approach is computationally expensive. In [15] a graph based GP (similar to CGP) was demonstrated on this problem, however the approach achieved a success rate of only 8% on the training portion (first 12 integers) of the sequence. Wilson and Heywood evolved recursive programs using Linear GP that solved up to order-3 Fibonacci sequences [16]. On average solutions were discovered in 2.12×10^5 evaluations, with a generalization rate of 83%. We evolve for both the first 12 and 50 numbers in the sequence and test for generality to 74 numbers (after which the value exceeds a long int).

Table 4 shows the success rate and number of evaluations required to evolve programs that produce the Fibonacci sequence (based on 287 runs). As with the squares problem, the fitness function is the number of correctly outputted numbers. The starting condition (either 0,1 or 1,2) and the length of the target sequence appear to have little influence on the success rate of time taken. The results show that sequences that are evolved to produce the first 50 numbers do better at generalizing to produce the first 74 numbers. However, the starting condition again makes no difference.

Sum of numbers. Here we wished to evolve a program that could sum an arbitrarily long list of numbers. At the n-th iteration, the evolved program should

Table 5. Evaluations required to evolve a program that can add a set of numbers

Average	Minimum	Maximum	Std. Deviation
6,922	2,27	2,9603	4,998

Table 6. Evaluations required to evolve a SMCGP program that can add a set of numbers of a given size. 100% of SMCGP experiments were successful. The % success rate for conventional CGP is also shown.

Size of set	Average	Minimum	Maximum	Std. dev	% CGP
2	50	50	50	0	100
3	618	54	3,248	566	80
4	1,266	64	9,334	1,185	95.8
5	1,957	116	9,935	1,699	48
6	2,564	120	11,252	2,151	38.1
7	3,399	130	17,798	2,827	0
8	4,177	184	17,908	3,208	0
9	5,138	190	18,276	3,871	0
10	5,918	201	22,204	4,401	0

be able to take n inputs and compute the sum of all the inputs. We devised this problem because we thought it would be difficult for genetic programming, but relatively easy for a technique such as neural networks. The premise being, that neural networks appear to perform well when combining input values and genetic programming seems to prefer feature selection on the inputs.

Input vectors consist of random sequences of integers. The fitness is defined as the absolute cumulative error between the output of the program and the expected sum of the values. We evolved programs which were evaluated on input sequences of 2 to 10 numbers. The function sets consists of the self modifying functions and just the ADD operator.

Table 5 summarizes the results this problem. All experiments were found to be successful, in that they evolved programs that could sum between 2 and 10 numbers (depending on the number of iterations the program is iterated). Table 6 shows the number of evaluations required to reach the nth sum (where n is from 2 to 10).

After evolution, the best individual for each run was tested to see how well it generalized. This test involved summing a sequence of 100 numbers. It was found that most solutions generalized, however in 0.07% of cases, they did not.

We also tested the ability of conventional CGP to sum a set of numbers. Here CGP could only be evolved for a given size of set of input numbers. The results (based on 500 runs) are also shown in table 6. It is revealed that CGP is able to solve this problem only for a smaller sets of numbers. This shows a clear benefit of the self-modification approach in comparison with the direct encoding.

5.2 Regression and Classification

Bioinformatics Classification. In this experiment, SMCGP is applied to the protein classification problem described in [17]. The task is to predict the location of a protein in a cell, from the amino acids in the particular protein. The entire dataset was used for the training set. The set consisted of 2,427 entries, with 19 variables each and 1 output. The function set for SMCGP includes all the mathematical operators in addition to the self modifying command set. The CGP function set contained just the mathematical operators (see section 4). For this type of problem, it is not clear that a self-modification approach would have advantages compared with classic CGP. Also, we added the number of iterations to the genotype so that the phenotype is iterated that many times before being executed on the training set.

Table 7 shows the summary of results for this problem (based on 100 runs of each representation). Both CGP and SMCGP perform similarly. The addition of the self modification operations does not appear to hinder evolvability - despite the increase in search space size.

Table 7. Results summary for the bio informatics classification problem

-	CGP	SMCGP
Average fitness (training)	66.81	66.83
Std. dev. fitness (training)	6.35	6.45
Average fitness (validation)	66.10	66.18
Std. dev. fitness (validation)	5.96	6.46
Avg. evaluations to best fitness (training)	7,679	7,071
Std. dev. evaluations to best fitness (training)	2,452	2,644
Avg. evaluations to best fitness (validation)	7,357	7,161
Std. dev. evaluations to best fitness (validation)	2,386	2,597

Powers Regression. A problem was devised that tests the ability of SMCGP to learn a 'modular' regression problem. The task is to evolve a program that, depending on the iteration, approximates the expression x^n where n is the iteration number. The fitness function applies x as integers from 0 to 20. The fitness is defined as the number of wrong outputs (i.e. lower is better).

The function set contains all the modification operators (section 4) and NOP, DADD, DSUB, NOP, DADD, DSUB, DMULT, DDIV, CONST, INDX and IN-COUNT from the numeric operations (section 4).

Programs are evolved to $n = 10$ and then tested for generality up to $n = 20$. As with other experiments, the program is evolved incrementally. Where it first tries to solves n=1 and if successful, is evaluated and evolved for n=1 and n=2 until eventually it is evaluated on n=1 to 10.

Table 8 shows the results summary for the powers regression problem. All 337 runs were successful. In this instance, there is an interesting difference between the two starting conditions. If the fitness function starts with $n = 1$ to 5, it is found that that fewer evaluations are required to reach $n = 10$. However, this

Table 8. Summary of results for the powers regression problem

Number of Initial Test Sets	Average Evaluations	Std Dev.	Percentage Generalize
1	687,156	869,699	60.4
5	527,334	600,800	55.6

leads to reduced generalization. Using a Kolmogorov-Smirnov test, it was found that the difference in the evaluations required is statistically significant (p=0.0).

6 Conclusions

We have examined and discussed a developmental form of Cartesian Genetic Programming called Self Modifying CGP and evaluated and compared it with classic CGP on a number of diverse problems. We found that it is more efficient than classic CGP at solving four of the test problems: Fibonacci sequence, sequence of squares, sum of inputs, and a power function. In addition it appears that it was able to obtain general solutions for all these problems, although full confirmation of this will require further analysis of evolved programs. On a fourth problem, classification it performed no worse than CGP despite a larger search space.

Other approaches to solving problems, such as Fibonacci, produce a computer program. Instead, at each iteration we produce a structure that produces the output. This could be considered as unrolling the loops (or recursion) in a program. In a related paper [18], we use this structure building approach to construct digital circuits. In future work, we will investigate SMCGP when used as a continuous program. We believe combining both approaches will be beneficial.

References

1. Banzhaf, W., Beslon, G., Christensen, S., Foster, J.A., Képès, F., Lefort, V., Miller, J.F., Radman, M., Ramsden, J.J.: From artificial evolution to computational evolution: A research agenda. Nature Reviews Genetics 7, 729–735 (2006)
2. Kampis, G.: Self-modifying Systems in Biology and Cognitive Science. Pergamon Press, Oxford (1991)
3. Spector, L., Stoffel, K.: Ontogenetic programming. In: Koza, J.R., Goldberg, D.E., Fogel, D.B., Riolo, R.L. (eds.) Genetic Programming 1996: Proceedings of the First Annual Conference, pp. 394–399. MIT Press, Stanford University (1996)
4. Gruau, F.: Neural network synthesis using cellular encoding and the genetic algorithm. Ph.D. dissertation, Laboratoire de l'Informatique du Parallélisme, Ecole Normale Supérieure de Lyon, France (1994)
5. Miller, J.F., Thomson, P.: A developmental method for growing graphs and circuits. In: Tyrrell, A.M., Haddow, P.C., Torresen, J. (eds.) ICES 2003. LNCS, vol. 2606, pp. 93–104. Springer, Heidelberg (2003)
6. Kumar, S., Bentley, P.: On Growth, Form and Computers. Academic Press, London (2003)

7. Harding, S.L., Miller, J.F., Banzhaf, W.: Self-modifying cartesian genetic programming. In: Thierens, D., Beyer, H.-G., et al. (eds.) GECCO 2007: Proceedings of the 9th annual conference on Genetic and evolutionary computation, vol. 1, pp. 1021–1028. ACM Press, London (2007)
8. Miller, J.F., Thomson, P.: Cartesian genetic programming. In: Poli, R., Banzhaf, W., Langdon, W.B., Miller, J., Nordin, P., Fogarty, T.C. (eds.) EuroGP 2000. LNCS, vol. 1802, pp. 121–132. Springer, Heidelberg (2000)
9. Vassilev, V.K., Miller, J.F.: The advantages of landscape neutrality in digital circuit evolution. In: Miller, J.F., Thompson, A., Thompson, P., Fogarty, T.C. (eds.) ICES 2000. LNCS, vol. 1801, pp. 252–263. Springer, Heidelberg (2000)
10. Yu, T., Miller, J.: Neutrality and the evolvability of boolean function landscape. In: Miller, J., Tomassini, M., Lanzi, P.L., Ryan, C., Tetamanzi, A.G.B., Langdon, W.B. (eds.) EuroGP 2001. LNCS, vol. 2038, pp. 204–217. Springer, Heidelberg (2001)
11. Koza, J.: Genetic Programming: On the Programming of Computers by Natural Selection. MIT Press, Cambridge (1992)
12. Huelsbergen, L.: Learning recursive sequences via evolution of machine-language programs. In: Koza, J.R., Deb, K., et al. (eds.) Genetic Programming 1997: Proceedings of the Second Annual Conference, pp. 186–194. Morgan Kaufmann, Stanford University (1997)
13. Nishiguchi, M., Fujimoto, Y.: Evolution of recursive programs with multi-niche genetic programming (mnGP). In: Evolutionary Computation Proceedings, 1998. IEEE World Congress on Computational Intelligence, pp. 247–252 (1998)
14. Agapitos, A., Lucas, S.M.: Learning recursive functions with object oriented genetic programming. In: Collet, P., Tomassini, M., Ebner, M., Gustafson, S., Ekárt, A. (eds.) EuroGP 2006. LNCS, vol. 3905, pp. 166–177. Springer, Heidelberg (2006)
15. Shirakawa, S., Ogino, S., Nagao, T.: Graph structured program evolution. In: Proceedings of the 9th annual conference on Genetic and evolutionary computation, pp. 1686–1693. ACM, London (2007)
16. Wilson, G., Heywood, M.: Learning recursive programs with cooperative coevolution of genetic code mapping and genotype. In: GECCO 2007: Proceedings of the 9th annual conference on Genetic and evolutionary computation, pp. 1053–1061. ACM Press, New York (2007)
17. Langdon, W.B., Banzhaf, W.: Repeated sequences in linear genetic programming genomes. Complex Systems 15(4), 285–306 (2005)
18. Harding, S., Miller, J.F., Banzhaf, W.: Self modifying cartesian genetic programming: Parity. In: CEC 2009 (2009) (submitted)

Automatic Creation of Taxonomies of Genetic Programming Systems

Mario Graff and Riccardo Poli

School of Computer Science and Electronic Engineering, University of Essex, UK
{mgraff,rpoli}@essex.ac.uk

Abstract. A few attempts to create taxonomies in evolutionary computation have been made. These either group algorithms or group problems on the basis of their similarities. Similarity is typically evaluated by manually analysing algorithms/problems to identify key characteristics that are then used as a basis to form the groups of a taxonomy. This task is not only very tedious but it is also rather subjective. As a consequence the resulting taxonomies lack universality and are sometimes even questionable. In this paper we present a new and powerful approach to the construction of taxonomies and we apply it to Genetic Programming (GP). Only one manually constructed taxonomy of problems has been proposed in GP before, while no GP algorithm taxonomy has ever been suggested. Our approach is entirely automated and objective. We apply it to the problem of grouping GP systems with their associated parameter settings. We do this on the basis of performance signatures which represent the behaviour of each system across a class of problems. These signatures are obtained thorough a process which involves the instantiation of models of GP's performance. We test the method on a large class of Boolean induction problems.

1 Introduction

A *taxonomy* is the coherent arrangement of elements into groups. For many sciences, the construction of a taxonomy has been an important step towards maturity. Taxonomies have many applications. In biology, for example, taxonomies of animals and plants are the starting point in understanding the biodiversity of a region. Also, taxonomies help to model a group of individuals as a single entity, thereby removing the need to analyse each member of the group separately.

The importance of taxonomies is also clear for Evolutionary Algorithms (EAs). For example, from a practitioner's point of view, an *algorithm taxonomy* may reduce the time-consuming task of finding the most suitable algorithm to solve a problem at hand. This is generally done by selecting an algorithm that has already been applied to a "similar" problem. If this algorithm fails to return a satisfactory result, one would want to look for an algorithm that, due to its characteristics, is sufficiently different from the first, e.g., an algorithm that belonged to a different group of the taxonomy. *Taxonomies of problems* are also useful since it is often possible to associate algorithms to problem classes. Knowing which taxonomic class a problem is in can then guide a practitioner towards good algorithms for solving it.

So, what taxonomies are available for EAs? There are some general taxonomies. In fact, every author of a book on EAs is forced to come up with some structure within which to

L. Vanneschi et al. (Eds.): EuroGP 2009, LNCS 5481, pp. 145–158, 2009.

present the material in an orderly fashion; for example see [3,14]. This structure can be seen as a taxonomy of algorithms. Typically these taxonomies are constructed on the basis of a set of characteristics which are deemed to best represents such algorithms. These often focus on each of the main components of an algorithm, e.g., the representation, the genetic operators, the number of parents used to produce an offspring, the number of offspring, the way the offspring is included in the population, etc. That is, they take a reductionist approach, which looks at similarities and differences among the components of an algorithm, *as perceived by the user*. Also, often the topmost levels in the taxonomy are determined by historical reasons.

While these taxonomies are useful, they are, rather naturally, subjective. Furthermore, while some of the most useful taxonomies in other disciplines are *based on behaviour*, not just form (think, for example, of the periodic table), these types of EA taxonomies are typically only based on "form", i.e., the components of an algorithm. So, they aren't designed to give answers to practitioners, who are really interested in determining whether an algorithm will solve their problems.

Of course, there are other EA taxonomies which are much more precise and useful. For example [4] proposed to use a tabular representation to describe genetic algorithms, scatter search and ant systems. In [13] a taxonomy of parallel genetic algorithms was proposed, which was based on the different ways in which a genetic algorithm can be parallelised. [15] categorises various EAs and other mathematical and AI search algorithms. In [7] a taxonomy of crossover operators for real-coded genetic algorithms was presented. The taxonomy consisted in 18 different crossover operators grouped into 3 classes. There have also been a very small number of taxonomic efforts of this kind in GP, e.g., [18]. Also a GP problem taxonomy was proposed in [12].

One feature of these taxonomies is that they capture only specific aspects of an algorithm or they are very specific to a special class of algorithms. The reason is combinatorial: not only there are many different EAs to consider, but also each may have from several to tens of parameters (e.g., 17 parameters were used in the GP system described in [10]) which can alter its behaviour, in some cases very significantly. If one considers the combinations of these parameters, then the set of algorithms with their different configurations immediately becomes astronomically large. Another feature of these taxonomies is that, like the ones described above, they are *manually built* and so they are subjective and are normally based on form rather than behaviour of algorithms.

Modelling the behaviour of algorithms is the realm of EA theory. There are a variety of tools that have been used to understand and categorise EAs. These range from Markov chains [16] to schema theories [11] to computational complexity [8]. These have produced important results which go some way in the direction of a taxonomy based on behaviours. It is possible, for example, to determine whether a particular EA will solve a particular problem in polynomial vs. non-polynomial time or whether a particular EA is a global optimiser (given enough time). However, results are either specific to a problem or of purely mathematical interest. Furthermore, they don't really provide a detailed EA taxonomy, in the sense that typically they only include two classes. So, mathematical models, too, struggle to give taxonomies that meet the needs of practitioners.

A step in the direction of automating the creation of taxonomies as well as turning them into practical tools has been made in [1,2] where several tens of problems where automatically grouped into a *problem taxonomy*. The idea was to run a particular EA where mating is controlled via a neighbourhood structure recording the average number of fitness evaluations required to solve each problem using each of a small number of neighbourhood structures. The resulting performance signatures (vectors) were then clustered hierarchically, thereby producing a taxonomy.

Here, we present a method to create taxonomies which has some similarity with the one proposed by Ashlock *et al.*, but with also some differences. Firstly, we focus on GP. Secondly, we are interested in producing a practical *algorithm taxonomy*, which we think may be more useful than a problem taxonomy. Like Ashlock's work, our approach is automated and objective. We apply it to the problem of grouping GP systems with their associated parameter settings. Like [1,2] we do this on the basis of the performance signatures and a hierarchical clustering algorithm. However, our performance signatures represent the behaviour of each system across a class of problems, not the behaviour of multiple versions of an algorithm over one problem. Also, our signatures are not the measurement of performance over multiple cases: they are obtained thorough a process which involves the instantiation of a mathematical models of GP's performance recently introduced in [6]. It is then this model which provides the signature for each algorithm.

The paper is organised as follows. In Sec. 2 we describe the clustering algorithm used to create our GP taxonomy. In Sec. 3 we review the models of GP's performance proposed in [6]. Sec. 4 presents a new procedure to instantiate such models. Alternative similarity measures for comparing different GP performance models and systems are presented in Sec. 5. In Sec. 6 we describe the parameters setting and the different GP systems used to test our approach. This is followed (in Sec. 7) by a description of the system taxonomies created when evaluating performance over a large class of Boolean induction problems. Some conclusions are given in Sec. 8.

2   Creating GP's Taxonomies

As we already mentioned, our technique to create taxonomies for GP is based on a *hierarchical clustering algorithm* (see [9]). This works as follows. Let us assume we are given a set of objects (performance signatures, in our case) which we wish to group. Let $\mathcal{M}$ be a square matrix where the element (i,j) represents the distance (or similarity) between the i-th and j-th objects. We derive from it a similarity measure $s(X,\mathcal{Y})$ between pairs of clusters, X and $\mathcal{Y}$ (more on this below). Then the clustering algorithm of [9] involves the following three steps: (a) First, each object is assigned to a separate cluster; (b) A new cluster is created by merging the two clusters which are closest based on the similarity measure $s(X,\mathcal{Y})$, thereby reducing the number of clusters by one; (c) Repeat step 2 until there is only one cluster left.

Naturally the behaviour of the algorithm is influenced by the similarity measure $s(X,\mathcal{Y})$ used. In this paper, we used the *average linkage* as similarity measure. That is

$$s(X,\mathcal{Y}) = \frac{\sum_{i \in X} \sum_{j \in \mathcal{Y}} \mathcal{M}(i,j)}{|X||\mathcal{Y}|} \tag{1}$$

where $|\cdot|$ denotes the number of elements in a cluster.

This algorithm has now one key component left to specify: the distance matrix $\mathcal{M}$. A meaningful $\mathcal{M}$ can be computed using different procedures. In this paper we associate to each GP system a vector $\mathbf{c}$ (a performance signature) and build $\mathcal{M}$ by computing the distance (or similarity) between all the possible pairs of vectors $\mathbf{c}$ (more on this later).

Of course, then the question becomes how to choose the size and components of $\mathbf{c}$. Again, there are different ways to define a mapping between GP systems (and parameter settings) and $\mathbf{c}$ vectors. For example, taking a dual approach to Ashlock's [1,2], we could select a group of problems of interest, and associate a different component of $\mathbf{c}$ to each problem. The component could just represent the performance of a particular GP system on the problem associated to that component. Each $\mathbf{c}$ vector would then represent the behaviour of a GP system over all problems in our set.

While this is a simple and effective way of proceeding when the set of problems we want to use to build a taxonomy of algorithms is small, it is not viable when one wants to consider large problem classes. This is unfortunate, since to draw general taxonomic conclusions about different GP systems we need to test them over a large set of problems. So, we decided against this approach and instead adopted a more complex, but also more general technique which consists in creating the $\mathbf{c}$ vectors using the coefficients of a model of GP behaviour. In particular, we will use a version of the performance model originally proposed in [6]. We review it in the next section.

3   Performance Model

The model proposed in [6] is a practical model of GP which has been specifically designed to focus on performance. Its applicability was tested on the class of rational symbolic regression problems with excellent results using as a performance measure the best fitness recorded in a run.

The model is based on the idea that GP's performance can be modelled using a re-representation of the fitness function. The simplest re-representation is as follows. Let $\Omega = \{p_1, p_2, \ldots\}$ be the ordered set of all programs obtained by recursively composing primitives from the primitive set and let f be a function over Ω. Then, one can represent the fitness function as the ordered set $\mathcal{F}(\Omega) = \{f_{p_1}, f_{p_2}, \ldots\}$ where $f_{p_i} = f(p_i)$. Using this tabular re-representation of the fitness function, in principle, one could compute the following linear approximation (model) of GP's performance $P(f) \approx a_0 + \sum_{p \in \Omega} a_p f_p$ where the a_p are coefficients, which, once again, in principle, could be obtained running a least square method on a suitable training set of (P, f) pairs. In reality, however, this approach is not viable because one needs to identify $|\Omega| + 1$ parameters to create the model. So, if the search space is large (and it typically is), very large training sets would be needed. It is easy, however, to fix this problem: instead of re-representing f using all the programs in Ω, we can use a subset $S \subseteq \Omega$. We can then use the approximation

$$P(f) \approx a_0 + \sum_{p \in S} a_p f_p \tag{2}$$

where we can control the trade-off between model accuracy and training set size by varying the cardinality of S.

In GP, the fitness function $f(p)$ typically states how similar the functionality of a program p is to a target functionality t. Often t is represented via a finite set of fitness cases. So, we can think of both t and p as being ℓ-dimensional vectors, $\mathbf{t}$ and $\mathbf{p}$, respectively, ℓ being the number of fitness cases. We can then define $f(p) = d(\mathbf{p}, \mathbf{t})$ where d is a similarity measure between vectors.[1]

Under these conditions, (2) transforms into:

$$P(\mathbf{t}) \approx a_0 + \sum_{\mathbf{p} \in S} a_\mathbf{p} \cdot d(\mathbf{p}, \mathbf{t}) \tag{3}$$

In order to instantiate this model, one needs: a training set of problems T, a validation set V (to test the generality of the model), a closeness measure $d : \mathbb{R}^\ell \times \mathbb{R}^\ell \mapsto \mathbb{R}$ and a set of program behaviours from which elements of S are drawn. T and V are composed by pairs $(P(\mathbf{t}), \mathbf{t})$ where $\mathbf{t}$ is a problem (represented by its target vector) and $P(\mathbf{t})$ is the performance of GP on problem $\mathbf{t}$. To obtain $P(\mathbf{t})$ a GP system is run multiple times and its performance is estimated by averaging. Note that while (3) was derived from (2) under the assumption that $f(p) = d(\mathbf{p}, \mathbf{t})$, we are not forced to use the same similarity measure d in both: in (3) we can choose a different d if this, for example, improves the model's accuracy.

A model selection algorithm is used to obtain S and the coefficients $a_\mathbf{p}$ from the training set T. This is a point where our approach deviates from [6] where a GA and ordinary least squares were used for this purposes, while here we use a much simpler, yet effective approach. In the next section we provide details on the model identification techniques we used.

4   Model Optimisation

We used the Least Angle Regression (LAR) algorithm (see [5] for details) as a procedure to decide which elements should be included in S given a larger set of prospective target behaviours Σ. We stop LAR after m steps, where m is the desired size for the set S, and we pick as elements of S the m elements from Σ chosen by the algorithm so far. In this way we are certain to retain in S elements of Σ with a high correlation with GP's performance, thereby increasing the accuracy of the model over alternative ways of choosing S.

Of course, now we are faced with the question of how to choose the elements of Σ. Since problems $\mathbf{t}$ and program behaviours $\mathbf{p}$ are discretised as vectors of size ℓ, we could pick the elements of Σ from the problem class itself. For small, discrete domains, one could use all the possible functions in the class of problems. If the number of problems is too big or infinite, one could draw a representative set of samples from the problem class. In this work we effectively used the latter approach. More specifically, we took $\Sigma = T$.

Although this procedure automates the process of selecting S, it still requires the user to specify the size of S. To free the user from this delicate duty, we decided to use a cross-validation technique on the training set to determine the best size for S. This works as follows. The training set is split into five sets of equal size. Four sets are joined together and are used to produce a model (i.e., these sets are now the training set) while the remaining set is used to assess its generalisation. That is, we use the created model to predict the performance of the GP system in the problems of the remaining set. The process is repeated 5 times, each time leaving out a different fifth of the training set T. As a result, we get a performance predictions for the problems in the training set.

---

[1] Typically, $d(\mathbf{p}, \mathbf{t}) = \sum_i |p_i - t_i|^k$, with $k = 1$ (sum of absolute errors) or $k = 2$ (RMS error).

We use these predictions and the actual GP's performance to measure the quality of models using the Relative Squared Error (RSE). The RSE is defined as $rse = \frac{\sum_i (P_i - \tilde{P}_i)^2}{\sum_i (P_i - \bar{P})^2}$ where i ranges over the set of test problems used to evaluate the accuracy of a model, P_i is the average performance recorded for problem i, $\tilde{P}_i$ is the performance predicted by the model, and $\bar{P}$ is the average performance over all runs. The objective is to obtain values of rse as close as possible to zero.[2]

Cross-validation was applied to models produced by LAR with $m = 1, 2 \ldots, |T|$ with the aim of identifying the value of m which provided the best generalisation. The process ends when the best generalisation on the training set has been reached. Note that while the process of identifying S from T *with a known m* involves first running LAR and then doing least squares, to save computation during the determination of the optimal m we did not performed the least square method after LAR. Instead we simply used the models produced by LAR.

To compute the performance model we need to define a closeness measure d. Obviously, the accuracy of the model depends on the closeness measure used. In preliminary work we tested different closeness measures. Here we adopt the one that gave the model with best quality, namely $d = (\mathbf{p} \cdot \mathbf{t})^2$ where $\cdot$ is the scalar product.

To sum up, we create a performance model of a GP system by using the LAR algorithm to select the set S, a cross-validation technique to set the size of S, and, finally, a least square method to compute the coefficients a_p.

5 Similarity Matrices

We are now in the position to compute the matrix $\mathcal{M}$ required to apply (1). We present three different methods to compute $\mathcal{M}$. Two of them use the coefficients a_p of the model obtained using two different similarity measures d, namely the angle and the Euclidean distance between vectors of coefficients a_p. The third method uses the mean and the standard deviation GP's performance on a training set to associate a pair (μ, σ) to each GP system. The matrix $\mathcal{M}$ is then constructed using the Euclidean distance between pairs of (μ, σ) tuples. These methods are described in detail below.

It is possible to compute the angle between two GP systems by first noticing that (3) represents a hyper-plane. To see this, we rewrite (3) in normal form using the scalar product between two vectors, namely $\mathbf{c} \cdot ((P(\mathbf{t}), d(\mathbf{p_1}, \mathbf{t}), \cdots, d(\mathbf{p}_{|S|}, \mathbf{t})) - \mathbf{x_0}) = 0$ where $\mathbf{x_0} = (a_0, 0, \ldots, 0)$ is a particular point on the hyper-plane, $\mathbf{c} = (-1, a_{\mathbf{p_1}}, \cdots, a_{\mathbf{p}_{|S|}})$, and a_p are the coefficients of the performance model. Then, we can measure the angle (or dissimilarity) between two GP systems as the angle between the hyper-planes representing these systems. If the systems are characterised by the vectors $\mathbf{c}' = (-1, a'_{\mathbf{p_1}}, \cdots, a'_{\mathbf{p}_{|S|}})$ and $\mathbf{c}'' = (-1, a''_{\mathbf{p_1}}, \cdots, a''_{\mathbf{p}_{|S|}})$, the angle between them is $\alpha = \arccos\left(\frac{\mathbf{c}' \cdot \mathbf{c}''}{||\mathbf{c}'|| ||\mathbf{c}''||}\right)$.

Naturally, for this idea to work, one needs to make sure that the coefficients of the two models are associated to the same programs $\mathbf{t}$. Because different models are built independently (using the procedure described in Sec. 4), this may not be the case. So, to avoid

---

[2] A value of rse close to 1 means that the model is as good (or bad) at predicting performance differences as the mean. A value of rse less than 1 means that the model predicts better than the mean, while $rse > 1$ implies worse predictions than the mean.

problems, we used the following strategy. Let S' and S'' be two sets associated to the performance models of two different GP systems, a and b. We start by instantiating two models, represented by the vectors $c'_a = (-1, a'_{\mathbf{p}_1}, \cdots, a'_{\mathbf{p}_{|S'|}})$ and $c'_b = (-1, b'_{\mathbf{p}_1}, \cdots, b'_{\mathbf{p}_{|S'|}})$, which were obtained by creating performance models of a and b using set S'. We repeat the procedure, obtaining vectors $c''_a = (-1, a''_{\mathbf{p}_1}, \cdots, a''_{\mathbf{p}_{|S''|}})$ and $c''_b = (-1, b''_{\mathbf{p}_1}, \cdots, b''_{\mathbf{p}_{|S''|}})$, respectively, but this time using S''. We then compute two angles for each pair of GP systems $\alpha_1 = \arccos\left(\mathbf{c}'_a \cdot \mathbf{c}'_b / \|\mathbf{c}'_a\| \|\mathbf{c}'_b\|\right)$ and $\alpha_2 = \arccos\left(\mathbf{c}''_a \cdot \mathbf{c}''_b / \|\mathbf{c}''_a\| \|\mathbf{c}''_b\|\right)$. From these we finally compute the *angle between two GP systems* as $\alpha = \frac{\alpha_1 + \alpha_2}{2}$. By applying the procedure described above to all pairs of GP systems under study we obtain the elements of matrix $\mathcal{M}$ one by one.

In the second method to compute $\mathcal{M}$ we first re-write (3) as a scalar product between two vectors:

$$P(\mathbf{t}) \approx (a_0, \ldots, a_{|S|}) \cdot (1, d(\mathbf{p}_1, \mathbf{t}), \ldots, d(\mathbf{p}_{|S|}, \mathbf{t})). \tag{4}$$

Clearly, the performance model in (4) depends only on a vector $\mathbf{c} = (a_0, \ldots, a_{|S|})$. We can therefore calculate the similarity between GP systems by measuring the Euclidean distance between their associated $\mathbf{c}$ vectors.

As for the case discussed in Sec. 5, again for this to work we must make sure corresponding components of the $\mathbf{c}$ vectors being compared are associated to the same elements of Σ. Here again we used the same approach, i.e., compute two distance measures and then average them.

The third distance matrix $\mathcal{M}$ is computed using the statistics of GP's performance on a set of test problems. This set is the union of the training set T and validation set V. We define $\mathbf{c} = (\mu, \sigma)$ where μ is the mean and σ is the standard deviation of GP's performance.

These statistics were collected by running each GP system in the conditions described in the following section.

6   Test Problems, GP Systems and Parameter Settings

To test our approach, we used two, fairly standard GP systems. Both systems use subtree crossover and subtree mutation. One system is essentially identical to the one used by Koza [10] the only significant difference being that we select crossover and mutation points uniformly at random, while [10] used a non-uniform distribution. The other system is TinyGP (see [14, Appendix]). The main difference between the two systems is that the Koza-style system is generational, while TinyGP uses a steady-state strategy.

Tab.1 shows the parameters for the GP systems. We only use combinations of crossover and mutation rates such that $p_{xo} + p_m \leq 100\%$[3] because in GP crossover and mutation are mutually exclusive operators. For the Koza-style GP system, besides the traditional roulette-wheel selection, we also tested tournament selection with a tournament size of 2. When using the terminal set $\mathcal{T}_1$, we considered all possible combinations of the parameters presented in Tab. 1 for all three types of GP system (Koza-style system with fitness proportionate and tournament selection and TinyGP). We, therefore, obtained 36 different GP systems. Besides these systems, we included another set of systems using as terminal

---

[3] The systems with a crossover rate of 90% use a 10% rate of reproduction.

Table 1. Parameters used in the GP experiments

Function set	$\{AND, OR, NAND, NOR\}$
Terminal set	$\mathcal{T}_1 = \{x_1, x_2, x_3, x_4\}, \mathcal{T}_2 = \mathcal{T}_1 \cup \{\bar{x}_1, \bar{x}_2, \bar{x}_3, \bar{x}_4\}$
Crossover rate p_{xo}	100%, 90%, 50%, and 0%
Mutation rate p_m	100%, 50%, and 0%
Maximum tree depth used in mutation	4
Selection Mechanism	Tournament (size 2) and Roulette-wheel
Population size	5000, 1000, and 500
Number of generations	50
Number of independent runs	100

set $\mathcal{T}_2$ and restricting the population size to be 1000. Therefore, in total we analysed 48 different GP systems. While this number is not very large, the variety of systems considered is sufficiently representative for the purpose of exemplifying our method for the automated creation of algorithm taxonomies in GP.

The training set T and validation set V are composed of Boolean induction problems with 4 inputs. We randomly selected 1,100 different Boolean functions from this set and for each one of them we counted the number of times the GP algorithm found a program that encoded the target functionality in 100 independent runs. We took as our performance measure the fraction of successful runs out of the total number of runs. Then these 1,100 randomly generated problems were divided into two sets: a training set T composed of 500 functions and the validation set V comprising the remaining 600 functions.

7   Taxonomies

Once the matrix $\mathcal{M}$ is available, we can apply the hierarchical clustering algorithm described in Sec. 2 to progressively group our 48 GP systems into larger and larger clusters. These clusters form a hierarchy including 48 levels (one for each cycle of the clustering algorithm). The hierarchy can naturally be represented using a *dendrogram* such as the one shown in Fig. 1.

While dendrograms are a useful tool, they are not a taxonomy. Firstly, there are far too many classes for a taxonomy (the dendrogram for n objects includes $n - 1$ classes). Secondly, the dendrograms produced in the presence of many objects (such as our 48 GP systems) are complex and difficult to digest. For example, it is difficult to appreciate the relative distance between clusters. So, how can we convert a dendrogram of (models of) GP systems into a taxonomy?

To transform dendrograms into taxonomies we first perform a data reduction exercise. We focus only on the 10 topmost clusters in the dendrogram starting from its root node. We then imagine that each of these 10 clusters is a node in a fully connected graph. We associate to each edge a weight defined by $s(\mathcal{X}, \mathcal{Y})$. We add to the resulting graph a further set of 48 nodes, one for each GP system, and we connect each of them to the node representing the (lowest level) cluster containing the corresponding GP system. We also associate a weight $s(\mathcal{X}, \mathcal{Y})$ to these new edges. This new weight was defined to be the result of $s(\mathcal{X}, \mathcal{X})$ when the GP system belongs to the class $\mathcal{X}$, thereby we can interpreted it as the distance within cluster $\mathcal{X}$. This graph is our GP taxonomy.

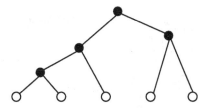

Fig. 1. Example dendrogram: hollow circles represent class members, solid ones represent clusters

Of course, an important objective is to be able to visualise this taxonomy, to see what we can learn from it. There are number of ways one can draw graphs. In this paper, we decided to use the neato package, which is part of the GraphViz library.

Neato uses the following strategy to position nodes. Nodes are interpreted as physical bodies connected by springs (which correspond to the edges between nodes in the graph). The user can set both the spring stiffness and the rest length. To produce a layout, the (virtual) physical system is initialised in some suboptimal configuration, which is then iteratively modified until the system relaxes into a state of minimal energy. Often this relaxation process leads to nicely drawn graphs where groups of nodes with either strong connections or many connections tend to be placed next to each other (since this effectively minimises the associated elastic energy).

To produce graphical representations of our taxonomies we gave neato the weighted graphs described above, setting each spring's rest-length to 10 times the weight of the corresponding edge. The stiffness of springs was set to be approximately inversely proportional to weights using the formula $\frac{1}{0.01+s(\mathcal{X},\mathcal{Y})}$.

Below we will show figures illustrating the results produced by neato. In each figure, edges between clusters are represented using dotted lines, while edges connecting GP systems to their mother cluster are drawn with solid lines. To label nodes we use the following nomenclature. Each system is represented by three terms: a) a term of the form P####, which represents the size of the population; b) a term of the form X####, which represents the crossover rate (as a percentage);[4] c) a term which indicates which terminal set was used. So, a node labelled P500_X50_T1 represents a system with a population of size 500, $p_{xo} = p_m = 0.5$ and terminal set $\mathcal{T}_1$. To distinguish different forms of selection and reproduction we use different graphical symbols to represent the corresponding nodes. Namely, the TinyGP systems are represented with ellipses, while the Koza-style GP systems with roulette-wheel section and tournament selection are represented using octagons and double octagons, respectively. Finally, since in many cases a cluster contains all the GP systems with a particular population size and terminal set, we represent such systems with a single node. In these cases, we drop element (b) from the node label.

Fig. 2 shows the taxonomy obtained when $\mathcal{M}$ is computed using the angles between GP performance models (interpreted as hyper-planes, as explained in Sec. 5). As one can see, the taxonomy includes a central cluster that is mainly populated by generational systems with small populations and tournament selection, which is surrounded by several satellite

---

[4] We did not include in the nomenclature the mutation rate because this can be inferred using the following rule: mutation is not used if $p_{xo} = 90\%$, otherwise its rate is given by $p_m = 1 - p_{xo}$.

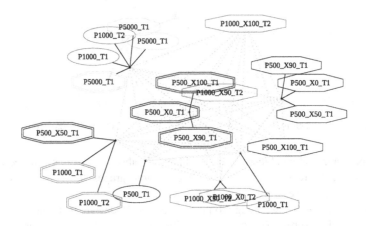

Fig. 2. Taxonomy created using the angles between pairs of performance models (see text). The octagon and double octagon shapes represent the Koza-style GP systems with roulette-wheel selection and tournament selection, respectively. Ellipses represent TinyGP systems.

clusters. What is striking about the satellites is that essentially all generational systems with roulette-wheel selection are on one side (the single octagons on the right), while the systems with tournament selection are on the other. It is also interesting to see that all the systems with a population of 5000 individuals have been clustered together (top left of the figure). Interestingly, that cluster also includes the TinyGP systems with a population size of 1000 individuals, suggesting that a steady state system can achieve with a smaller population what a generational system achieves with a larger one. Looking at the distance between the clusters we can also infer that the steady-state systems with tournament selection are at the antipodes w.r.t. the Koza-style systems with roulette-wheel selection, and that the Koza-style systems with tournament selection sit somehow in between them.

Fig. 3 shows the taxonomy created using the Euclidean distance between performance models. There are some similarities and differences w.r.t. Fig. 2. For example, the steady-state systems are closer to Koza-style systems with tournament selection than to Koza-style systems with roulette-wheel selection, suggesting that the selection method and reproduction strategy are key taxonomic features. Also, the generational systems with tournament selection and small populations are, by and large, again clustered together (in this case to the right of the figure). Furthermore, all the systems with a population of 5000 individuals are grouped together, although in this case the distance within the cluster is much reduced indicating that such systems have very similar behaviours. Another difference is that in Fig. 3 there is a group composed of TinyGP systems and Koza-style system with tournament selection — all of them with a population of 1000 individuals.

Fig. 4 shows the taxonomy obtained when using the Euclidean distance between vectors of performance statistics.[5] The taxonomy obtained is very different from the previous

---

[5] It is worth to mention that the distances within all the clusters are very small making most of the nodes in each cluster overlap. For clarity, we forced neato to remove these overlaps. So, the distances between the systems and their cluster is not to scale in the figure.

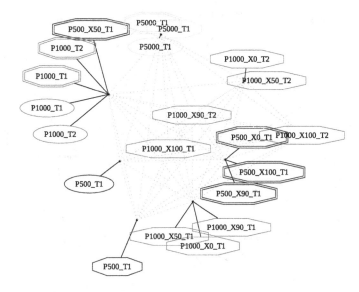

Fig. 3. Taxonomy created using the Euclidean distance between performance models. Octagon=Koza's GP w/ roulette selection, double octagon=Koza's GP w/ tournaments, ellipse=TinyGP.

taxonomies and it is of harder interpretation. One of the most notable differences is that this taxonomy does not group the systems with a population of 5000 individuals. Furthermore, both Fig. 2 and Fig. 3 indicated that systems with a population size of 5000 and the Koza-style systems with a population size of 500 are very distant performance-wise, while this information is not made explicit in the taxonomy shown in Fig. 4. This suggests that models of GP performance capture more information than simple raw statistics. Raw statistics (particularly mean performance) are constrained by the limits imposed by the no-free lunch [17]. They are unable to discover if two systems have distinct specialisation niches or whether they show similar average performance because they succeed and fail to the same degree on the same problems. The performance models presented above can. These are, therefore, better suited for the construction of algorithm taxonomies.

If we look at the taxonomies in Figs. 2 and 3 as a whole, we can see a fairly clear overall picture of the behaviour of GP systems over the class of Boolean problems.[6] In particular we find that behaviours tend to be relatively little influenced by the choice of crossover and mutation rates, while changes in the selection and reproduction schemes can give substantial performance differences. This in turn suggests that adjusting crossover and mutation rates is particularly useful if one has already found a good GP system for a problem (or group of problems) and now wants to get the maximum out of that system. Changing selection strategy, instead, may give better chances of success if the user is not satisfied

---

[6] While our taxonomies were built using a subset of all Boolean functions with 4 inputs, the accuracy of our performance model on the validation set V suggests that the model is in fact general. So, taxonomies based on it should represent reasonably well the behaviour of the 48 GP systems studied across the whole class of Boolean functions.

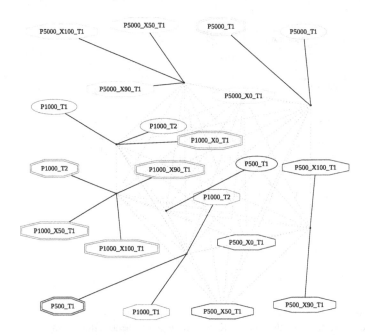

Fig. 4. Taxonomy created using Euclidean distances between vectors of empirical statistics of GP performance. Octagon=Koza's GP w/ roulette selection, double octagon=Koza's GP w/ tournaments, ellipse=TinyGP.

with a particular GP system and wants to find a better alternative. Furthermore, it is clear from the taxonomies that as the population size increases, also the choice of selection and reproduction scheme become less important in determining GP's behaviour. This suggests that another important taxonomic factor in relation to performance on Boolean induction problems is population size.

Before concluding this section we would like to mention that although we used figures to explain the structure of the taxonomies produced with our method and the relations between different GP systems, all our observations are also supported by the raw distance between the clusters formed. Unfortunately, due to lack of space we cannot present tables containing the similarities between the clusters and within each cluster.

8   Conclusions

In this paper we have presented an automatic procedure to produce taxonomies in GP. The procedure is based on performance models of GP systems. We use the models' parameters as signatures which are then clustered via a hierarchical algorithm to create dendrograms. The dendrograms are further processed to identify the key groups and their relationships. These are formalised into a weighted graph, which really represents our taxonomies. Plotting the graph with standard tools produces a highly informative representation of taxonomies from which a variety of lessons can be learnt. Since our taxonomies are based on performance, these lessons are particularly useful to guide practitioners.

Two lessons were particularly obvious in all taxonomies: population size and selection scheme are more important factors than crossover and mutation rates in determining performance on Boolean problems. So, the former should be changed first if one is not happy with the performance provided by a GP system, while the latter could be changed later to finely optimise an already satisfactory system.

Although in this contribution we have only presented taxonomies in relation to Boolean induction problems, preliminary results show that this procedure produces meaningful taxonomies of algorithms also for the class of rational symbolic regression problems (the class of functions used originally in [6]). In the future, we will explore this in depth. In addition, we hope to apply our taxonomic approach to some classes of real-world applications. Finally, while in this paper we explicitly focused on GP, nothing prevents the application of the method to other EAs. In the future, we intend to explore this avenue.

References

1. Ashlock, D., Bryden, K., Corns, S.: On taxonomy of evolutionary computation problems. In: Proceedings of the 2004 IEEE Congress on Evolutionary Computation, Portland, Oregon, pp. 1713–1719. IEEE Press, Los Alamitos (2004)
2. Ashlock, D.A., Bryden, K.M., Corns, S., Schonfeld, J.: An updated taxonomy of evolutionary computation problems using graph-based evolutionary algorithms. In: Yen, G.G., Wang, L., Bonissone, P., Lucas, S.M. (eds.) Proceedings of the 2006 IEEE Congress on Evolutionary Computation, pp. 403–410. IEEE Press, Los Alamitos (2006)
3. Back, T., Fogel, D.B., Whitley, D., Angeline, P.J.: Mutation operators. In: Baeck, T., Fogel, D.B., Michalewicz, Z. (eds.) Evolutionary Computation 1 Basic Algorithms and Operators, ch. 32, pp. 237–255. Institute of Physics Publishing, Bristol (2000)
4. Calégari, P., Coray, G., Hertz, A., Kobler, D., Kuonen, P.: A taxonomy of evolutionary algorithms in combinatorial optimization. J. Heuristics 5(2), 145–158 (1999)
5. Efron, B., Hastie, T., Johnstone, I., Tibshirani, R.: Least angle regression. Annals of Statistics (2004)
6. Graff, M., Poli, R.: Practical model of genetic programming's performance on rational symbolic regression problems. In: O'Neill, M., Vanneschi, L., Gustafson, S., Esparcia-Alcázar, A., Falco, I.D., Cioppa, A.D., Tarantino, E. (eds.) EuroGP 2008. LNCS, vol. 4971, pp. 122–133. Springer, Heidelberg (2008)
7. Herrera, F., Lozano, M., Sánchez, A.M.: A taxonomy for the crossover operator for real-coded genetic algorithms: An experimental study. Int. J. Intell. Syst. 18(3), 309–338 (2003)
8. Jansen, T., Wegener, I.: The analysis of evolutionary algorithms - a proof that crossover really can help. Algorithmica 34(1), 47–66 (2002)
9. Johnson, S.: Hierarchical clustering schemes. Psychometrika 32(3), 241–254 (1967)
10. Koza, J.R.: Genetic Programming. The MIT Press, Cambridge (1992)
11. Langdon, W.B., Poli, R.: Foundations of Genetic Programming. Springer, Heidelberg (2002)
12. Martin, P.: Building a taxonomy of genetic programming. In: Spector, L., Goodman, E.D., Wu, A., Langdon, W.B., Voigt, H.-M., Gen, M., Sen, S., Dorigo, M., Pezeshk, S., Garzon, M.H., Burke, E. (eds.) Proceedings of the Genetic and Evolutionary Computation Conference (GECCO 2001), p. 182. Morgan Kaufmann, San Francisco (2001)
13. Nowostawski, M., Poli, R.: Parallel genetic algorithm taxonomy. In: Jain, L.C. (ed.) Proceedings of the Third International conference on knowledge-based intelligent information engineering systems (KES 1999), pp. 88–92. IEEE, Los Alamitos (1999)
14. Poli, R., Langdon, W.B., McPhee, N.F.: A field guide to genetic programming (2008), http://lulu.com http://www.gp-field-guide.org.uk (With contributions by Koza, J.R)

15. Poli, R., Logan, B.: The evolutionary computation cookbook: Recipes for designing new algorithms. In: Proceedings of the Second Online Workshop on Evolutionary Computation, Nagoya, Japan (March 1996)
16. Vose, M.D.: The simple genetic algorithm: Foundations and theory. MIT Press, Cambridge (1999)
17. Wolpert, D.H., Macready, W.G.: No free lunch theorems for optimization. IEEE Transactions on Evolutionary Computation 1(1), 67–82 (1997)
18. Zhang, B.-T.: A taxonomy of control schemes for genetic code growth. In: The Workshop on Evolutionary Computation with Variable Size Representation at ICGA 1997, July 20 (1997)

Extending Operator Equalisation:
Fitness Based Self Adaptive Length Distribution for Bloat Free GP

Sara Silva[1] and Stephen Dignum[2]

[1] CISUC, Department of Informatics Engineering, University of Coimbra,
Polo II - Pinhal de Marrocos, 3030-290 Coimbra, Portugal
`sara@dei.uc.pt`
[2] School of Computer Science and Electronic Engineering, University of Essex,
Wivenhoe Park, Colchester, CO4 3SQ, UK
`sandig@essex.ac.uk`

Abstract. Operator equalisation is a recent bloat control technique that allows accurate control of the program length distribution during a GP run. By filtering which individuals are allowed in the population, it can easily bias the search towards smaller or larger programs. This technique achieved promising results with different predetermined target length distributions, using a conservative program length limit. Here we improve operator equalisation by giving it the ability to automatically determine and follow the ideal length distribution for each stage of the run, unconstrained by a fixed maximum limit. Results show that in most cases the new technique performs a more efficient search and effectively reduces bloat, by achieving better fitness and/or using smaller programs. The dynamics of the self adaptive length distributions are briefly analysed, and the overhead involved in following the target distribution is discussed, advancing simple ideas for improving the efficiency of this new technique.

Keywords: Genetic Programming, Bloat, Operator Equalisation.

1 Introduction

The search space of Genetic Programming (GP) is virtually unlimited and programs tend to grow larger during the evolutionary process. Code growth is a natural result of genetic operators in search of better solutions, but it has been shown that beyond a certain program length the distribution of fitness converges to a limit [1]. Bloat can be defined as an excess of code growth without a corresponding improvement in fitness. Several theories explaining bloat, and many different bloat control methods, have been proposed in the literature. For a review see [2,8]. Here we use the terms *length* and *size* interchangeably, both meaning the number of nodes of a tree based individual.

Crossover bias [3,4,6] is the most recent theory concerning bloat. It explains code growth in tree-based GP by the effect that standard subtree crossover has

L. Vanneschi et al. (Eds.): EuroGP 2009, LNCS 5481, pp. 159–170, 2009.

on the distribution of tree sizes in the population. In short, the average length of programs in the mating pool does not differ from that of the resultant child population after the application of standard subtree swapping crossover. However, the length distribution of the child population, under normal GP experimental conditions, is biased towards smaller programs. Smaller programs will be unable to obtain reasonable fitness and will be ignored by selection, hence increasing the average size of programs in the mating pool for the succeeding generation. Some of the most popular bloat control methods rely on the usage of length limits to prevent the larger individuals from entering the population. Although this effectively keeps the population from growing too large, it was found that length limits actually speed code growth in the beginning of the run because they promote the proliferation of the small unfit individuals [5]. Counterintuitively, bloat may be better counteracted by preventing the *smaller* individuals from entering the population. The findings from this research suggest that GP experimentation should bias search, during a run, towards program lengths that have been found to provide relatively better fitness.

Operator equalisation [7] is a recent bloat control technique that allows accurate control of the program length distribution during a GP run. By filtering which individuals are allowed in the population, it can easily bias the search towards smaller or larger programs. This technique has previously achieved promising results with different predetermined target length distributions, using a static number of length classes, or bins, constrained by a conservative program length limit. Here we improve operator equalisation by giving it the ability to automatically determine and follow the ideal length distribution for each stage of the run, and removing the fixed maximum program length constraint. Updated every generation, the target length distribution is based on the fitness measured on each length class of the previous generation. This creates a fast moving bias towards the areas of the search space where the fittest programs are, avoiding the small unfit individuals resulting from the crossover bias. The removal of the fixed maximum program length allows the dynamic creation of new length classes in runtime. Starting with the number of bins strictly necessary for the initial population, a new bin is created whenever needed to accommodate the new best-of-run individual, a criterion inspired by previous work on dynamic limits [2,8].

The next section describes the operator equalisation technique, both previous work and new improvements. Section 3 specifies the problems and settings used to test the new technique, while Section 4 reports and discusses the results. Section 5 concludes.

2 Operator Equalisation

Developed alongside the crossover bias theory [3,4,5,6] (see Sect. 1), a new method for bloat control has recently been proposed, called operator equalisation [7]. It is capable of accurately controlling the distribution of sizes inside the population, easily biasing the search towards smaller or larger programs at runtime. Next we describe the original operator equalisation technique, followed by the improvements we propose.

2.1 Previous Work

The original operator equalisation technique controls the distribution of sizes inside the population by probabilistically accepting each newly created individual based on its size. The probabilities of acceptance are calculated considering a predetermined target distribution. A maximum allowed program length, and the desired number of length classes, or bins, are used to calculate the width of the bins. Except for the probabilities of acceptance, all these elements remain static throughout the run.

Initially, the probability of acceptance for each bin is calculated to be directly proportional to the desired frequency of individuals on the corresponding target bin. After each generation the current distribution is used to update the probabilities of acceptance so that the target is approximated at a certain rate (see [7] for details). Naturally, the probabilistic nature of the acceptance procedure allows large discrepancies between the target and current distributions. For example, if crossover presents large numbers of programs of a certain size in one generation the corresponding bins will overflow in comparison with the target. This will, however, be corrected with a punitive acceptance probability in the next generation.

Operator equalisation was tested on a 10-variate symbolic regression problem, and on the 10-bit even parity problem, using different target distributions. All the targets were easily reached, and some distributions proved more efficient for one or the other problem, but none provided the best results in both problems.

2.2 Determining the Ideal Length Distribution

Provided with a simple and effective method for biasing the search towards the desired program lengths, the question immediately arises: what is the best length distribution? Preliminary steps were taken in this direction [7] and evidence has shown that (1) it depends on the problem and (2) it depends on the stage of the run. Remember the search should be biased towards program lengths that have better fitness (Sect. 1), so our target length distribution simply follows fitness.

For each bin, we calculate the average fitness of the individuals within. Our target is directly proportional to these values. Bins with higher average fitness will accept more individuals, because that is where search is proving to be more successful. Formalizing, the desired number of individuals for each bin b, t_b, is calculated as $t_b = round(n \times \bar{f}_b / \sum_i \bar{f}_i)$, where $\bar{f}_i$ is the average fitness in the bin with index i, $\bar{f}_b$ is the average fitness of the individuals in b, and n is the number of individuals in the population. The rounded t_b values may not actually add to the exact number of individuals in the population, but that is irrelevant as the target does not have to be exactly matched.

Updated every generation, this dynamic target can easily and quickly lead the search towards the most promising program lengths. It avoids the small unfit individuals resulting from the crossover bias, as well as the excessively large individuals that do not provide better fitness than the smaller ones already found. In short, our target is capable of self adapting to any problem and any stage of the run.

2.3 Following the Self Adaptive Target Distribution

Like the original operator equalisation technique, the new implementation filters which individuals are allowed in the population, based on the size of the individual and on the target length distribution. However, in the new technique the acceptance of each individual is not probabilistic, but instead relies on a deterministic acceptance procedure. Figure 1 shows two pieces of pseudocode: on the left, the cycle that generates a new population; on the right, the details of the acceptance procedure.

generation of a new population:

```
n_accepted = 0
while n_accepted < pop_size
  select parents
  apply genetic operator
  for each child i
    l = length of i
    b = bin that holds length l
    if accept(i,b)
      n_accepted = n_accepted + 1
```

accept(individual i, bin b):

```
accept = false
if b exists
  if b is not full
  or i is the new best-of-bin
    accept = true
else
  if i is the new best-of-run
    create new bin b
    accept = true
```

Fig. 1. Pseudocode of operator equalisation: on the left, the cycle that generates a new population; on the right, the details of the acceptance procedure

Unlike the original technique, the new operator equalisation does not use a maximum limit on program length. Therefore, the number of bins to use is not specified. The run begins with the number of bins strictly necessary for the initial population. As individuals grow larger, their corresponding bins may still not exist and can be created during the run. A new bin will only be created for a new best-of-run individual, i.e., one with better fitness than any other individual found so far. This criterion is highly inspired on previous work on dynamic limits [2,8], a technique that has proven to be very effective in bloat control. Like the target, the maximum permissible program length is thus self adapted along the run. A bin is full when it has reached the number of individuals specified in the target. From then on, it can only increase its capacity to accept another individual if this individual is the new best-of-bin, meaning it has better fitness than any other already in the bin. Any newly created bin will automatically be full once it accepts the first individual.

Clearly, the evaluation of the individuals as part of the filtering process introduces a major computational overhead, and this will be addressed again in Sects. 4.2 and 4.4. Nevertheless, this allows a clever overriding of the target distribution, by allowing the addition of more individuals to the bins where the search is having a greater degree of success. Furthermore, the removal of the limit on program length permits a shift of attention to more promising areas of the search space, by dynamically opening new bins when it is advantageous in terms of fitness. In this new version of the operator equalisation technique, good individuals do not go unnoticed. Instead, they are used to quickly increase the search bias towards the right lengths without even having to wait for the next generation.

3   Experiments

Problems. Four different problems were chosen to test the new operator equalisation technique: Symbolic Regression of the quartic polynomial ($x^4 + x^3 + x^2 + x$, with 21 equidistant points in the interval -1 to $+1$), Artificial Ant on the Santa Fe food trail, 5-Bit Even Parity, and 11-Bit Boolean Multiplexer. The function sets are the same as in [9]. The regression problem uses no random constants. In all four problems, fitness was calculated such that lower values represent better fitness.

Techniques. Three different variants of operator equalisation were tested, using bin widths of 1, 5, and 10 ("Op Eq 1", "Op Eq 5", and "Op Eq 10"). These were compared with three other techniques: a no-limits approach ("No Limits") were individuals grow completely free of restrictions, the popular Koza approach ("Koza") of limiting tree depth to the fixed value of 17 [9], and a successful dynamic depth approach ("Dyn Depth") where the maximum allowed tree depth is initially set with a low value, and only increased when needed to accommodate a new best-of-run individual [2,8].

Parameters. A total of 30 runs were performed with each technique for each problem. All the runs used populations of 1000 individuals allowed to evolve for 50 generations. Most of the remaining parameters follow the settings indicated in [9] and [2,8], to facilitate the comparison between the techniques. Tree initialization was performed with the Ramped Half-and-Half method [9] with a maximum initial depth of 6 (the same value was used to initialize the dynamic depth). Selection for reproduction used tournaments of size 7. The reproduction (replication) rate was 0.1, no mutation was used, and crossover adopted a uniform selection of crossover points. Selection for survival was not elitist.

Tools. All the experiments were performed using a modified version of GPLAB[1] Statistical significance of the null hypothesis of no difference was determined with pairwise Kruskal-Wallis non-parametric ANOVAs at $p = 0.01$.

4   Results and Discussion

A new technique can be regarded as successful when it provides better results than what had been achieved so far. But what are the criteria to decide whether some given results are better than others? We show how misleading some plots can be, by presenting an optimistic view, followed by a pessimistic view, of the results achieved by the new operator equalisation technique. The differences are striking. Then we present what we believe is a fair and realistic view of the results. We close

---

[1] A Genetic Programming Toolbox for MATLAB, http://gplab.sourceforge.net

the section with a brief analysis of the evolutionary dynamics of the rejections and length distributions. All the claims of *significant* differences in the results are supported by statistical evidence as described in Sect. 3 (*Tools*).

4.1   The Optimistic View

Figure 2 shows the evolution of the best fitness along the run for the Regression (a) and Artificial Ant (b) problems. Ignore the small inset plots for now. This is the most traditional way of presenting such results: fitness versus generations. In this case we plot the median values of the 30 runs.

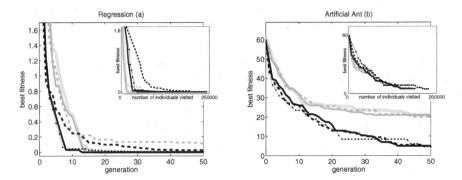

Fig. 2. Best fitness versus generations. (Insets: Best fitness versus number of individuals visited.) Legend: ▬▬ No Limits ▬▬▬ Koza ▬ ▬ ▬ Dyn Depth ▬▬▬ Op Eq 1 ▪ ▪ ▪ Op Eq 5 ⋯•⋯ Op Eq 10

Looking at the results presented this way, the new operator equalisation technique seems to be very successful indeed, for all three variants. Take a look at the inset plots of Fig. 3 for a snapshot of the remaining problems, Parity and Multiplexer. For all problems, operator equalisation is able to improve fitness faster and in most cases reaches significantly better values than the non-equalising techniques.

4.2   The Pessimistic View

When describing how to make the population follow the target length distribution by filtering the individuals that enter the new generation (Sect. 2.3) we mentioned that a major computational overhead was being introduced because of the need to evaluate the individuals before deciding whether to accept or reject them. In each generation of the new operator equalisation technique, many more individuals are created and evaluated than in the non-equalising techniques.

Figure 3 shows an alternative way of representing the evolution of the best fitness along the run, for the Parity (a) and Multiplexer (b) problems. Fitness is not plotted against the generations, but against the number of individuals visited[2] so far. For each rejected individual, another one must be visited. Again

---

[2] Individuals visited are individuals created, evaluated, or simply reused.

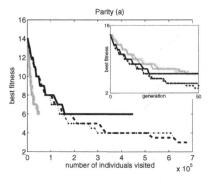
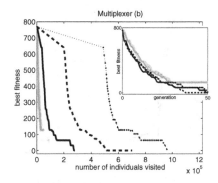

Fig. 3. Best fitness versus number of individuals visited. (Insets: Best fitness versus generations.) Legend: ▬▬ No Limits ▬▬ Koza ▬ ▬ ▬ Dyn Depth ▬▬ Op Eq 1 ▪ ▪ ▪ Op Eq 5 ⋯⋯ Op Eq 10

we plot the median values of the 30 runs. The inset plots provide a view of the optimistic version of these results, for comparison. The inset plots of Fig. 2 show the pessimistic view for the Regression and Artificial Ant problems.

Looking at Figs. 2 and 3 we are tempted to conclude that the good results of the new operator equalisation only come at the expense of unacceptably high computational effort (except maybe for the Artificial Ant problem). However, in Sect. 4.4 we will take a deeper look at the distribution of the rejected individuals and the way it relates to the self-adapting targets. This analysis will reveal a much brighter scenario where simple modifications to the current implementation of the operator equalisation technique can dramatically improve its efficiency.

4.3   The Realistic View

Going back to the subject of what defines a good result, we should stress that fitness is not the only important element. Most GP users also want *simple* solutions, as they are easier to understand and tend to generalize better [10]. So the goal is to find short programs that provide accurate results. This is, after all, what bloat is all about: the relationship between size and fitness.

Figure 4 shows, for all problems considered, the best fitness plotted against the program length, median values of the 30 runs. These are somewhat unconventional plots, but they provide a clear view of how varied the bloat dynamics is among the different problems and the different techniques. Depending on how fast the fitness improves with the increase of program length, the lines in the plot may point downward (south), or they may point to the right (east). Lines pointing south represent a rapidly improving fitness with little or no code growth. Lines pointing east represent a slowly improving fitness with strong code growth. Lines pointing southwest (bottom left) represent improvements in fitness along with a reduction of program length. We want our lines to point as south (and west) as possible.

As shown in figure 4, for all except the Regression problem all the variants of operator equalisation exhibit a more downwards tendency than the other

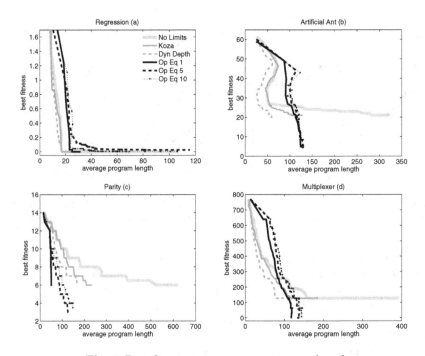

Fig. 4. Best fitness versus average program length

techniques. Even if this tendency is not apparent in the earlier stages of the run, it quickly becomes very marked and decisively leads the search towards short and accurate solutions. The dynamic depth approach is the only one that would stand a slight chance of equaling these results, if given more running time. Even in the Regression problem, all the techniques present statistically similar fitness values, although the operator equalisation variants use significantly longer programs than the Koza and dynamic depth approaches. It is common in symbolic regression problems for programs to get larger with very small fitness improvements, bloat being caused mostly by unoptimized, not inviable, code. However, the increased program length observed with the equalisation technique seems to be of a different nature, as explained at the end of the next section.

4.4   Dynamics of Rejections and Length Distributions

In the next three figures the left plot (a) refers to the operator equalisation technique with bin width 1 in the Regression problem, and the right plot (b) refers to operator equalisation with bin width 10 in the Parity problem. Figure 5 shows the number of rejected individuals by length class and generation. Figure 6 shows the desired number of individuals by length class and generation, i.e., the evolution of the target distribution of program lengths along the run. Figure 7 shows the evolution the actual distribution of program lengths along the run.

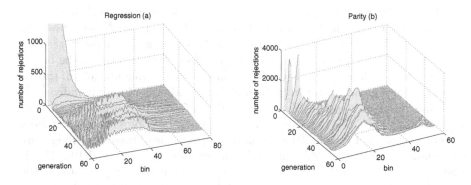

Fig. 5. Rejections. Regression uses bin width 1 and Parity uses bin width 10.

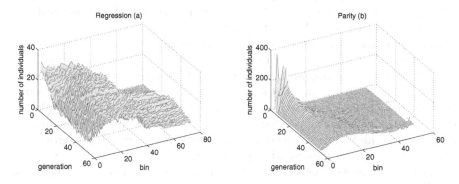

Fig. 6. Target distributions. Regression uses bin width 1 and Parity uses bin width 10.

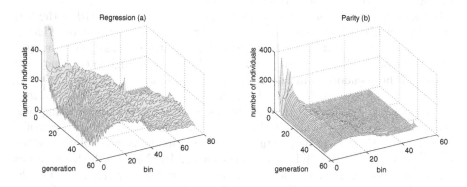

Fig. 7. Actual distributions. Regression uses bin width 1 and Parity uses bin width 10.

Each value is the mean over the 30 runs. There is considerable variance between Regression runs, but not much in the Parity case.

Beginning with the Regression problem, one interesting fact is immediately apparent on Fig. 5a. The "number of rejections" axis does not go beyond the value

1000 for visualization purposes, but the number of rejected individuals in the conspicuous peak reaches higher than 10 thousand. If completely drawn, this peak would be roughly 10 times higher than shown. This reveals that in the Regression problem there is a concentration of rejections of small individuals in the beginning of the run. More specifically, with a bin width of 1, roughly 16% of the total number of rejections happen in the first 5 generations and refer to individuals with sizes between 1 and 10. With bin widths of 5 and 10 this percentage rises dramatically to 75–76%. If drawn, these peaks would be approximately 350 times higher than shown! In the Multiplexer problem (not shown) these percentages are also relatively high (10%, 23%, and 40% for bin widths of 1, 5, and 10). This is a good indication that the high number of visited individuals represented in the pessimistic views of Sects. 4.1 (insets) and 4.2 (main plots) may not be so dramatic. After all, such small individuals are not very computationally expensive. However, for the Artificial Ant and Parity problems the initial peaks only contain 0.7–1.9% of the rejections. Figure 5b shows the modest initial rejection peaks for the Parity problem with a bin width of 10.

Apart from the initial peaks, there is a concentration of rejections that seems to follow a path along the generations (Fig. 5, both plots). As we can see in Fig. 6, for both problems this roughly falls on the path formed by the moving target distribution of program lengths. The same happens for the remaining problems (not shown). One could suspect that, since most rejections happen on bins that already exist, many individuals are accepted into newly created bins outside the boundaries of the target. This is, however, incorrect. When comparing the actual length distributions on Fig. 7 with the desired targets on Fig. 6 we see they are close matches, except for the undesired peak of the actual length distribution in the Regression problem (Fig. 7a, peak reaches the value 130). The same happens for all problems with all operator equalisation variants: a close match between actual and desired distributions, except for some traces of the highest initial rejection peaks.

The knowledge that most rejections happen in existing bins within the target, along with the fact that the target is, for the most part, not overridden, provides the perfect setting for improving the operator equalisation technique. A single minor modification can provide a major breakthrough in efficiency. We go back to the pseudocode in Fig. 1, Sect. 2.3, and in the acceptance procedure (on the right) remove the line "or i is the new best-of-bin". Remove this, and the need to evaluate all the individuals attempting to enter existing bins disappears. The bins will accept the individuals while there's room available, and will reject them once they're full. Only the individuals who fall outside the boundaries of the target will have to be evaluated. And because the target is not overridden anyway, the actual length distribution will barely suffer any modifications. To understand the full impact that this simple modification can have on the efficiency of operator equalisation, we have measured the percentage of rejections that actually fall *outside* the boundaries of the target. The numbers are presented in Table 1.

An additional observation can be made regarding the rejection plot of the Regression problem (Fig. 5a), i.e., the jagged look of the rejection pattern. This

Table 1. Percentage of rejections falling outside the target distribution

	Bin width		
	1	5	10
Regression	7.6%	0.5%	0.003%
Artificial Ant	5.0%	4.1%	4.6%
Parity	13.1%	4.9%	4.0%
Multiplexer	3.8%	2.8%	1.8%

characteristic is almost absent, but still detectable, in the target length distribution (Fig. 6a). The Parity problem with bin width 1 exhibits an even stronger jagged surface (not shown), with absolutely no rejections for the even length bins. However, the target for this problem is also completely jagged (not shown). This makes total sense, as the function set of the Parity problem does not allow the creation of even sized individuals, so the mean fitness of those bins is actually null, and no individuals are expected there. In the Regression problem, despite the mixed arity function set used, there seems to be some difficulty in creating individuals of even size. But because they *do* appear in the population, the mean fitness of those bins is not null, and this is reflected in the relatively smooth surface of the target distribution. Even more interesting, as we can see in Fig. 7a the *actual* distribution of program lengths observed during the run is equally smooth, meaning that the operator equalisation is actually inserting these difficult individuals in the population. Without operator equalisation both the target and actual distributions of program lengths are jagged (not shown). In both cases many more odd sized individuals are created than even sized ones, but with operator equalisation the most requested bins fill up sooner and start increasing their number of rejections, while the least requested bins still have room available. This is what causes the jagged pattern of rejections. Without equalisation the most common sizes fill the population leaving no room for the difficult ones. By granting the rare individuals a place in the population, operator equalisation is actually increasing the genotypic diversity, that indeed reaches the highest levels of all the tested techniques (not shown). On the other hand, this increased diversity is accompanied by the proliferation of inviable code (not shown), as if introns were being used to fill the nodes needed to build even sized individuals. This could explain why the jagged pattern is more pronounced for the small sizes, where there is not enough room for introns. It could also explain the atypical proliferation of inviable code in the Regression problem [2,8], and ultimately the poor results achieved by the equalisation technique.

5   Conclusions

We have improved the operator equalisation technique by giving it the ability to automatically determine and follow the ideal length distribution for each stage of the run, unconstrained by a fixed maximum limit. The results were approached from several different points of view, and the main conclusion is that, despite introducing a considerable overhead on computational effort, the new technique effectively reduces bloat, by achieving better fitness and/or using smaller programs in most of the problems considered.

A brief analysis of the dynamics of rejections and length distributions has revealed that (1) in some problems there is a very high concentration of rejections of small individuals in the beginning of the run, (2) most of the remaining rejections are located on the path of the target distribution, and not on unexplored areas of the search space, and (3) the actual length distribution closely follows the target. These observations have dismissed the pessimistic view about the efficiency of the new operator equalisation technique, and provided a simple idea for greatly improving it.

Acknowledgments. Sara Silva acknowledges grant SFRH/BPD/47794/2008 from Fundação para a Ciência e a Tecnologia, Portugal.

References

1. Langdon, W.B., Poli, R.: Foundations of Genetic Programming. Springer, Heidelberg (2002)
2. Silva, S.: Controlling bloat: individual and population based approaches in genetic programming. Ph.D thesis, Dep. Informatics Engineering, Univ. Coimbra (2008)
3. Poli, R., Langdon, W.B., Dignum, S.: On the limiting distribution of program sizes in tree-based genetic programming. In: Ebner, M., O'Neill, M., Ekárt, A., Vanneschi, L., Esparcia-Alcázar, A.I. (eds.) EuroGP 2007. LNCS, vol. 4445, pp. 193–204. Springer, Heidelberg (2007)
4. Dignum, S., Poli, R.: Generalisation of the limiting distribution of program sizes in tree-based genetic programming and analysis of its effects on bloat. In: Thierens, D., et al. (eds.) Proceedings of GECCO 2007, pp. 1588–1595. ACM Press, New York (2007)
5. Dignum, S., Poli, R.: Crossover, sampling, bloat and the harmful effects of size limits. In: O'Neill, M., Vanneschi, L., Gustafson, S., Esparcia Alcázar, A.I., De Falco, I., Della Cioppa, A., Tarantino, E. (eds.) EuroGP 2008. LNCS, vol. 4971, pp. 158–169. Springer, Heidelberg (2008)
6. Poli, R., McPhee, N.F., Vanneschi, L.: The impact of population size on code growth in GP: analysis and empirical validation. In: Keijzer, M., et al. (eds.) Proceedings of GECCO 2008, pp. 1275–1282. ACM Press, New York (2008)
7. Dignum, S., Poli, R.: Operator equalisation and bloat free GP. In: O'Neill, M., Vanneschi, L., Gustafson, S., Esparcia Alcázar, A.I., De Falco, I., Della Cioppa, A., Tarantino, E. (eds.) EuroGP 2008. LNCS, vol. 4971, pp. 110–121. Springer, Heidelberg (2008)
8. Silva, S., Costa, E.: Dynamic limits for bloat control in genetic programming and a review of past and current bloat theories. Genet. Program. Evolvable Mach. (January 13, 2009), http://www.springerlink.com/content/k001162572j4vh70, doi:10.1007/s10710-008-9075-9
9. Koza, J.R.: Genetic programming – on the programming of computers by means of natural selection. MIT Press, Cambridge (1992)
10. Rosca, J.P.: Generality versus Size in Genetic Programming. In: Koza, J.R., et al. (eds.) Proceedings of GP 1996, pp. 381–387. MIT Press, Cambridge (1996)

Modeling Social Heterogeneity with Genetic Programming in an Artificial Double Auction Market

Shu-Heng Chen and Chung-Ching Tai

AI-ECON Research Center, Department of Economics,
National Chengchi University, Taiwan
chchen@nccu.edu.tw, chungching.tai@gmail.com

Abstract. Individual differences in intellectual abilities can be observed across time and everywhere in the world, and this fact has been well studied by psychologists for a long time. To capture the innate heterogeneity of human intellectual abilities, this paper employs genetic programming as the algorithm of the learning agents, and then proposes the possibility of using population size as a proxy parameter of individual intelligence. By modeling individual intelligence in this way, we demonstrate not only a nearly positive relation between individual intelligence and performance, but more interestingly the effect of decreasing marginal contribution of IQ to performance found in psychological literature.

1 Introduction

Inequality has long been an important issue in economics since Vilfredo Pareto's research in wealth distribution. Pareto probed into this issue and thought that one of the sources of the inequality comes from the heterogeneity in measured and unmeasured abilities and skills in the labor pool [1]. Pareto termed this fact "Social Heterogeneity," which constitutes an important part of his economic thinkings.

"Human society is not homogenous; it is made up of elements which differ more or less, not only according to the very obvious characteristics such as sex, age, physical strength, health, etc., but also according to less observable, but no less important, characteristics such as intellectual qualities, morals, diligence, courage, etc. " (Pareto [2], Chapter II, 102)

Unfortunately, the significance of the heterogeneity in individual capabilities did not attract much attention from economists until late twentieth century.

In 1994, Herrnstein and Murray published their controversial book entitled *The Bell Curve: Intelligence and Class Structure in American Life*, which discusses group differences in IQ and their socioeconomic consequences [3]. Regardless of the sources of differences between IQ, *The Bell Curve* and subsequent research using sibling data has revealed the importance of intelligence on various socioeconomic performances [4] [5]. Ever since then, IQ has been employed to

L. Vanneschi et al. (Eds.): EuroGP 2009, LNCS 5481, pp. 171–182, 2009.

explain statistically the causes of the inequalities in income and wealth between nations [6] [7], or been explicitly put into the growth model as a proxy of human capital to model GDP growth [8] [9].

While IQ has been used to explain macroeconomic variables, not much attention was paid to model the influence of intelligence on individual behavior in economic literature.[1] However, we know from psychological literature that intelligence is crucial to people's learning capability [11], that brings about an important question: Are there proper ways to model individuals with heterogeneous intelligence?

This paper aims to serve as a preliminary effort to model heterogeneous-intelligence individuals in an agent-based double auction market. The purpose of Agent-based modeling is to construct heterogeneous-agent systems. In these models, algorithms from computational intelligence can be used to portray agents' learning behavior. Although agents could be endowed with "intelligence" to learn and adapt, however, to the best of our knowledge, none of the existing agent-based economic models takes into account the degree of intelligence as a factor of learning behavior.

In this paper, we employ Genetic Programming (GP) as agents' learning algorithms. Furthermore, we choose the parameter of population size as the proxy variable of IQ. A series of simulations will be reported, and the results will be compared to what we have known about IQ from psychological studies.

This paper is organized as follows: Research questions will be elaborated in section 2. Section 3 depicts the experimental design, including market mechanism, trading strategies, and experiment settings. Results, evaluations, and analysis of experiment are presented in section 4. Section 5 is the conclusion.

2   IQ, Learning, and Economic Performance

Intelligence is an important and widely investigated issue in psychological studies, and psychologists have developed various IQ tests and models to measure and explain human intelligence. One of the facts that interests us is the relation between intelligence and learning. Gottfredson defines intelligence as a term used to describe people's capability of learning:

> "Intelligence is a very general mental capability that, among other things, involves the ability to reason, plan, solve problems, think abstractly, comprehend complex ideas, *learn quickly* and *learn from experience*." (italics added, see [11])

As what can be seen from the definition, it suggests that a positive relation should exist between intelligence and the learning capability. As a model which

---

[1] Rydval and Ortmann's research is the closest one. They compared the influences of IQ and monetary incentives on human subjects' performances, and found that most of the differences in performances can be attributed to differences in IQ instead of the amount of monetary payoffs. See [10].

seeks to find the parameter representing individual's intelligence, there is no doubt that it should be able to exhibit such a positive relation.

Nevertheless, two additional questions are worthy of investigation in the quest of the parameter for intelligence. First, in a series of human-agent competition studies, researchers found that human subjects did learn, but most people were defeated by software trading programs, only few of them performed comparably to software agents [12] [13] [14]. This leaves us a question: How much intelligence should a learning agent have to defeat other well-designed software trading strategies?

Second, as the psychological literature points out, high intelligence does not always contribute to high performance–the significance of intelligence performance is more salient when the problems are more complex. Also, it appears that intelligence exhibits decreasing marginal contribution in terms of performances.[2] Can our model generate phenomena consistent with these observations?

In the next section, we will introduce an agent-based double auction market where GP's population size is taken as a proxy variable for intelligence. By doing so, we then try to answer the questions mentioned above.

3   Experimental Design

Experiments in this paper were conducted in AIE-DA (Artificial Intelligence in Economics - Double Auction) platform which is an agent-based discrete double auction simulator with build-in software agents.

3.1   Market Mechanism

AIE-DA is inspired by the Santa Fe double auction tournament held in 1990, and in this study we adopted the same token generation process as Rust et al.'s design [18]. Our experimental markets consist of four buyers and four sellers. Each of the traders can be assigned a specific strategy–either a designed trading strategy or a GP agent.

During the transactions, traders' identities are fixed so they cannot switch between buyers and sellers. Each trader has four units of commodities to buy or to sell, and can submit only once for one unit of commodity at each step in a trading day. Every simulation lasts 7,000 trading days, and each trading day consists 25 trading steps. AIE-DA is a discrete double auction market and adopts AURORA trading rules such that at most one pair of traders are allowed to make transaction at each trading step. The transaction price is set to be the average of the winning buyer's bid and the winning seller's ask.

At the beginning of each simulation, each trader will be randomly assigned a trading strategy or as a GP agent. Traders' tokens (reservation prices) are also

---

[2] [15] demonstrates that the correlation between intelligence and performance increases when the tasks are made more complex. As to the decreasing marginal value of intelligence, please see [16] and [17].

randomly generated with random seed 6453. Therefore, each simulation starts with a new combination of traders and a new demand and supply schedule.[3]

3.2 Trading Strategies and GP Agents

In order to test the ability of GP agents, we developed several trading strategies from the double auction literature as GP agents' competitors. They are: **Kaplan**, **Ringuette**, and **Skeleton** modified from Rust et al.'s tournament [18]; **ZIC** from Gode and Sunder [19]; **ZIP** from Cliff and Bruten [20]; **Markup** from Zhan and Friedman [21]; **Gjerstad-Dickhaut (GD)** from Gjerstad and Dickhaut [22]; **BGAN** from Friedman [23]; **Easley-Ledyard (EL)** from Easley and Ledyard [24]; **Empirical** strategy is inspired by Chan et al. [25], and it works in the same way as Friedman's BGAN but develops its belief by constructing histograms from opponents' past shouted prices.[4]

Although most of the strategies were created for the purpose of studying price formation processes, we still sent them to the "battlefield" because they can represent, to a certain degree, various types of trading strategies which can be observed in financial market studies.

GP agents in this study adopt only standard crossover and mutation operations, by which it means no election, ADFs nor other mechanisms are implemented. At the beginning of every trading day, each GP trader randomly picks a strategy from his/her population of strategies and uses it through the whole day. The performance of each selected strategy is recorded, and if a specific strategy is selected more than once, a weighted average will be taken to emphasize later experiences.

GP traders' strategies are updated–with selection, crossover, and mutation– every N days, where N is called the "select number." To avoid the flaw that a strategy is deserted simply because it was not selected, we set N twice the size of the population so that theoretically each strategy has the chance to be selected twice. Tournament selection is implemented and the size of the tournament is 5, however big the size of the population is. We also preserve the elite to the next generation, and the size of the elite is 1. The mutation rate is 5%, in which 90% of this operation is tree mutation.

In order to examine the validity of using population sizes as GP traders' intelligence, a series of experiments were conducted, GP traders' population sizes were set at 5, 20, 30, 40, 50, 60, 70, 80, 90, and 100 respectively. Such a sampling enables us to scrutinize the issues posted in section 2.

---

[3] Considering the vast number of combinations and permutations of traders, we did not try out all possible trader combinations. Instead, 300 random match-ups were created for each series of experiment.

[4] Named by or after their original designers, these strategies were modified to accommodate our discrete double auction mechanism in various ways. They were modified according to their original design concepts as possible as we can. As a result, they might not be 100% the same as they originally are.

4 Results and Discussions

In this section, we evaluate traders' performances with a profit-variation point of view. Profit ability is measured in terms of individual efficiencies.[5] In addition to profits, a strategy's profit stability is also taken into account because in double auction markets, the variation of profits might be considered in human trading strategies, which are determined by human's risk attitudes [26]. Here we procure variations of strategies by calculating the standard deviation of each strategy's individual efficiencies.

4.1 Learning Capabilities of GP Agents

In investigating into GP traders' learning capability, we simply compare GP agents with designed strategies collected from the literature. We are interested in the following questions:

1. Can GP traders defeat other strategies?
2. How many resources are required for GP traders to defeat other strategies?

GP traders with population size of 5, 20, and 50 are sampled to answer these questions.[6] Figure 1 is the result of this experiment. Here we represent GP traders of population size 5, 20, and 50 with P5, P20, and P50 respectively. We have the following observations from Figure 1:

- No matter how big the population is, GP traders can gradually improve and defeat other strategies.
- GP traders can still improve themselves even under the extreme condition of a population of only 5.[7] Figure 2 shows the evolution of average complexity of GP strategies. In the case of P5, the average complexity almost equals to 1 at the end of the experiments, meaning that GP traders could still gain superior advantages by constantly updating their strategy pools composed of very simple heuristics. In contrast with P5, in the case of bigger population, GP develops more complex strategies as time goes by.
- What is worth noticing is that GP might need a period of time to evolve. The bigger the population, the fewer generations are needed to defeat other strategies. In any case, it takes hundreds to more than a thousand days to achieve good performances for GP traders.

---

[5] In order to evaluate the performance of each strategy, we adopted the notion of *individual efficiency*. Considering the inequality in each agent's endowment due to random matching of strategies as well as random reservation prices, individual efficiency is calculated as the ratio of one's actual profits to his/her theoretical surplus, which is the sum of the differences between one's intramarginal reservation prices and the market equilibrium price.

[6] The corresponding select number were set at 10, 40, and 100 respectively. Briefly speaking, the number of selection is the evaluation cycle for each GP generation.

[7] The fact that the tournament size is also 5 means that strategies in the population might converge very quickly.

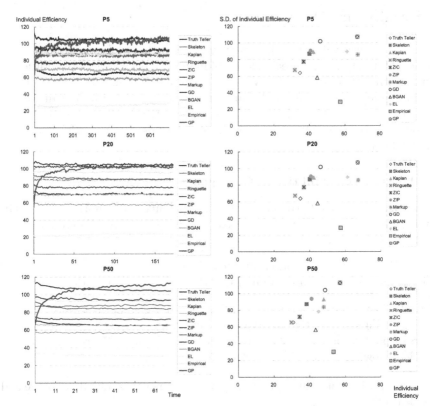

Fig. 1. Comparison of GP Traders with Designed Strategies. From the top to the bottom rows are comparisons when GP traders' population sizes are 5, 20, and 50 respectively. (a) The left panels of each row is the time series of individual efficiencies. (b) The right panels of each row is the profit-variation evaluation on the final trading day. The horizontal axis stands for their profitability (individual efficiency), and the vertical axis stands for the standard deviation of their profits.

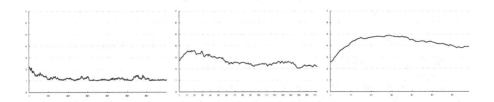

Fig. 2. Evolution of average GP complexity when the population sizes are 5, 20, and 50 respectively (from the left panel to the right panel)

- Figure 1 also shows the results in a profit-variation viewpoint. Other things being equal, a strategy with higher profit and less variation is preferred. Therefore, one can draw a frontier connecting the most efficient trading

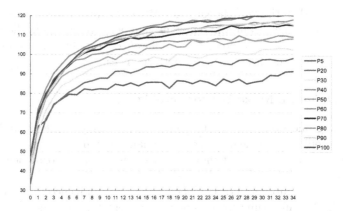

Fig. 3. GP traders' performances at different "intelligence" levels. The horizontal axis is generation; the vertical axis is the profit level attained by GP traders.

strategies. Figure 1 shows that GP traders, although with more variation in profits in the end, always occupy the ends of the frontier.

The result of this experiment shows that learning GP traders can outperform other (adaptive) strategies, even if those strategies may have a more sophisticated design.

4.2   Intelligence and Learning Speed

Psychologists tell us that intelligence of human beings involves the ability to "learn quickly and learn from experiences" [11]. To investigate the influence of individual intelligence on learning speed, we think of a GP trader's population size as a proxy of his/her IQs. Is this parameter able to generate behavioral outcomes consistent with what psychological research tells us?

Figure 3 delineates GP traders' learning dynamics with a more complete sampling. Roughly speaking, we can see that the bigger the population size, the less time GP traders need to perform well. In other words, GP traders with higher intelligence tend to learn faster and gain more wealth consequently.

However, if we are careful enough, we may also notice that this trend is not as monotone as we might have thought. It seems that there are three groups of learning dynamics in this figure. From P5 to P30, there exists manifest positive relation between "intelligence" and performance. P40 and P50 forms the second group: they are not very distinguishable, but both of them are better than traders with lower "intelligence". The most inexplainable part is P60 to P100. Although this group apparently outperform traders with lower "intelligence", the inner-group relation between "intelligence" and performance is quite obscure.

For a better understanding of this phenomenon, a series of nonparametric statistical tests were performed upon these simulation results. The outcomes of these tests are presented in part A of Table 1. Pairwise Wilcoxon Rank Sum

Tests show that when the "intelligence" levels are low, small differences in intelligence may result in significant differences in final performances. On the contrary, among those who have high intelligence, differences in intelligence do not seem to cause any significant discrepancy in performances.

4.3   Can Diligence Make UP for Intelligence?

One interesting questions about intelligence and learning is that if we let those with lower intelligence to learn longer in financial markets, are they able to discover better strategies so that they can have competent performances as those with high IQs?

To test this, we simply let GP traders with lower intelligence to evolve longer. Table 2 is the learning time for different intelligent GP traders. For example, traders with population size 5 have only 1/20 of the intelligence of those whose population size is 100, therefore, we let traders with IQ 5 evolve 20 times longer than traders with IQ 100.

Part B of Table 1 demonstrates the results of this experiments. We can observe that if the difference of intelligence is not so large, it is possible to make up for the deficiency of endowed ability by hard working. However, diligence can only partially offset such deficiency when the difference of intelligence is large. Take GP traders with IQ 5 as an example, they can catch up with traders with IQ 40 if they work eight times longer. Nevertheless, when facing traders with IQ 100, they cannot reduce the gap even by working twenty times longer. This result seems to deny the hypothesis that traders with low intelligence can fairly achieve appreciable performances as smarter ones in double auction markets.

4.4   A Test for Fluid Intelligence

In section 4.2, we've seen that GP with different IQs improve with different speed. However, we can also observe from Figure 3 that the learning dynamics of GP traders are quite similar, except for the magnitude of initial leaps. This may suggest that GP traders with higher IQs perform better than those with lower IQs simply because they can improve more at the begining, while they don't improve a lot in the rest of the simulations, just like those with lower IQs.

This brings us a question: according to psychological theory, *fluid intelligence* measures the ability to problem-solving and learning in new situations, so if GP traders with higher IQs can perform better in a fixed market environment, they should be able to do the same thing when facing constantly changing new environments. The next experimental result demonstrates GP traders' performances when the market demand and supply changes at the begining of every generation.[8]

Part C of Table 1 demonstrates the results of this experiments. We can see from the table that a similar pattern emerges as those in Part A and Part B.

---

[8] To be more specific, we reassigned market participants' reservation prices randomly at the begining of every generation.

Table 1. Wilcoxon Rank Sum Tests for GP traders' performances on individual efficiencies. "*" symbols significant results under 10% significance level; "**" symbols significant results under 5% significance level.

		P5	P20	P30	P40	P50	P60	P70	P80	P90	P100
A	P5	X									
	P20	0.099*	X								
	P30	0.010**	0.328	X							
	P40	0.002**	0.103	0.488	X						
	P50	0.000**	0.009**	0.129	0.506	X					
	P60	0.000**	0.000**	0.003**	0.034**	0.130	X				
	P70	0.000**	0.000**	0.015**	0.121	0.355	0.536	X			
	P80	0.000**	0.000**	0.003**	0.036**	0.131	1.000	0.558	X		
	P90	0.000**	0.000**	0.011**	0.079*	0.250	0.723	0.778	0.663	X	
	P100	0.000**	0.000**	0.000**	0.002**	0.009**	0.284	0.093*	0.326	0.150	X
B	P5	X									
	P20	0.571	X								
	P30	0.589	0.288	X							
	P40	0.170	0.060*	0.442	X						
	P50	0.090*	0.020**	0.236	0.834	X					
	P60	0.004**	0.001**	0.019**	0.159	0.207	X				
	P70	0.066*	0.015**	0.191	0.671	0.848	0.280	X			
	P80	0.016**	0.003**	0.062*	0.333	0.422	0.656	0.577	X		
	P90	0.043**	0.010**	0.144	0.552	0.714	0.384	0.845	0.658	X	
	P100	0.001**	0.000**	0.008**	0.062*	0.091*	0.736	0.150	0.444	0.201	X
C	P5	X									
	P20	0.085*	X								
	P30	0.011**	0.410	X							
	P40	0.000**	0.028**	0.131	X						
	P50	0.001**	0.093*	0.379	0.620	X					
	P60	0.000**	0.013**	0.096*	0.799	0.460	X				
	P70	0.000**	0.060*	0.265	0.704	0.882	0.503	X			
	P80	0.000**	0.004**	0.029**	0.475	0.223	0.645	0.250	X		
	P90	0.000**	0.007**	0.050*	0.663	0.357	0.851	0.376	0.745	X	
	P100	0.000**	0.000**	0.000**	0.022**	0.005**	0.038**	0.004**	0.101	0.053*	X

Table 2. Learning Span of GP traders

IQ	Generations	IQ	Generations
5	699	60	57
20	174	70	49
30	115	80	42
40	86	90	37
50	69	100	34

This result therefore serves as a support for the suitability of modeling fluid intelligence with GP traders' population sizes.

5   Conclusion

The purpose of this paper is to raise the issue of heterogeneity in individual cognitive capacity since most agent-based economic or financial models do not deal with it. In this paper, we propose a method to model individual intelligence in agent-based double auction markets. We then run a series of experiments to validate our results according to what psychological studies have shown to us.

Preliminary experimental results in this paper show that it is viable to use population size as a proxy of intelligence for GP traders. In general, the results are consistent with psychological findings–a positive relation between intelligence and learning performance, and a decreasing marginal contribution of extra intelligence. Our study therefore shows that, by employing Genetic Programming as the learning algorithm, it is possible to model both the individual learning behavior and the innate heterogeneity of individuals at the same time.

The results of this study remind us a possibility that there is another facet to connect human intelligence and artificial intelligence. Artificial intelligence not only can be used to model intellectual behavior individually, but is able to capture social heterogeneity through a proper parameterization as well.

Acknowledgements

The authors are grateful to an anonymous referee for very helpful suggestions. The research support in the form of NSC grant no. NSC 95-2415-H-004-002-MY3 is gratefully acknowledged.

References

1. Nielsen, F.: Economic inequality, Pareto, and sociology: The route not taken. American Behavioral Scientist 50(5), 619–638 (2007)
2. Pareto, V.: Manual of Political Economy. In: Schwier, A.S., Page, N. (eds.) Manual of Political Economy. A. M. Kelley, New York (1971) (Original work published 1906)

3. Herrnstein, R.J., Murray, C.: The Bell Curve: Intelligence and Class Structure in American Life. Free Press, New York (1994)
4. Murray, C.: Income Inequality and IQ. AEI Press, Washington(1998), http://www.aei.org/books/filter.all,bookID.443/bookdetail.asp
5. Murray, C.: IQ and income inequality in a sample of sibling pairs from advantaged family backgrounds. American Economic Review 92(2), 339–343 (2002)
6. Lynn, R., Vanhanen, T.: IQ and the Wealth of Nations. Praeger, Westport (2002)
7. Lynn, R., Vanhanen, T.: IQ and Global Inequality. Summit Publishers, Washington (2006)
8. Jones, G., Schneider, W.J.: Intelligence, human capital, and economic growth: A Bayesian averaging of classical estimates (BACE) approach. Journal of Economic Growth 11, 71–93 (2006)
9. Ram, R.: IQ and economic growth: Further augmentation of Mankiw-Romer-Weil model. Economic Letters 94, 7–11 (2007)
10. Rydval, O., Ortmann, A.: How financial incentives and cognitive abilities affect task performance in laboratory settings: An illustration. Economic Letters 85, 315–320 (2004)
11. Gottfredson, L.S.: Mainstream science on intelligence: An editorial with 52 signatories, history, and bibliography. Intelligence 24(1), 13–23 (1997)
12. Das, R., Hanson, J.E., Kephart, J.O., Tesauro, G.: Agent-human interactions in the continuous double auction. In: Proceedings of the 17th International Joint Conference on Artificial Intelligence (IJCAI). Morgan-Kaufmann, San Francisco (2001)
13. Taniguchi, K., Nakajima, Y., Hashimoto, F.: A report of U-Mart experiments by human agents. In: Shiratori, R., Arai, K., Kato, F. (eds.) Gaming, Simulations, and Society: Research Scope and Perspective, pp. 49–57. Springer, Heidelberg (2004)
14. Grossklags, J., Schmidt, C.: Software agents and market (in)efficiency–A human trader experiment. IEEE Transactions on System, Man, and Cybernetics: Part C, Special Issue on Game-theoretic Analysis & Simulation of Negotiation Agents, IEEE SMC 36, 56–67 (2006)
15. Christal, R.E., Tirre, W., Kyllonen, P.: Two for the money: Speed and level scores from a computerized vocabulary test. In: Lee, G., Ulrich, T. (eds.) Proceedings, Psychology in the Department of Defense. Ninth Annual Symposium (USAFA TR 8-2), U.S. Air Force Academy, Colorado Springs (1984)
16. Detterman, D.K., Daniel, M.H.: Correlations of mental tests with each other and with cognitive variables are highest for low-IQ groups. Intelligence 13, 349–359 (1989)
17. Hunt, E.: The Role of intelligence in Modern Society. American Scientist, 356–368 (July/August 1995)
18. Rust, J., Miller, J.H., Palmer, R.: Characterizing effective trading strategies: Insights from a computerized double auction tournament. Journal of Economic Dynamics and Control 18, 61–96 (1994)
19. Gode, D., Sunder, S.: Allocative efficiency of markets with zero-intelligence traders: Market as a partial substitute for individual rationality. Journal of Political Economy 101, 119–137 (1993)
20. Cliff, D., Bruten, J.: Zero is not enough: On the lower limit of agent intelligence for continuous double auction markets. Tech. Rep. no. HPL-97-141, Hewlett-Packard Laboratories (1997), http://citeseer.ist.psu.edu/cliff97zero.html
21. Zhan, W., Friedman, D.: Markups in double auction markets. Journal of Economic Dynamics and Control 31, 2984–3005 (2007)

22. Gjerstad, S., Dickhaut, J.: Price formation in double auctions. Games and Economic Behavior 22, 1–29 (1998)
23. Friedman, D.: A simple testable model of double auction markets. Journal of Economic Behavior and Organization 15, 47–70 (1991)
24. Easley, D., Ledyard, J.: Theories of price formation and exchange in double oral auction. In: Friedman, D., Rust, J. (eds.) The Double Auction Market-Institutions, Theories, and Evidence, Addison-Wesley, Reading (1993)
25. Chan, N.T., LeBaron, B., Lo, A.W., Poggio, T.: Agent-based models of financial markets: A comparison with experimental markets. MIT Artificial Markets Project, Paper No. 124 (September 1999),
   http://citeseer.ist.psu.edu/chan99agentbased.html
26. Kagel, J.: Auction: A survey of experimental research. In: Kagel, J., Roth, A. (eds.) The Handbook of Experimental Economics. Princeton University Press, Princeton (1995)

Exploring Grammatical Evolution for Horse Gait Optimisation

James E. Murphy, Michael O'Neill, and Hamish Carr

University College Dublin, Belfield,
Dublin 4, Ireland
{james.murphy,m.oneill,hamish.carr}@ucd.ie
http://ncra.ucd.ie

Abstract. Physics-based animal animations require data for realistic motion. This data is expensive to acquire through motion capture and inaccurate when estimated by an artist. Grammatical Evolution (GE) can be used to optimise pre-existing motion data or generate novel motions. Optimised motion data produces sustained locomotion in a physics-based model. To explore the use of GE for gait optimisation, the motion data of a walking horse, from a veterinary publication, is optimised for a physics-based horse model. The results of several grammars are presented and discussed. GE was found to be successful for optimising motion data using a grammar based on the concatenation of sinusoidal functions.

Keywords: Grammatical Evolution, physics-based animation, gait optimisation, quadrupedal locomotion, Fourier analysis.

1   Introduction

A well-constructed physics-based animal model can produce physically realistic animations given good quality motion data. This data can be expensive to measure and unreliable to estimate. We propose to take some potentially inaccurate motion data and optimise it using an evolutionary algorithm approach.

Animal motion data is sometimes published in biomechanical and veterinary literature. Data can also be gleaned from sequential high-speed photographs of an animal in motion [1]. Although this data can be extracted and formatted for use with a physics-based model, it will not automatically produce stable locomotion. The data must be optimised for use with a specific model.

In this paper we present an approach to gait optimisation which utilises the Grammatical Evolution (GE) evolutionary computation technique. Rather than simply performing a parameterised optimisation on the data, which is known to work well, we propose a Fourier analysis based approach. By representing an animal's gait as a summation of sinusoidal functions, we optimise a more minimal set of parameters, as we successively concatenate sinusoidal functions of differing amplitude, phase and frequency. This representation is compact and mimics the sinusoidal nature of muscle movement. It also gives the evolutionary

L. Vanneschi et al. (Eds.): EuroGP 2009, LNCS 5481, pp. 183–194, 2009.

process more freedom to evolve than a parameterised optimisation, including the potential to retarget gait cycles from one animal to another.

We discuss how GE is used to optimise a walk gait for a physics-based horse model. Using a horse simulation application as the fitness function, gait cycles are generated and assessed. The gait cycle representation and the manner in which the simulation application acts as fitness function is described in Section 3. We compare the results of a variety of grammar types and speed-up strategies in Section 4. Section 5 concludes the paper with a brief discussion of the presented results. First of all, we present a brief overview of some related work.

2   Related Work

A gait is a pattern in which an animal moves using its limbs. The four main natural gaits of a horse are the walk, trot, canter and gallop. During locomotion, a sequence of muscles, determined by the current gait pattern, are contracted periodically to produce movement in the bones. The rotating bones cause the hoof to push off the ground surface, thrusting the animal forwards. The cyclical nature of a gait allows it to be quantified and described in terms of gait cycles.

A gait cycle begins when a foot contacts the ground and ends when that same foot contacts the ground again. The fraction of the cycle in which the foot is in contact with the ground is called the duty factor. The most distinguishing feature of a gait cycle is its footfall sequence, as the order in which an animal's feet impact the ground differs between gaits. Gaits are also described in terms of stride length and stride frequency [2].

The gait an animal will utilise, when travelling at a particular velocity, can be predicted based on the dynamic similarity principles [3]. This theory is based on the dimensionless Froude number, which is a function of the animal's velocity, height of hip from the ground and acceleration due to gravity. The dynamic similarity hypothesis states that when different mammals are travelling at equal Froude numbers, their gait patterns will be dynamically similar. This is exploited in our fitness function as will be described in Section 3.2.

Reproducing animal locomotion through animation is a well-studied topic in computer graphics and other disciplines such as robotics. Physics-based animal models can realistically reproduce animal motion given a well-constructed model and motion controller. An animal model is constructed as a series of interconnected rigid bodies. The rigid bodies represent the animal's bones and the connections between these bones are the joints. A physics engine ensures that the bones react in a physically realistic manner to gravity, friction, collisions, applied forces and torques [4]. Forces and torques applied to the bones produce linear and angular accelerations. As the bones are connected together by joints, forces applied to a single rigid body may propagate throughout the rigid body system. To produce motion in the model, a motion controller applies torques of specific magnitude about each of the joints in a limb, with precise timing, to produce a desired bone rotation. The development of animal models and motion controllers has been studied in the computer animation field for many years.

Genetic Algorithms are used to generate quadruped robot gaits in [5,6,7]. The fitness function in [5] is of interest as it is concerned with finding a gait that uses minimal energy, while covering a required distance at a specified speed. This can be applied to gait generation for quadruped animations as they state that the optimal gaits produced for the robots are comparable to those expected of a real-life animal travelling at the same Froude number. Each flavour of evolutionary algorithm will vary in performance depending on the problem domain. We apply a relatively new type of evolutionary algorithm called Grammatical Evolution (GE) to gait optimisation. GE is one of the most popular forms of grammar-based Genetic Programming (GP) due to the convenience by which a user can specify and modify the grammar, whilst ignoring the task of designing specific genetic search operators. It has been successfully employed to financial prediction [9], but has never been applied to gait generation and optimisation.

The GE search adapts principles from molecular biology. GE is distinct from other evolutionary algorithms as it uses a variable length binary or integer string to derive solutions from a Backus Naur Form grammar. In GE, the evolutionary algorithm's genetic operators are applied to the strings (genotypes) rather than problem domain solutions (phenotypes). Potential complexities of the phenotype are inconsequential. This makes GE a good choice for gait optimisation because the horse gait problem domain is large and complex, as will be discussed in the following section. For further information on GE please refer to [10].

3   GE for Gait Optimisation

Our goal is to optimise gait data and explore how motion variations in a joint affect an animal's gait. GP allows us to explore these motion variations as a gait is optimised. The advantage of GP over other model/structure learning methods, such as Neural Networks, is that output structures are generally in a human-understandable format. We can examine motions and identify patterns which may allow us to generate motion for morphologies for which we have no data.

Measured data, and experience of manually tuning motion data for animation purposes, provide a general template of how each joint should move for realistic motion of the model. Information on an animal's musculature and joint limits can be implicitly included through constraints imposed on the evolving structures. We wish to compare results of multiple optimisations utilising differing styles and levels of constraint on evolutionary freedom. Grammar-based GP, specifically GE, is ideal for this purpose as various constraint methods can be easily incorporated into a grammar and rapidly performance tested.

During the GE process, each phenotype produced from a grammar is passed to a simulation application as motion data. This application, acting as fitness function, assesses each phenotype for a few gait cycles. The GE process proceeds until an optimal solution is found. Technical details of the GE set-up are beyond the scope of this paper however, the GE parameters used for each run are presented in Table 1. All results in Section 4 are generated using GEVA [11].

Table 1. GE parameters used for every run

Parameter	Value
Generations	50
Population	75
Max. wrapping	3
Replacement	generational
Elite size	7
Selection	Tournament (3)
Initialisation	RampedFullGrow
Max. depth	10
Grow prob.	0.5
Crossover prob.	0.9
Crossover point	fixed
Mutation prob.	0.02

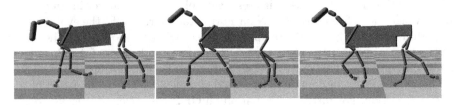

Fig. 1. Sequential screen-shots of the physics-based horse model walking

Sequential screen-shots of the simulation application, implemented using Open Dynamics Engine [8], are shown in Figure 1. Simulation application details are beyond the scope of this paper however, the most important aspect of the simulation application is the gait cycle representation. It is based on the observation that, as muscles tend to relax and contract in a sinusoidal manner, real-life joint-angle data can be decomposed into a sum of sinusoidal functions through Fourier analysis. Motion data for each joint is therefore represented as a sum of these sinusoids, referred to as the sum-of-sines representation. It is a compact and elegant way of representing cyclical data such as a single cycle of a gait, which is used repeatedly to produce sustained locomotion. Details of our grammars and fitness function follow in Section 3.1 and Section 3.2 respectively.

3.1   Gait Optimisation Grammar

The grammars must allow for construction of syntactically correct motion data in the sum-of-sines format. If the goal is to optimise real-life animal gait data, that data must be incorporated into the grammar. The greatest consideration in the grammar construction is the degree to which phenotypes are allowed differ from the seed data. The grammar should produce motion data for each joint in the model, represented as a summation of sinusoidal functions, in the form:

$$\langle amp \rangle * sin(\langle freq \rangle * 2 * PI * time + \langle phase \rangle) \tag{1}$$

Motion data, extracted from plots in [2], is decomposed into its component sinusoidal functions through Fourier analysis. Fourier terms whose amplitude is below some arbitrarily chosen threshold value are discarded, leaving a more compact summation of sinusoidal functions approximating the motion data.

This representation could be simply optimised by manipulating the variable values of its constituent sinusoidal functions. The values of the *amp* and *phase* parameters in each function can be optimised within some defined range. As the extracted motion, or seed, data dictates the frequencies of the functions we can optimise, the range of potential solutions is constrained. To provide greater flexibility, our GE grammars produce phenotypes by concatenating sinusoidal functions to the seed data from a fuller range of frequencies, not just those dictated by the minimal form of the Fourier analysis. Depending on the grammar, phenotypes can be constrained to remain close to the seed data or allowed to deviate. Each of the generated phenotypes must be assessed in the animal simulation application by means of a fitness function.

3.2 Gait Optimisation Fitness Function

Our fitness function is based on energy efficiency and gait predictions based on the dimensionless Froude number, which is calculated as follows:

$$Fr = v/\sqrt{g * h} \qquad (2)$$

Where Fr is the Froude number, v is velocity, g is gravity and h is height of the animal's hip from the ground. Dynamic similarity theory states that an animal travelling at a particular Froude number will share gait characteristics with other animals travelling at the same Froude number. This implies that if we have gait information for a single animal moving at a range of Froude numbers, we can predict the gaits of other animals. Such data is published in [3].

A gait is optimised to move the model at a particular Froude number. From that Froude value we calculate velocity and predict the phase difference between limbs, stride frequency, stride length and duty factors.

These predictions are used in the fitness function to score the phenotypes. The phase difference and stride frequency are set in the application based on the Froude number argument. Only the joint-angle motion data is generated. An optimal generated gait moves the model with the velocity, duty factor and stride length values predicted by the dynamic similarity theory. Energy efficiency of a gait is also a factor in the fitness score. In nature, animal morphology and joint-angle motion has generally evolved to use the minimum energy to travel a desired distance at a desired velocity. The fitness function therefore rewards those phenotypes that use minimal energy.

Each of the fitness components has an associated weight. Each component's contribution to the fitness score is a function of its weight and a measure of the error from its predicted value. The better the gait, the lower the score. Using this technique, a perfect score of 0 should never occur. The energy component is a weighted sum of the model's total energy use, averaged per cycle. As the model must expend energy to move, even the most optimal gaits will have a positive fitness value, as will be seen in the following section.

4   Experiments and Results

To explore the use of GE for gait optimisation, grammars which utilise seed data are investigated as well as free-style grammars which do not. In the case of these unconstrained, free-style grammars, the goal is not to produce aesthetically realistic gaits, but rather create novel movement and test the capabilities of our multivariable fitness function. It is apparent from experimentation, that grammars providing parameterised optimisation of seed data perform well. A future goal of our research is to retarget optimal gait data from one animal to another as outlined in [12]. Our experiments have shown that a simple parameterised optimisation of the seed data does not have the scope to alter motion data significantly enough to allow retargeting to animals with different morphology to that of the source. This motivates our exploration of grammars which provide the flexibility to evolve gaits from one animal to another.

The results presented in this paper are divided into three categories. In Section 4.1, we describe a grammar which optimises data by concatenating sinusoidal functions to the seed data. This grammar is compared with parameterised optimisation approaches. In Section 4.2, the investigation of the concatenating functions grammar continues with two attempts to speed-up the evolutionary process through generational manipulation of the grammar and fitness function. Finally, in Section 4.3, grammars that do not use seed data are presented.

4.1   Parameterised Optimisation Comparison

For our parameterised optimisation, each term in the compact summation of sinusoidal functions data approximation is represented as a triple (amplitude, frequency and phase). Each of these values is optimised within a range of 25% of itself. While this approach performs well on data which has been measured from an animal with the same morphology as our model, as in the presented case, it is constrained to produce a limited set of motion, unsuitable for interspecies gait retargeting. We aim to develop a grammar which can at least match and hopefully surpass the parameterised approach, while having the scope to completely diverge from the seed data during our retargeting experiments.

The grammar presented in Figure 2 optimises the seed data by adding (or subtracting) sine and cosine functions of differing amplitude and frequency to the seed data, which is itself in the sum-of-sines form. Generated motion may deviate from the seed data through addition of functions of differing frequency.

The frequencies of the appendable sinusoidal functions and the seed data summations have a range of 1 to 8Hz. Fourier analyis of the seed data shows that higher frequency functions have amplitudes less than our arbitrarily chosen threshold value of 1. These functions are considered less influential to the overall movement and are discarded for compactness. The grammar also constrains

```
<prog> ::= <fcurve0> <newline> <fcurve1> <newline> ... <fcurve11>

<fcurve0> ::= <curve0> | <curve0> + <funcs>
...
<fcurve11> ::= <curve11> | <curve11> + <funcs>

<funcs> ::= <funcs> <op> <funcs>
          | <function>
          | <med_amp_var>

<op> ::= + | -

<function> ::= <low_amp_var> * sin( <low_freq_var> * 2 * PI * t )
             | <low_amp_var> * cos( <low_freq_var> * 2 * PI * t )
             | <med_amp_var> * sin( <med_freq_var> * 2 * PI * t )
             | <med_amp_var> * cos( <med_freq_var> * 2 * PI * t )
             | <hi_amp_var> * sin( <hi_freq_var> * 2 * PI * t )
             | <hi_amp_var> * cos( <hi_freq_var> * 2 * PI * t )

<low_freq_var> ::= 1 | 2
<med_freq_var> ::= 3 | 4
<hi_freq_var> ::= 5 | 6 | 7 | 8

<low_amp_var> ::= 0 | 0.25 | 0.5 | ... | 20
<med_amp_var> ::= 0 | 0.1 | 0.2 | ... | 4
<hi_amp_var> ::= 0 | 0.05 | 0.1 | ... | 1

<curve0> ::= 6.97+7.7*sin(1*2*PI*t+-1.07)+2.56*sin(2*2*PI*t+2.97) ...
...
<curve11> ::= ...
```

Fig. 2. An illustrative example of a grammar based on the concatenation of sinusoidal functions to the seed data. Note the seed data at the bottom of the grammar. (Omitted terms represented by '...'.).

appended functions of a particular frequency to have an amplitude of a specific range. This range is based on observations from the Fourier analysis.

The best fitness plot in Figure 3, shows the numerical (parameterised) optimisation starting off worst. Gradually it improves and achieves a similar optimal solution score to the concatenating functions grammar. The overall winner in terms of best fitness is a grammar which uses a combination of parameterised optimisation and concatenating functions. The sinusoidal functions are added to the seed data, whose parameters are optimised in parallel.

4.2   Concatenating Functions Speed-Up Attempts

To improve the performance of the concatenating functions grammar, two tests are presented in which the fitness function and grammar dynamically changes. For the varying fitness function test, the fitness weights are changed from generation to generation, as shown in Table 2. It was hoped that this would speed up the evolutionary process, prevent the process from becoming stuck at local minima and produce a more well rounded solution, i.e. one which optimises aspects of velocity, duty factor and stride length equally.

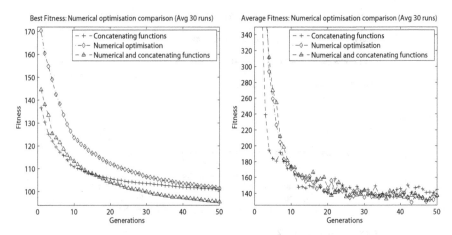

Fig. 3. A numerical optimisation approach is contrasted with the concatenating functions grammar and a hybrid of the two approaches. Best and average fitness (averaged, 30 runs) are presented on the left and right respectively. *(Note difference in scale.).*

Table 2. Fitness function weights during the fitness function variation run

Fitness measure	Gen. 0-10	Gen. 11-20	Gen. 21-30	Gen. 31-40	Gen.41-50
Distance	1	2	1	1	1
Duty factor	1	1	2	1	1
Stride length	1	1	1	2	1
Energy	1	1	1	1	1

The results presented in Figure 4 do not show any speed-up. The change in fitness function does seem to drive the evolution forward in some situations however, in this instance, the change in fitness function causes a plateau in the best fitness score from generation 10-20. The large spike in the corresponding average fitness plot indicates that the model has a high distance error value at generation 10. The increase in the distance-scalar's weight causes temporary chaos. The process recovers and quickly proceeds to an optimal solution.

The second test involves restricting joint motion on a generational basis. The sequence in which joints are given freedom is presented in Table 3. While again this approach does not speed up the evolutionary process, it is clear from the average fitness plots in Figure 4 that the varying joint freedom grammar produces very stable gaits from the earliest generations. The large starting values apparent in most of the average fitness plots are the result of the unviable phenotypes passed to the simulation application, usually at the start of the evolutionary process. The motion data can cause the model to wildly gyrate its limbs or provide such a boisterous gait that the model flips over. These bad phenotypes are awarded a very high score (corresponding to the worst fitness possible). By initially restricting the model's degrees of freedom, production of the bad phenotypes appears minimised. The results are summarised in Table 4.

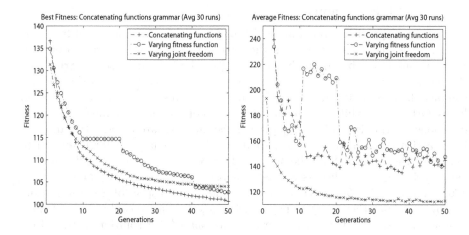

Fig. 4. The concatenating functions grammar with alternating fitness and grammar strategies. Best and average fitness (averaged, 30 runs) are presented on the left and right respectively. *(Note difference in scale)*.

Table 3. Generational joint freedom. Only joints with a ✓ are free to move and evolve motion data for each generation range. All other joints remain static.

Joint	Gen. 0-10	Gen. 11-20	Gen. 21-30	Gen. 31-50
Scapula (fore)	✓	✓	✓	✓
Shoulder (fore)	✓	✓	✓	✓
Elbow (fore)	-	✓	✓	✓
Carpal (fore)	-	-	✓	✓
Fetlock (fore)	-	-	-	✓
Hip (hind)	✓	✓	✓	✓
Stifle (hind)	-	✓	✓	✓
Tarsal (hind)	-	-	✓	✓
Fetlock (hind)	-	-	-	✓
Proximal (neck)	-	✓	✓	✓
Mid (neck)	-	-	✓	✓
Atlas (head/neck)	-	-	-	✓

Table 4. Best and average fitness with standard deviations (averaged, 30 runs), of the concatenating functions grammar with the fitness function and grammar variations

Strategy	Avg. Best	Std. Dev.	Avg. Avg.	Std. Dev.
None	100.333	8.5231	140.9203	43.1542
Varying fitness weights	102.7	5.3572	154.5484	50.3666
Varying joint freedom	104.0333	7.8892	112.2726	8.4968

4.3 Free-Style Grammars

In contrast to previous grammars which contain seed data, the use of free-style grammars is also investigated. As the grammar, shown in Figure 5, does not contain constraints based on the animal's joint limits and muscle distribution, the motions resulting from this free grammar vary greatly across the 30 runs completed. In some instances, the model moves utilising only its front or hind limbs. Other runs exhibit a sequence of sudden hops to move the model. On a few occasions, motion is produced by placing the limbs squarely under the animal's body and using a high-frequency, small amplitude, back and forth motion to "vibrate" the animal along the surface. The fact that these very different gait cycles achieve similar fitness scores demonstrates a flaw in the multivariable fitness function. It appears that improvements in one aspect of the fitness score can overshadow other components, which suggests a more sophisticated fitness function is required. Currently, the ultimate shape of a solution using a free-style grammar may be randomly determined early in the evolutionary process. An interactive evolutionary computation technique could be employed in the early generations to guide the process towards a realistic motion.

```
<prog>    ::= <fcurve> <newline> ... <fcurve>

<fcurve>  ::= <expr>

<expr>    ::= <expr> <op> <expr>
            | (<expr> <op> <expr>)
            | <pre-op> (<expr> * t)
            | <var>

<op>      ::= + | - | / | *
<pre-op>  ::= sin | cos
```

Fig. 5. An illustrative example of a free grammar. The t variable is required by our simulation application so that generated motion data may be a function of time. The *fcurve* terms, for each joint presented in Table 3, are omitted and represented by '...'.

A free-style sum of sinusoidal functions technique is also tested. The grammar is similar to the concatenating functions grammar in Figure 2, except that the functions are not concatenated to any seed data. Our knowledge that animal gait data can be decomposed into a summation of sinusoidal functions, each having parameters which fall within particular frequency, amplitude and phase values, is incorporated into the grammar. In contrast to the free grammar in Figure 5, information about the animal's musculature is implicitly included through the specific frequency and amplitude ranges of the sinusoids. While a small number of generated gaits are visually unrealistic, the majority are comparable to the real-life motion of a horse. While not as good as those grammars which include seed data, there is potential for improvement given a more sophisticated fitness function and increased population and generation values.

Figure 6 shows the free and sinusoidal grammar scoring comparably to the concatenating functions grammar in terms of fitness. This illustrates the pitfalls

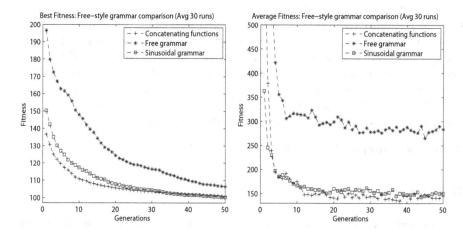

Fig. 6. Free and free-style sinusoidal grammars compared with the concatenating functions grammar. Best and average fitness (averaged, 30 runs) are presented on the left and right respectively. *(Note difference in scale.)*

of using a multivariable fitness function and few motion constraints. Out of the free grammar's 30 runs, very different "optimal" solutions score similarly. It demonstrates that if realism is the goal, seed data, or constraints built into the grammar based on observations of animal motion, are required.

5   Discussion and Conclusions

GE allows us to rapidly explore different approaches to gait generation. Phenotypes are produced in a human-readable format which assists understanding of gait motion. Grammars also allow us to construct gait representations other than the presented sum-of-sines, necessary for non-cyclical gait transition motions.

Table 5 shows that the concatenating functions grammar improves upon the parameterised approach in terms of fitness score, albeit by a small margin. The overall winner is a combination of the two. The concatenating sinusoidal functions approach is found to be a compact method of representing and optimising

Table 5. Overall best and average fitness scores (averaged, 30 runs) achieved by each grammar alongside their respective standard deviations

Grammar	Avg. Best	Std. Dev.	Avg. Avg.	Std. Dev.
Numerical & concatenating functions	95.3	5.2729	138.8273	57.2904
Concatenating functions	100.333	8.5231	140.9203	43.1542
Sinusoidal grammar	100.4333	6.0039	150.2595	44.2394
Numerical only	101.3333	8.5715	132.6506	37.6833
Free	106.3	19.05	279.1248	81.3519

gait data. By optimising seed data whilst appending new sinusoidal functions, we have the flexibility to retarget to other morphologies whilst maintaining realism.

The multivariable nature of our fitness function allows for significant motion variance between similarily scoring phenotypes. It may be benificial to use a multi-objective optimisation approach in future. With further refinement of the grammars and fitness function, this GE method of gait optimisation can be utilised in our gait retargeting solution. In future we also hope to use a grammar-based approach to produce sophisticated balance and directional systems.

Acknowledgments. Thanks to IRCSET and IBM.

References

1. Muybridge, E.: Horses and Other Animals in Motion. Dover Publications, New York (1985)
2. Back, W., Clayton, H.M.: Equine Locomotion. Harcourt Publishers (2001)
3. Alexander, R.M., Jayes, A.S.: A Dynamic Similarity Hypothesis for the Gaits of Quadrupedal Mammals. Journal of Zoology (London) 201, 135–152 (1983)
4. Parent, R.: Computer Animation: Algorithms and Techniques. The Morgan Kaufmann Series in Computer Graphics and Geometric Modeling. Morgan Kaufmann, San Francisco (2002)
5. Kiguchi, K., Kusumoto, Y., Watanabe, K., Izumi, K., Fukuda, T.: Energy-Optimal Gait Analysis of Quadruped Robots. In: Artificial Life and Robotics, vol. 6, pp. 120–125. Springer, Japan (2002)
6. Garder, L.M., Høvin, M.E.: Robot Gaits Evolved by Combining Genetic Algorithms and Binary Hill Climbing. In: Genetic and Evolutionary Computation Conference 2006, pp. 1165–1170. ACM, New York (2006)
7. Xu, K., Chen, X., Liu, W., Williams, M.: Legged Robot Gait Locus Generation Based on Genetic Algorithms. In: Proc. International Symposium on Practical Cognitive Agents and Robots, vol. 213, pp. 51–62. ACM, New York (2006)
8. Smith, R.: Open Dynamics Engine v0.5 User Guide (2006)
9. Brabazon, A., O'Neill, M.: Biologically Inspired Algorithms for Financial Modelling. Springer, Heidelberg (2006)
10. O'Neill, M., Ryan, C.: Grammatical Evolution. Genetic Programming Series. Kluwer Academic Publishers, Dordrecht (2003)
11. GEVA: Grammatical Evolution in Java, http://ncra.ucd.ie/geva
12. Murphy, J.E., Carr, H., O'Neill, M.: Grammatical Evolution for Gait Retargeting. In: Proc. Theory and Practice of Computer Graphics 2008. Eurographics, pp. 159–162 (2008)

There Is a Free Lunch for Hyper-Heuristics, Genetic Programming and Computer Scientists

Riccardo Poli and Mario Graff

School of Computer Science and Electronic Engineering, University of Essex, UK
{rpoli,mgraff}@essex.ac.uk

Abstract. In this paper we prove that in some practical situations, there is a free lunch for hyper-heuristics, i.e., for search algorithms that search the space of solvers, searchers, meta-heuristics and heuristics for problems. This has consequences for the use of genetic programming as a method to discover new search algorithms and, more generally, problem solvers. Furthermore, it has also rather important philosophical consequences in relation to the efforts of computer scientists to discover useful novel search algorithms.

1 Introduction

Hyper-heuristics are often defined as "heuristics to choose heuristics" [4]. A *heuristic* is considered as a rule-of-thumb or "educated guess" that reduces the search required to find a solution. The difference between meta-heuristics and hyper-heuristics is that the former operate directly on the problem search space with the goal of finding optimal or near-optimal solutions. The latter, instead, operate on the heuristics search space (which consists of all the heuristics that can be used to solve a target problem). The goal then is finding or generating high-quality heuristics for a problem, for a certain group of instances of a problem, or even for a particular instance. Put another way, hyper-heuristics are search algorithms that don't directly solve problems, but, instead, search the space of solvers, searchers, meta-heuristics or heuristics that can then solve the problems of interest.

The situation is depicted in Figure 1 where a hyper-heuristic (the point drawn in thick line in the leftmost circle in the figure) is sampling a subset of elements in the space of heuristics (the middle circle) in order to find a good solver for a problem, represented as a search space endowed with a fitness function (the rightmost circle in the figure). Each solver sampled by the hyper-heuristic needs to be executed in order to evaluate the quality of such a solver. This, in its turn, involves the sampling of a subset of elements of the space of solutions. Since fitness values are associated to these, information percolates up the hierarchy to guide the hyper-heuristic. Note that there are many hyper-heuristics in the space of hyper-heuristics. Each may sample the space of heuristics differently, which in turn may lead to sampling the solution space differently.

Genetic Programming (GP) [3,15,21] has been very successfully used as a hyper-heuristic. For example, GP has evolved competitive SAT solvers [1,2,8,14], evolutionary algorithms [16,17], bin packing algorithms [5,6,22], particle swarm optimisers [7,19,20], and travelling salesman problem solvers [11,12,13,18].

L. Vanneschi et al. (Eds.): EuroGP 2009, LNCS 5481, pp. 195–207, 2009.
© Springer-Verlag Berlin Heidelberg 2009

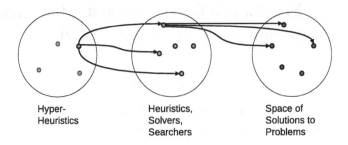

| Hyper-
Heuristics | Heuristics,
Solvers,
Searchers | Space of
Solutions to
Problems |

Fig. 1. A hyper-heuristic exploring the space of search algorithms (or more generally solvers) for a particular problem

The question we want to investigate in this paper is whether or not the No-Free-Lunch (NFL) theory of Wolpert and Macready [26] and its more modern reformulations have implications for hyper-heuristics and GP (when used to invent problem solvers and other search algorithms). In principle, there would seem to be no a priori reason why NFL would not apply to hyper-heuristics in the sense that NFL is said to apply to all forms of search. However, nobody has actually ever explored the specific applicability of NFL to meta-search (i.e., hyper-heuristics). In this paper we explore under what conditions NFL may apply to hyper-heuristics and what implications the availability or otherwise of a "free lunch" may have on researchers interested in developing search algorithms, including GP.

The paper is organised as follows. In Section 2 we cover the basics of NFL. In Section 3 we consider the implications of NFL for hyper-heuristic search. Section 4 shows that free lunches are possible for hyper-heuristics. Exactly what this means is explained in Section 5. We then look at the consequences this has in a broader context in Section 6. We consider higher-order hyper-heuristic search in Section 7. Finally, we provide some conclusions in Section 8.

2   No Free Lunch

Informally speaking, the NFL theory [26] states that, when evaluated over all possible problems, all algorithms are equally good or bad irrespective of our evaluation criteria.

In the last decade there have been a variety of results that have refined and specialised NFL (see for example [25] for a comprehensive review and [24] for a discussion on its implications). One important result states that if one selects a set of fitness functions that are *closed under permutation* (more on this below) then the expected performance of any search algorithm over that set of problems is constant, i.e., it does not depend on the algorithm we choose [23] nor the chosen performance measure. In formulae, if $\mathcal{F}$ is a set of fitness functions closed under permutation, we have that

$$\sum_{f \in \mathcal{F}} P(f, a_1) = \sum_{f \in \mathcal{F}} P(f, a_2) \tag{1}$$

for any pair of pair of (non-resampling) search algorithms a_1 and a_2 and for any performance measure P. Furthermore, [23] showed that the connection between closure and

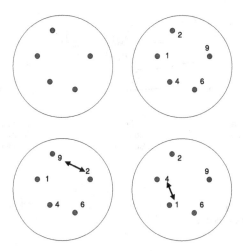

Fig. 2. A sample search space (top left), a problem (i.e., an assignment of fitness to the elements of the search space (top right), and two sample permutations of the fitness function (bottom)

f1	2		f1	1		f1	4		f1	6		f1	9	
f2	1		f2	2		f2	1		f2	1		f2	1	
f3	4		f3	4		f3	2		f3	4		f3	4	... (120) .
f4	6		f4	6		f4	6		f4	2		f4	6	
f5	9		f5	9		f5	9		f5	9		f5	2	

Fig. 3. A fitness function seen as a vector and its permutations

NFL is an "if and only if" one. That is, it is also the case that two arbitrary algorithms will have identical performance over a set of functions only if that set of functions is closed under permutation.

What does it mean for a set of functions to be closed under permutation? The NFL is limited to finite search spaces. A fitness function is an assignment of fitness values to the elements of the search space, as exemplified graphically in Figure 2 (top). A permutation of a fitness function is simply a rearrangement of the fitness values originally allocated to the objects in the search space. Examples of permutations are shown at the bottom of Figure 2. If we enumerate the elements of a search space according to some scheme, we can then represent a fitness function as a vector that stores the fitness associated to each point of the space. For example, if we enumerate the five points in Figure 2 (top left) in anti-clockwise order, the fitness function in Figure 2 (top right) can be represented as indicated in the leftmost vector in Figure 3. In this case, permuting a fitness function simply means shuffling the elements of the vector representing it. A subset of permutations of our sample fitness function are shown in Figure 3. A set of problems/fitness functions is closed under permutation if, for every function in the set, all possible shuffles of that function are also in the set. If all the possible 120

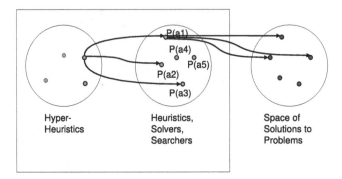

Fig. 4. In the space of hyper-heuristics (exploring the space of search algorithms/solvers for a particular problem), fitness is determined by how well each algorithm solves the original problem

permutations (including the identity permutation) of the elements of the leftmost vector in Figure 3 were considered, then the resulting set of fitness functions would be closed under permutation.

3   Implications of No-Free Lunch for Hyper-Heuristics

So, let us see exactly what the applicability of NFL would imply in the case of hyper-heuristics. Firstly, let us consider what guides the search of hyper-heuristics.

Hyper-heuristics require some feedback from the problem domain in order for them to have any hope to discover good heuristics. As illustrated in Figure 4, they are guided by a performance measure, $P(a)$, that indicates how good each heuristic a (for "algorithm") is in relation to the problem at hand. Note that while in practice $P(a)$ is computed (in fact, it can often only be estimated) only for the algorithms actually sampled by the hyper-heuristic, from a logical point of view we can imagine that all possible algorithms in the heuristics space have a precomputed performance measure. In this situation, we can forget about the original problem space, and, instead, look at the algorithm space as an ordinary search space endowed with an objective measure. That is, we can focus on the rectangular region marked out in Figure 4.

In order to make further progress, we need to generalise and formalise our treatment a little. Let $\mathcal{P}$ be a set of problems for which we want to find a good solver via hyper-heuristic search. The set $\mathcal{P}$ may include a single problem instance (as exemplified in Figures 1 and 4), a set of problem instances, or even a whole class of problems (as illustrated in Figure 5). However, for simplicity, let us imagine that all such problems share the same solution space Ω and that Ω is finite. If, for example, we were interested in solving SAT problems, we imagine that $\mathcal{P}$ is a set of SAT instances with the same number of variables, n, so that $\Omega = \{0, 1\}^n$ for all problems in $\mathcal{P}$.

What distinguishes a problem from another is the assignment of fitness to the elements of Ω. Let $\mathcal{G}$ be the set of fitness functions associated to the problems in $\mathcal{P}$. $\mathcal{G}$ corresponds to the rectangular area to the right of Figure 5. We can think of the elements of $\mathcal{G}$ graphically as in the figure or as vectors/tables as indicated in Section 1.

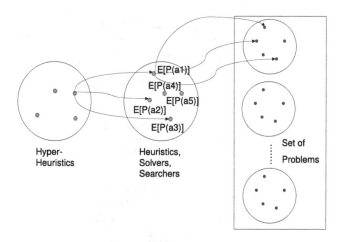

Fig. 5. A hyper-heuristic exploring the space of search algorithms (or more generally solvers) for a particular set of problems

As we discussed above, a hyper-heuristic is a search algorithm that explores the space of search algorithms. Let us call this latter space Λ.[1] Λ corresponds to the circle in the middle of Figure 5. In order to make decisions as to which heuristics in Λ to explore and in which order, a hyper-heuristic needs to be guided by some objective function that evaluates how well each of the heuristic it samples solves the problems in $\mathcal{P}$. Let us look at this more closely.

Given a problem with fitness function $g \in \mathcal{G}$, a searcher/heuristic will sample Ω to achieve some goal. Often the goal is to find good values of fitness and we would like the heuristic to find a member of Ω that has the largest (or smallest, if minimising) value in g using the smallest possible number of trials. However, one may want to adopt other performance measures. Let P be one such performance measure. Because we don't have just one problem, but a set of problems, $\mathcal{P}$, the performance measure $P(a)$ for an algorithm $a \in \Lambda$ will need to be applied to all such problems and then averaged to produce an estimation of how good an element of the space of searchers, Λ, is. Naturally, in the presence of stochasticity, P will be noisy. For the purpose of this argument, however, we will imagine that we can associate to each element a of Λ the exact expected value of $P(a)$, $E[P(a)]$, where the expectation is taken over all problems in $\mathcal{P}$, i.e., all functions in the set $\mathcal{G}$. As we indicated above, once the performance of each algorithm/solver is available, we can forget about the problem space, and, instead, look at the algorithm space as an ordinary search space endowed with a "fitness measure" $E[P(a)]$, as illustrated in Figure 6 (top).

Of course, as we know from the NFL theory summarised in Section 2, if the set of fitness functions, $\mathcal{G}$, characterising our problem set, $\mathcal{P}$, is closed under permutation, then all non-resampling search algorithms will show identical average performance over $\mathcal{G}$, i.e., $\sum_{p \in \mathcal{P}} P(a, p) = $ constant. So, $\forall a : E[P(a)] = $ constant.

---

[1] Naturally, Λ may depend on the problems in $\mathcal{P}$, the solution space Ω, the adopted representation for search behaviours, etc. We will not investigate these details here.

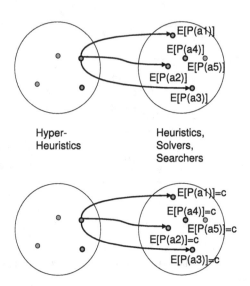

Fig. 6. The search as perceived from the point of view of a hyper-heuristic (top). If the problem set is closed under permutation, the "fitness function" over the heuristic search-space (Λ) is flat (bottom). In this case, searching for hyper-heuristics is futile, since every single one of them is equally good.

This has a simple and important implication. A hyper-heuristic using $E[P(a)]$ (for any P) as its fitness measure over Λ will find that the fitness landscape is flat! This situation is illustrated in Figure 6 (bottom). So, the hyper-heuristic has nothing to guide it. In fact, no hyper-heuristic can have any guidance whatsoever. Therefore, a hyper-heuristic that chooses a random solver from Λ is as good as any other. In other words, *hyper-heuristic search makes no sense if the set of problems for which we want to find a solver is closed under permutation.*

4   Free Lunches for Hyper-Heuristics

How likely is the situation depicted above? Igel and Toussaint [9] were able to show that if one considers all possible sets of functions with a given domain and a given co-domain, in most conditions *the sets that are closed under permutation, and, hence, over which NFL applies, represent a tiny fraction of the whole.* From the point of view of hyper-heuristic search, this would suggest that sets of problems $\mathcal{P}$ for which there is no point in using hyper-heuristics are in fact quite rare, which is encouraging. However, this is no guarantee that NFL will not be applicable for a particular set of problems.

Proving whether or not a set of fitness functions $\mathcal{G}$ is closed under permutation may be difficult in general. However, there are cases where it is trivial to see that $\mathcal{G}$ cannot be closed under permutation via basic counting argument. For example, if $\mathcal{P}$ includes only one problem instance (which is sometimes the case in hyper-heuristic search), and the fitness function associated to that problem instance, f, is non-flat (i.e., there is at least one point $x \in \Omega$ with higher than average fitness, which is typically the case), then $\mathcal{P}$ cannot be closed under permutation. This is because with f there is always a $y \in \Omega$, with $y \neq x$, such that $f(y) \neq f(x)$. It is, therefore, possible to find a way to shuffle f to produce a new fitness function $\tilde{f}$ in which the fitness originally associated to x is now assigned to y, that is $\tilde{f}(y) = f(x)$. But then, since $f(x) \neq f(y)$, we have that $\tilde{f}(y) \neq f(y)$, so $\tilde{f}$ is distinct from f. Since $\mathcal{P}$ is a singleton set, so is $\mathcal{G}$. However, $f \in \mathcal{G}$ and, so, $\tilde{f} \notin \mathcal{G}$, which implies $\mathcal{G}$ is not closed under permutation. So, *when a hyper-heuristic approach is applied to finding a solver for a specific problem, there can be a free lunch.*

This counting argument can be made more general. If we know that one problem has at least n distinct fitness values $f_1, \ldots, f_n$ at points $x_1, \ldots, x_n$, then it is possible to shuffle the assignments of fitness values to $x_1, \ldots, x_n$ in at least $n!$ ways. If our problem set $\mathcal{P}$ contains fewer that $n!$ problems, then it cannot be closed under permutation and there is a free lunch for hyper-heuristic search. It stands to reason that these conditions are easily met by any practical problem set, since a) the size, n, of the co-domain for realistic fitness functions will be relatively large (i.e., the fitness function can return a variety of different values), and b) practical problem sets can never be too large because one needs to calculate the performance of a in each of the problems which is usually done by running a on each problem thereby the time required for this procedure would limit the size of $\mathcal{P}$. So, the condition $|\mathcal{G}| < n!$ should easily be satisfied.[2] So, *in practice, when a hyper-heuristic approach is applied to finding a solver for a not-too-large set of problems, there will likely be a free lunch.*

5   What's a Free Lunch?

So far we have not looked very closely at what having a "free lunch" means. One might think, for example, that having a free lunch means that for any performance measure and any two algorithms, one algorithm is superior to the other. Or perhaps having a free lunch might mean that for any performance measure, there is at least one algorithm that is superior to some other. Unfortunately, neither interpretation is correct.

As we indicated above, [23] showed that two arbitrary algorithms have identical performance (irrespective of the chosen performance measure) over a set of functions if and only if that set is closed under permutation. This result was proven by designing a performance measure such that if NFL held over a set of functions $\mathcal{F}$ that is not closed under permutation, it would then be possible to identify an algorithm whose average performance over $\mathcal{F}$ is inferior to that of some other algorithm, leading to a contradiction. So, another way to look at this result is to say that if the set of functions $\mathcal{F}$ is *not* closed under permutation, then *there exists at least one performance measure under which at least one algorithm has better (worse) than average performance* at optimising functions from $\mathcal{F}$.

So, the fact that NFL does not apply to hyper-heuristic search when the set of fitness functions $\mathcal{G}$ associated to a problem set $\mathcal{P}$ is not closed under permutations means that

---

[2] In fact this condition is always satisfied in the case of Boolean induction problems with $k > 1$ inputs, since there can be at most 2^{2^k} different Boolean functions in $|\mathcal{G}|$ and, normally, there are $2^k + 1$ different possible fitness values, but $2^{2^k} < (2^k + 1)!$ for $k > 1$.

the search for superior search algorithms is meaningful, for at least some performance measures. We know nothing about which and how many algorithms exist that perform better/worse than average. We know nothing about which and how many performance measures exist for which this is possible. All we know is that answers to these questions exist. Finding them may be a totally different matter.

While this may sound discouraging, in fact testing whether any particular performance measure is suitable is not difficult. All we need to do is find two points in the search space for which the performance measure gives different outputs. We suspect in many practical situations this may just require testing a handful of points.

6   Implications for Computer Scientists and Other Searchers

Until now we have focused on the situation where we are evaluating one particular hyper-heuristic. As indicated in Figures 4–6, however, such a hyper-heuristic is just one element in the space of hyper-heuristics. In this section we want to move one level up in the "search hierarchy" to explore important questions such as: Are there better (non-resampling) hyper-heuristics than others? Are computer scientists designing hyper-heuristics wasting their time?

Let us start by considering a special (but typical) case. Let us imagine that we have fixed the set of problems $\mathcal{P}$ we want to solve and that we are interested in finding solvers for these problems that maximise $E[P(a)]$ for a given performance measure P. That is, we are not looking at all possible such measures, but just one particular P that we think works well for our objectives. For example, P might just be the success rate for an algorithm or the best fitness found in n fitness evaluations. As we noticed above, $E[P(a)]$ can be seen as a fitness function over the elements of Λ. We also noticed that if $\mathcal{G}$ is closed under permutation, such a fitness function is flat. Note that even if the set is not closed under permutation, the fitness function on Λ may look flat. It will not be flat only if, additionally, P is a suitable performance measure (i.e., one will show differences in performance).

If we enumerate Λ we can store the values of $E[P]$ associated to each heuristic in Λ in a vector h. If we fix every detail of our search (performance measure, set of problems, fitness measure over the solution space Ω, etc.), our hyper-heuristic is simply faced with one problem which is to find the maximum value in h. This is a very common situation in hyper-heuristics. The question is: which hyper-heuristic should one use for the job?

If $\mathcal{P}$ is not closed under permutation, and we have chosen a performance measure that will show differences in performance, then h is non-flat, and so there are better elements of Λ than others. Because of this, the singleton set $\{h\}$ is not closed under permutation (see argument in Section 4). As a result, there exists at least one hyper-heuristic that is better than average for at least one performance measure. In other words, if we consider problem sets that are not closed under permutation, looking for good hyper-heuristics is not a waste of time. Connecting this observation with the fact (highlighted in Section 4) that sets of problems which are closed under permutation may in practice be a rarity, we can conclude that *there may be a free lunch for computer scientist developing hyper-heuristics.*

While, clearly, not every user of hyper-heuristics is directly involved in developing new ones, the modification of hyper-heuristics and the tuning of their parameters is very common. For example, users are likely to try to optimise the rates of application of the hyper-heuristic's search operators (e.g., with some sort of trial and error approach) in an attempt to obtain a hyper-heuristic that is able to locate elements of Λ associated with good h performance values using the smallest number of samples from Λ. This *tuning and tweaking is a form of search, where, we, the humans, are the searchers.* This process is guaranteed to be a waste of time if h is flat. However, as we noted above, this situation is unlikely and, so, *there may be a free lunch for users tuning hyper-heuristics.*

7   Higher-Order Hyper-Heuristics

Looking into the future a little, we may imagine that at a not-too-distant point, we will have enough computer power to automate the search for hyper-heuristics via hyper-hyper-heuristics. Is there hope for (future) hyper-hyper-heuristics? Does NFL hold for hyper-hyper-heuristics?

Clearly hyper-hyper-heuristics will search the same space currently searched by computer scientists who try to design good hyper-heuristics and users who tune their hyper-heuristics before using them. So, the search will be hopeless whenever human search is hopeless and *vice versa*. In other words, *in the right conditions[3], there is a free lunch for hyper-hyper-heuristics.*

Looking further into the future, one may wonder what would happen if we considered hyper-hyper-····-heuristics. The question is not purely theoretical, since hyper-hyper-heuristics are just round the corner, and researchers interested in designing them will in fact find themselves exploring the space of hyper-hyper-hyper-heuristics. Does it make sense to explore that space or does NFL hold there? As we have seen above, if the set of problems $\mathcal{P}$ is closed under permutation, then all heuristics are equally good, and, so, there is no point in using hyper-heuristics. The reason is that every time a hyper-heuristic samples the space of heuristics, it gets the same performance value back. Therefore, irrespective of what performance measure we may have in mind for hyper-heuristics, they will all appear to be equally good. That is the hyper-heuristics fitness landscape is flat too. It follows from this, that the same is true for higher order hyper-heuristics too. To sum up, *if we are working with a set of problems that is closed under permutation, there is no point in using hyper-hyper-····-heuristics.*

The converse is not necessarily true though. That is, even if the set $\mathcal{P}$ is not closed under permutation, in order for hyper-hyper-····-heuristics to make sense, we need to make sure at every level in the chain an objective measure is used that reveals performance differences at the level immediately below. While we suspect that in practice this may not be difficult to achieve (see discussion at the end of Section 5), there is no guarantee that such performance measures could be found without incurring in substantial overheads.

Can we imagine situations more general than this? Actually, yes, and in fact these may be quite important for the future of hyper-heuristics. Consider the case depicted in Figure 7 where a hyper-heuristic is asked to explore not one space of solvers with their associated performance values, but many such spaces (those in the rectangular region in the middle of the figure). For simplicity, let us imagine that these spaces share the

---
[3] See Section 4.

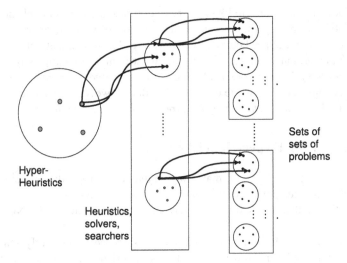

Fig. 7. A hyper-heuristic required to explore multiple spaces of search algorithms where each space has potentially a different set of problems associated to it

same structure and include the same algorithms/solvers but that they are characterised by different assignments of fitness (i.e., performance) to their points. In essence, from the point of view of the hyper-heuristic, these are different problems. Note that each heuristic space has potentially a different set of problems associated to it (one of the rectangles on the right of Figure 7).

Why would one want to consider this scenario? For example, we might want to evaluate how good our hyper-heuristic is at finding solvers for different classes of problems. In this case, each problem class induces a different h function over the space of heuristics, Λ. Another important case in where one is given only one set of problems (in this case all the rectangles on the right of Figure 7 would be identical), but we want to test whether our hyper-heuristic can work well with different performance measures P (e.g., best end of runs fitness, success rate, etc.). Then each performance measure would produce a different h over Λ. Alternatively, one might want the hyper-heuristic to work well both across problem classes and performance measures. Finally, one might even want to test a hyper-heuristic when searching different classes of heuristics (e.g., genetic algorithms, particle swarm optimisers and tabu search).

In all these cases, instead of having to deal with one fitness function over Λ, h, we will have a set of such functions, $\mathcal{H}$. In principle such a set could be closed under permutation. However, following arguments similar to the ones we have put forward for heuristics, it is easy to convince oneself that in many practical situations there will be a free lunch for this type of search and for its higher-order variants.

8 Conclusions

Hyper-heuristics are an important class of problem-solving techniques that find solvers for problems or classes of problems rather than attempting to solve the problems themselves. The last few years have seen an increased effort and many successes in applying GP as a hyper-heuristic.

In this paper we have shown that NFL does not automatically apply to hyper-heuristic search. NFL is guaranteed to apply to hyper-heuristics and higher-order hyper-heuristics only if it applies to the lowest levels in the search hierarchy, i.e., if the set of problems of interest is closed under permutation. When this is not the case, and, as we have argued, this happens in many realistic situations, there is a free lunch for hyper-heuristic techniques provided that, at each level in the search hierarchy, heuristics are evaluated using performance measures that reveal the differences present at the level immediately below.

Naturally, the fact that NFL results may not hold for hyper-heuristic search does not imply that the existing hyper-heuristic methods are good. This may need to be proved by other means. The non-applicability of NFL, however, means that it is worthwhile to try to come up with new and more powerful hyper-heuristic algorithms, including those based on GP.

Having reviewed what this paper has achieved, let us also briefly reconsider what we have left out. Throughout the paper we have made three important assumptions (deriving directly from the original NFL theory). We discuss them below.

Firstly, we used expectations of performance measures $E[P(a)]$ to assess the performance of an algorithm across problems, of a hyper-heuristic across algorithms, of a hyper-hyper-heuristic across hyper-heuristics and so on. While this makes sense in many cases, in principle one could come up with different ways of judging the worth of a searcher over a set of searches. For example, one might be more interested in higher order statistics, tails of distributions or extreme values of performance. In this case, NFL cannot tell much about what happens in search. So, we don't have this nice tool to explore under what circumstances there may be a free lunch for hyper-heuristics. What remains true is that there still is no point in using hyper-heuristics if the aggregate performance measure (whatever it is) computed over a set of problems produces identical results for all search algorithms. This is because, in this case, NFL would apply to the higher levels of the search hierarchy, rendering the search for better hyper-heuristics and higher-order hyper-heuristics futile.

Secondly, we implicitly assumed that problems are drawn from problem sets either by deterministic enumeration or probabilistically by sampling $\mathcal{P}$ with uniform probability. However, there are cases where not all the problems in $\mathcal{P}$ are equally likely (or important). In these cases the expected performance of an algorithm over the problem set $\mathcal{P}$ is given by

$$E[P(a)] = \sum_{f \in \mathcal{G}} p(f)P(f,a) \tag{2}$$

where $p(f)$ is a function that indicates with which probability each problem is presented to a searcher. If $p(f)$ is uniform then, as we have seen in Section 2, NFL holds if and only if $\mathcal{G}$ is closed under permutation. The question is: what happens if $p(f)$ is non-uniform? As proved in [10], NFL may still hold, but rather stringent conditions must be met on the form of $p(f)$. A function and its permutations are characterised by a unique histogram of fitness values. For NFL to hold, $\mathcal{G}$ must be closed under permutation *and* all the functions in $\mathcal{G}$ with a particular fitness histogram must have identical probability of being drawn from $\mathcal{G}$. While it is clearly trivial to extend the theory presented in this paper to cover the non-uniform case, these additional constraints make it even less likely that NFL will apply to any set of problems presented to a hyper-heuristic in a practical situation. So, hyper-heuristics have an even broader applicability on non-uniform problem domains.

Thirdly, we did not consider re-sampling (some algorithms are worse than others even in the presence of sets of problems that are closed under permutation simply because they waste more time revisiting points). In future research we will explore what implications re-sampling has for hyper-heuristic search.

References

1. Bader-El-Den, M., Poli, R.: Generating SAT local-search heuristics using a GP hyper-heuristic framework. In: Proceedings of Evolution Artificielle (October 2007)
2. Bader-El-Den, M.B., Poli, R.: A GP-based hyper-heuristic framework for evolving 3-SAT heuristics. In: Thierens, D., Beyer, H.-G., Bongard, J., Branke, J., Clark, J.A., Cliff, D., Congdon, C.B., Deb, K., Doerr, B., Kovacs, T., Kumar, S., Miller, J.F., Moore, J., Neumann, F., Pelikan, M., Poli, R., Sastry, K., Stanley, K.O., Stutzle, T., Watson, R.A., Wegener, I. (eds.) GECCO 2007: Proceedings of the 9th annual conference on Genetic and evolutionary computation, vol. 2, p. 1749. ACM Press, New York (2007)
3. Banzhaf, W., Nordin, P., Keller, R.E., Francone, F.D.: Genetic Programming – An Introduction; On the Automatic Evolution of Computer Programs and its Applications. Morgan Kaufmann, San Francisco (1998)
4. Burke, E., Kendall, G., Newall, J., Hart, E., Ross, P., Schulenburg, S.: Hyper-heuristics: an emerging direction in modern search technology. In: Glover, F.W., Kochenberger, G.A. (eds.) Handbook of metaheuristics, ch. 16, pp. 457–474. Kluwer Academic Publishers, Boston (2003)
5. Burke, E.K., Hyde, M.R., Kendall, G.: Evolving bin packing heuristics with genetic programming. In: Runarsson, T.P., Beyer, H.-G., Burke, E.K., Merelo-Guervós, J.J., Whitley, L.D., Yao, X. (eds.) PPSN 2006. LNCS, vol. 4193, pp. 860–869. Springer, Heidelberg (2006)
6. Burke, E.K., Hyde, M.R., Kendall, G., Woodward, J.: Automatic heuristic generation with genetic programming: evolving a jack-of-all-trades or a master of one. In: Thierens, D., Beyer, H.-G., Bongard, J., Branke, J., Clark, J.A., Cliff, D., Congdon, C.B., Deb, K., Doerr, B., Kovacs, T., Kumar, S., Miller, J.F., Moore, J., Neumann, F., Pelikan, M., Poli, R., Sastry, K., Stanley, K.O., Stutzle, T., Watson, R.A., Wegener, I. (eds.) GECCO 2007: Proceedings of the 9th annual conference on Genetic and evolutionary computation, vol. 2, pp. 1559–1565. ACM Press, New York (2007)
7. Di Chio, C., Poli, R., Langdon, W.B.: Evolution of force-generating equations for PSO using GP. In: Manzoni, S., Palmonari, M., Sartori, F. (eds.) AI*IA Workshop on Evolutionary Computation, Evoluzionistico GSICE 2005, University of Milan Bicocca, Italy, September 20 (2005)
8. Fukunaga, A.: Automated discovery of composite SAT variable selection heuristics. In: Proceedings of the National Conference on Artificial Intelligence (AAAI), pp. 641–648 (2002)
9. Igel, C., Toussaint, M.: On classes of functions for which no free lunch results hold. Information Processing Letters 86(6), 317–321 (2003)
10. Igel, C., Toussaint, M.: A no-free-lunch theorem for non-uniform distributions of target functions. Journal of Mathematical Modelling and Algorithms 3, 313–322 (2004)
11. Keller, R.E., Poli, R.: Cost-benefit investigation of a genetic-programming hyperheuristic. In: Monmarché, N., Talbi, E.-G., Collet, P., Schoenauer, M., Lutton, E. (eds.) EA 2007. LNCS, vol. 4926, pp. 13–24. Springer, Heidelberg (2008)

12. Keller, R.E., Poli, R.: Linear genetic programming of metaheuristics. In: Thierens, D., Beyer, H.-G., Bongard, J., Branke, J., Clark, J.A., Cliff, D., Congdon, C.B., Deb, K., Doerr, B., Kovacs, T., Kumar, S., Miller, J.F., Moore, J., Neumann, F., Pelikan, M., Poli, R., Sastry, K., Stanley, K.O., Stutzle, T., Watson, R.A., Wegener, I. (eds.) GECCO 2007: Proceedings of the 9th annual conference on Genetic and evolutionary computation, p. 1753. ACM Press, New York (2007)
13. Keller, R.E., Poli, R.: Linear genetic programming of parsimonious metaheuristics. In: Proceedings of IEEE Congress on Evolutionary Computation (CEC) (September 2007)
14. Kibria, R.H., Li, Y.: Optimizing the initialization of dynamic decision heuristics in DPLL SAT solvers using genetic programming. In: Collet, P., Tomassini, M., Ebner, M., Gustafson, S., Ekárt, A. (eds.) EuroGP 2006. LNCS, vol. 3905, pp. 331–340. Springer, Heidelberg (2006)
15. Koza, J.R.: Genetic Programming: On the Programming of Computers by Natural Selection. MIT Press, Cambridge (1992)
16. Oltean, M.: Evolving evolutionary algorithms for function optimization. In: Chen, K. (ed.) The 7th Joint Conference on Information Sciences, September 2003, vol. 1, pp. 295–298. Association for Intelligent Machinery (2003)
17. Oltean, M.: Evolving evolutionary algorithms using linear genetic programming. Evolutionary Computation 13(3), 387–410 (Fall 2005)
18. Oltean, M., Dumitrescu, D.: Evolving TSP heuristics using multi expression programming. In: Bubak, M., van Albada, G.D., Sloot, P.M.A., Dongarra, J. (eds.) ICCS 2004. LNCS, vol. 3037, pp. 670–673. Springer, Heidelberg (2004)
19. Poli, R., Di Chio, C., Langdon, W.B.: Exploring extended particle swarms: a genetic programming approach. In: Beyer, H.-G., O'Reilly, U.-M., Arnold, D.V., Banzhaf, W., Blum, C., Bonabeau, E.W., Cantu-Paz, E., Dasgupta, D., Deb, K., Foster, J.A., de Jong, E.D., Lipson, H., Llora, X., Mancoridis, S., Pelikan, M., Raidl, G.R., Soule, T., Tyrrell, A.M., Watson, J.-P., Zitzler, E. (eds.) GECCO 2005: Proceedings of the 2005 conference on Genetic and evolutionary computation, vol. 1, pp. 169–176. ACM Press, New York (2005)
20. Poli, R., Langdon, W.B., Holland, O.: Extending particle swarm optimisation via genetic programming. In: Keijzer, M., Tettamanzi, A.G.B., Collet, P., van Hemert, J., Tomassini, M. (eds.) EuroGP 2005. LNCS, vol. 3447, pp. 291–300. Springer, Heidelberg (2005)
21. Poli, R., Langdon, W.B., McPhee, N.F.: A field guide to genetic programming (2008), http://lulu.com http://www.gp-field-guide.org.uk (With contributions by Koza, J.R.)
22. Poli, R., Woodward, J., Burke, E.K.: A histogram-matching approach to the evolution of bin-packing strategies. In: Proceedings of the IEEE Congress on Evolutionary Computation, Singapore (2007) (accepted)
23. Schumacher, C., Vose, M.D., Whitley, L.D.: The no free lunch and problem description length. In: Proceedings of the Genetic and Evolutionary Computation Conference (GECCO), pp. 565–570. Morgan Kaufmann, San Francisco (2001)
24. Valsecchi, A., Vanneschi, L.: A study of some implications of the no free lunch theorem. In: Giacobini, M., Brabazon, A., Cagnoni, S., Di Caro, G.A., Drechsler, R., Ekárt, A., Esparcia-Alcázar, A.I., Farooq, M., Fink, A., McCormack, J., O'Neill, M., Romero, J., Rothlauf, F., Squillero, G., Uyar, A.Ş., Yang, S. (eds.) EvoWorkshops 2008. LNCS, vol. 4974, pp. 633–642. Springer, Heidelberg (2008)
25. Whitley, D., Watson, J.P.: Complexity theory and the no free lunch theorem. In: Burke, E.K., Kendall, G. (eds.) Search Methodologies: Introductory Tutorials in Optimization and Decision Support Techniques, ch. 11, pp. 317–339. Springer, US (2005)
26. Wolpert, D., Macready, W.: No free lunch theorems for optimization. IEEE Transactions on Evolutionary Computation 1(1), 67–82 (1997)

Tree Based Differential Evolution

Christian B. Veenhuis

Berlin University of Technology, Germany
veenhuis@googlemail.com

Abstract. In recent years a new evolutionary algorithm for optimization in continuos spaces called Differential Evolution (DE) has developed. DE turns out to
need only few evaluation steps to minimize a function. This makes it an interesting candidate for problem domains with high computational costs as for instance
in the automatic generation of programs. In this paper a DE-based tree discovering algorithm called Tree based Differential Evolution (TreeDE) is presented.
TreeDE maps full trees to vectors and represents discrete symbols by points in a
real-valued vector space providing this way all arithmetical operations needed for
the different DE schemes. Because TreeDE inherits the 'speed property' of DE, it
needs only few evaluations to find suitable trees which produce comparable and
better results as other methods.

1 Introduction

One aspect of Computational Intelligence is to learn or discover trees for a wide field
of applications. The most famous algorithm for tree discovery is (tree-based) Genetic
Programming [2], where trees representing programs, mathematical expressions and so
on are discovered.

In 1995 a new evolutionary algorithm for optimization called Differential Evolution
(DE) was introduced by Storn and Price in [7]. The main difference to other evolutionary algorithms is that the mutation process is guided by other population members and
not purely random. As stated in [9] DE is a fast optimizer in terms of 'needed number
of evaluation steps' to minimize a function.

Up to now, only few tree discovering algorithms are taking advantage of Differential
Evolution. In [3] the concept of Grammatical Evolution is combined with DE leading to
Grammatical Differential Evolution (GDE). The vectors of GDE are fixed length lists
of numbers of production rules from a Backus-Naur Form grammar. These production
rules are sequentially executed creating this way a tree.

A tree discovering algorithm using the Swarm Intelligence paradigm was introduced
in [1]. There, Abbass et al. proposed AntTAG, an Ant Colony Optimization method
using Tree Adjunct Grammars. For this, they used a pre-defined set of elementary trees
which are assembled by the ants to produce better trees. They applied AntTAG successfully to a typical symbolic regression problem. According to [6] a difficulty of this
approach is that the set of elementary trees needs to be adapted even for other symbolic
regression problems.

In this paper a DE-based tree discovering algorithm called Tree based Differential
Evolution (TreeDE) is presented. In TreeDE all vectors represent full trees as implicit

L. Vanneschi et al. (Eds.): EuroGP 2009, LNCS 5481, pp. 208–219, 2009.
© Springer-Verlag Berlin Heidelberg 2009

data structure. Furthermore, discrete symbols are represented by points in a real-valued vector space. Thus, it becomes possible to provide the arithmetical operations needed for the schemes.

This paper is organized as follows. Section 2 introduces the Differential Evolution algorithm. The extension of DE to become the Tree based Differential Evolution algorithm is explained in section 3. Some experimental results are given in section 4. Finally, in section 5 some conclusions are drawn.

2   Differential Evolution

Storn and Price introduced in [7,8,9] a real-valued vector based evolutionary algorithm for optimization in continuos spaces they called Differential Evolution (DE). The main idea employed is not to mutate vector components by simply replacing their values by random values. Instead, two population mates are randomly selected whose weighted difference is added to a third randomly selected population member creating this way a mutant vector. Then, either a two-point crossover ([7]) or a multi-point crossover ([9]) is performed between the mutant vector and the current population member being considered. At the end, the new created offspring vector (trial vector) replaces the current considered vector in the next generation, iff its fitness is better. Otherwise, the trial vector is discarded.

Notation. Let D be the dimension of the problem (i.e., the dimension of the search space $\mathbb{R}^D$), NP the number of vectors and X_G the set of vectors $X_G = \{ x_{1,G}, \cdots, x_{NP,G}\}$ of generation G. Each vector $x_{i,G} \in X_G$ is an element of the search space ($x_{i,G} \in \mathbb{R}^D$).

Typically, an upper and lower bound is defined for each component j: $(x)_j^{\mathbf{L}}$ (lower bound) and $(x)_j^{\mathbf{U}}$ (upper bound). Thus, for all indices $j \in \{1, \cdots, D\}$ we have that $(x_{i,G})_j \in [(x)_j^{\mathbf{L}}, (x)_j^{\mathbf{U}}]$.

The best vector of generation G (global best) is denoted by $x_{best,G}$. To compute the objective value (or fitness) of a vector, a fitness function $F : \mathbb{R}^D \to \mathbb{R}$ is used.

Initialization. Firstly, the generation counter G is set to 0. Then, the initial population X_0 is created by randomly creating NP vectors $x_{i,0}$. For this, each component $(x_{i,0})_j$ of each vector is taken from $U((x)_j^{\mathbf{L}}, (x)_j^{\mathbf{U}})$.

Iteration. The iteration works as follows. Firstly, for each vector of the population a certain number of population mates are selected randomly (typically 3 or 5). The number of these selected mates depends on the scheme used to compute a mutant vector v_i. A well-known standard scheme using 3 population mates is shown in Eq. (1).

$$v_i = x_{r_1} + F(x_{r_2} - x_{r_3}) \tag{1}$$

All selected mates need to be mutual different and also to be different from the current vector x_i. Secondly, the mutant vector v_i is computed using a scheme. Thirdly, a crossover is performed between the mutant and the current vector leading to a trial vector u_i. Finally, this trial vector replaces the current vector, iff its fitness is better. In pseudo-code this looks as follows:

1. **for all** vectors $x_{i,G} \in X_G$ **do**:
   1.1. Determine indices $r_1, r_2, r_3, \cdots \sim U(1, NP)$ such that $i \neq r_1 \neq r_2 \neq r_3 \neq \cdots$
   1.2. Compute mutant vector v_i with a scheme, e.g., Eq. (1)
   1.3. Create trial vector u_i via crossover between current and mutant vector (with $r \sim$ $U(0,1)^D$ being a randomized vector determined for each u_i and $d \sim U(1,D)^{NP} \subset$ $\mathbb{N}^{NP}$ a vector containing a randomly selected index for each population member):

$$u_i = (\cdots, (u_i)_j = \begin{cases} (v_i)_j \, , \, r_j \leq CR \vee j = d_i \\ (x_i)_j \, , \, r_j > CR \wedge j \neq d_i \end{cases}, \cdots)^T$$

   The randomly selected index d_i ensures that at least one vector component will be crossovered.
   1.4. Decide which one goes to next generation:

$$x_{i,G+1} = \begin{cases} u_i \quad , \, F(u_i) < F(x_{i,G}) \\ x_{i,G} \, , \, \text{otherwise} \end{cases}$$

2. $G \leftarrow G + 1$

There are three user-defined parameters which are application-dependent: (a) the mutation factor F as a real-valued constant from $[0,2]$ (but Storn also suggests in [9] that $F > 0$), (b) the crossover probability $CR \in [0,1]$ which is applied to each vector component separately and (c) the population size NP. In [9] the authors suggest to use the following parameters as a first try: $F = 0.5$, $CR = 0.1$ and $5 \cdot D \leq NP \leq 10 \cdot D$.

Variants. There are several well-known schemes for computing the mutant vectors. In this work the following schemes were used as introduced in [8] (names are overtaken):

DE/rand/1   : $v_i = x_{r_1} + F(x_{r_2} - x_{r_3})$
DE/rand/2   : $v_i = x_{r_1} + F(x_{r_2} + x_{r_3} - x_{r_4} - x_{r_5})$
DE/best/1   : $v_i = x_{best} + F(x_{r_2} - x_{r_3})$
DE/best/2   : $v_i = x_{best} + F(x_{r_2} + x_{r_3} - x_{r_4} - x_{r_5})$
DE/rand-to-best   : $v_i = x_{r_1} + \lambda(x_{best} - x_{r_1}) + F(x_{r_2} - x_{r_3})$
DE/current-to-rand : $v_i = x_i + \lambda(x_{r_1} - x_i) + F(x_{r_2} - x_{r_3})$
DE/current-to-best : $v_i = x_i + \lambda(x_{best} - x_i) + F(x_{r_1} - x_{r_2})$

Parameter F has the same meaning as introduced in the iteration part. The λ parameter controls the greediness of the appropriate scheme. Storn recommends in [8] to set $\lambda = F$.

3   Tree Based Differential Evolution

The scope of Tree based Differential Evolution (TreeDE) is to discover trees in tree spaces of, e.g., program-trees which can be interpreted to control robots or other entities, decision-trees for classification purposes, parse-trees representing mathematical expressions for symbolic regression tasks, operator-trees representing image processing filters and so on. For simplification all trees, whose nodes represent some kind of symbols are called symbol-tree hereafter.

Symbol-trees are built of non-terminal symbols (e.g., SIN, ADD, IF, $\geq$) and terminal symbols (e.g., X,MOVE,1,CLASS1). Non-terminal symbols have subordinated children which are used, e.g., as operands or actions. Terminal symbols have no children and are used as values, classes, actions and so on.

The algorithmic core of TreeDE is the Differential Evolution algorithm as described in section 2. In the following sections it is shown how the DE algorithm is extended to become the TreeDE method.

3.1 Tree Representation

In DE the representation of solutions is a D-dimensional real-valued vector. TreeDE replaces this vector representation by a static tree.

Let $T_{L,A}$ be an ordered rooted full A-ary tree with directed edges and a fixed number of L levels, i.e., a tree with exactly one root node and every internal node has exactly A children, whereby these children are ordered from left to right. Because it's a full tree, all nodes of the last level are in existence and no internal node is missing (often this type of tree is also called complete tree, perfect tree or perfect A-ary tree).

In Figure 1 the left part depicts a symbol-tree representing the mathematical expression $x^2 + 1$. This is a full tree $T_{3,2}$ with 3 levels and an arity of 2.

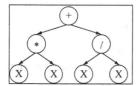 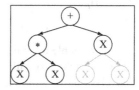

Fig. 1. Symbol-trees for $x^2 + 1$ (left) and $x^2 + x$ (right)

Definition 1 (Full Tree Space). *The set of all full trees $T_{L,A}$ with L levels and an arity of A is denoted with $\mathbb{T}_{L,A}$, i.e., $T_{L,A} \in \mathbb{T}_{L,A}$. If $[t_i]_{L,A} := T_{L,A}$ is an alternative notation of a full tree, then t_i represents the ith node of that tree.*

The total number of nodes of a full tree $T_{L,A} \in \mathbb{T}_{L,A}$ can be computed by Eq. (2).

$$N_{nodes}(L,A) = \sum_{i=0}^{L-1} A^i = A^L - 1 \ , \ (L \geq 0, A \geq 1) \tag{2}$$

Although structurally full trees are used, it is possible to represent trees of different sizes semantically. On the right part of Figure 1 a symbol-tree representing the mathematical expression $x^2 + x$ is shown. Here, one of the internal nodes represents a terminal symbol instead of a non-terminal one. An interpreter of this symbol-tree will ignore the following nodes (gray-colored). This way, the number of levels L constrains only the maximum depth of a tree.

3.2 Mapping Trees to Vectors

TreeDE uses full trees for the DE individuals $x_{i,G}$. Because all trees of the population have the same number of levels and the same arity, they also have the same structure

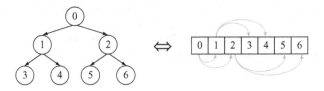

Fig. 2. Mapping a full tree to an equivalent array

and the same number of nodes. It is well-known that trees with a static and defined structure can be represented equivalently as array or vector. The size of the vector can be computed as in Eq. (2). In Figure 2 a tree $T_{3,2} \in \mathbb{T}_{3,2}$ is shown. The numbers within the nodes are the indices. The right part in Figure 2 shows the indexing scheme, i.e., how the nodes are arranged as an array or vector: One counts from the first to the last level starting at 0 and on each level from left to right.

The index-numbers of the children of a node within a vector can be computed as in Eq. (3). Here, p means the index-number of the parent-node within the vector, for which the child-node is searched. The index-number of the searched child-node is given with c (counted from left to right).

$$I_{child}(p,c) = A \cdot p + c \qquad (3)$$

With $I_{child}(p,c)$ it is possible to 'traverse' the vector as if it would be a tree. This is shown in Figure 2 on the right part with gray-colored arrows. To recognize a leaf (i.e., a node of the last level) of a tree within a vector one can use Eq. (4).

$$LEAF(p) = \begin{cases} 1, & I_{child}(p,1) > N_{nodes}(L,A) - 1 \\ 0, & otherwise \end{cases} \qquad (4)$$

There are also other possibilities to represent full trees. In [5] Poli and McPhee introduced a 2-dimensional indexing scheme for representing full trees. They index a node of a tree by a tupel composed of the level and the position within this level. The aim of the concept proposed in this paper is to allow any existing DE implementation to be used for the TreeDE approach. Thus, the simple 1-dimensional indexing scheme as presented above is used, because it reflects the vector representation of DE in a direct way. Additional to the structural representation of the tree, the representation of a symbol as content of a node needs to be realized as described in the next section.

3.3   Symbol Representation

DE individuals are moving in a real vector space ($x_{i,G} \in \mathbb{R}^D$). Thus, the vector addition and the scalar multiplication as needed for the different DE schemes are well-defined.

In TreeDE trees are used whose nodes have to represent symbols. But to use the schemes, the operations $+, -$ and multiplication with a real number have to be realized on the level of trees as well as on the level of single nodes.

Tree level. Let $[t_i]_{L,A} = T_{L,A}$ be an alternative notation of a tree as defined in Def. (1). Then, the operations $+$ and $-$ are realized simply by adding / subtracting all nodes at the same position of two trees $T_{L,A}, U_{L,A} \in \mathbb{T}_{L,A}$ as shown in Eq. (5).

$$T_{L,A} \pm U_{L,A} = [t_i]_{L,A} \pm [u_i]_{L,A} := [t_i \pm u_i]_{L,A} \tag{5}$$

Multiplication with a real number $k \in \mathbb{R}$ is realized by multiplying all nodes of a tree $T_{L,A} \in \mathbb{T}_{L,A}$ with k as shown in Eq. (6).

$$k \cdot T_{L,A} = k \cdot [t_i]_{L,A} := [k \cdot t_i]_{L,A} \tag{6}$$

Node level. Of what type or structure is a single node t_i of a tree $T_{L,A}$? If nodes would directly represent discrete symbols, how to define the operations $+,-$ and multiplication with a real number? Usually, expressions as, e.g., $SIN + ADD$ or $1.2 \cdot IF$ don't make sense. Therefore, in TreeDE, nodes represent discrete symbols by using points of a symbol vector space.

A single symbol is defined as a tuple $SYM(i) = (ID(i), ARITY(i))$ with its ID (e.g., SIN) and its ARITY (the number of expected children-nodes or operands). The ordered list of used symbols for a given application is $S := (SYM(1), \ldots, SYM(N_{sym}))$, with N_{sym} being the number of symbols. All symbols representing terminals are placed to the end of S. The index number of the first symbol representing a terminal is denoted with I_{term} $(1 \le I_{term} \le N_{sym})$. For an application with an ADD operator and the terminals X and Y we would get $S = ((ADD,2),(X,0),(Y,0))$ with $N_{sym} = 3$ and $I_{term} = 2$.

Definition 2 (Symbol Vector Space). *The symbol vector space $\mathbb{S} = \mathbb{R}^{N_{sym}}$ contains for each symbol $SYM(i)$ in S a standard basis vector, i.e., the dimension of the symbol vector space is the number of used symbols N_{sym}. The first symbol $SYM(1)$ gets the first dimension, the second symbol $SYM(2)$ gets the second dimension and so on. A vector of $\mathbb{S}$ is called symbol vector $\sigma = (\sigma_1, \ldots, \sigma_{N_{sym}})^T \in \mathbb{S}$.*

In Figure 3 three symbol vector spaces are shown. Symbol vector spaces with that low dimensions have a low usefulness, but they show how they are constructed.

To get the symbol ID of a node, the symbol vector has to be transformed as shown in Eq. (7). For the internal nodes of the tree, $\mathcal{T}_{internal}$ is used to determine a symbol. For internal nodes non-terminals as well as terminals are allowed. The symbols for leaves are determined with $\mathcal{T}_{leaf}$, because at this place only terminals make sense. To support the vector traversal process (see section 3.2), the number of expected operands of a given symbol can be determined by $\mathcal{T}_{arity}$.

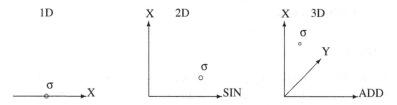

Fig. 3. Exemplary symbol vector spaces of 1, 2 and 3 dimensions. The labels denote the symbols assigned to the axes. The symbol vectors represent: 1D) $\mathcal{T}_{internal}(\sigma) = X$, $\mathcal{T}_{leaf}(\sigma) = X$,   2D) $\mathcal{T}_{internal}(\sigma) = SIN$, $\mathcal{T}_{leaf}(\sigma) = X$ and   3D) $\mathcal{T}_{internal}(\sigma) = X$, $\mathcal{T}_{leaf}(\sigma) = X$.

$$\mathcal{T}_{internal}(\sigma) = ID(\ \arg\max_{c \in \{1,...,N_{sym}\}}\{\sigma_c\}\ )$$
$$\mathcal{T}_{leaf}(\sigma) = ID(\ \arg\max_{c \in \{I_{term},...,N_{sym}\}}\{\sigma_c\}\ )$$
$$\mathcal{T}_{arity}(\sigma) = ARITY(\ \arg\max_{c \in \{1,...,N_{sym}\}}\{\sigma_c\}\ ) \tag{7}$$

A symbol vector represents that symbol which is assigned to the symbol vector component with the highest value. For instance, in the above ADD, X, Y example the symbol vector $(1.3, 0.3, 0.8)^T$ represents the ADD symbol, because the first component is the maximum. Symbol vector $(0.3, 0.1, 0.8)^T$ would represent the Y symbol. In case there are more symbol vector components with the maximum, the first symbol is chosen (this fact is not reflected by Eq. (7)). For example, the symbol vector $(0.3, 0.8, 0.8)^T$ represents the X symbol.

A node t_i of a tree $T_{L,A}$ is a symbol vector of dimension N_{sym}: $t_i \in \mathbb{S}$. Because the symbol vector space is a real vector space, the operations $+, -$ and multiplication with a real number are now defined on the level of nodes, too.

The concept of representing a symbol by a point in a higher dimensional space was also used by Page, Poli and Langdon in [4]. They used the outcome of a truth table as bitstring to represent a symbol (e.g., 1000 for AND, 1110 for OR, etc.). For the DE approach this can not be used in a comfortable way, if you want to maintain the arithmetical properties. Although you can imagine a $+$ and $-$ operation for bitstrings, the meaning of a scalar multiplication is not that obvious. Another question is how to define these operations while fulfilling the vector space axioms.

3.4   The Vector Space $\mathbb{T}_{L,A}(\mathbb{S})$

Now that the structure of the used trees as well as the structure of the nodes are defined, the full symbol tree space can be defined. Trees out of the full symbol tree space are capable of representing symbol-trees in a non-discrete way.

Definition 3 (Full Symbol Tree Space). *A full symbol tree space is a full tree space* $\mathbb{T}_{L,A}$, *where each node* t_i *of each full tree* $[t_i]_{L,A} \in \mathbb{T}_{L,A}$ *is a symbol vector, i.e.,* $t_i \in \mathbb{S}$. *The full symbol tree space is denoted as* $\mathbb{T}_{L,A}(\mathbb{S})$.

$\mathbb{T}_{L,A}(\mathbb{S})$ together with the operations in equations (5) and (6) is a vector space over $\mathbb{R}$. Thus, all presented DE schemes in section 2 are still valid, if the vectors $x_{i,G}$ are replaced by full symbol trees.

3.5   Algorithm

TreeDE consists of two phases: (1) initialization and (2) iteration. With N_{iter} denoting the number of iterations, the main algorithm can be described as follows:

1. Initialize population X_0 by randomly creating NP full symbol trees $x_{i,0}$.
2. Iterate population N_{iter} times

In the following sections both phases are described in more detail.

Initialization. In DE a lower $((x)_j^L)$ and upper $((x)_j^U)$ bound is used for the vector components j of the individuals. TreeDE uses for all components the same bounds which are denoted as x_L (lower bound) and x_U (upper bound).

The arity A for the used trees $T_{L,A}$ is computed as shown in Eq. (8). The highest arity of all used symbols is A.

$$A = \max\{ARITY(i)|\forall i \in \{1,\ldots,N_{sym}\}\} \tag{8}$$

The trees of TreeDE are mapped to the vectors of DE. For this, the dimension of the vectors needs to be computed. This can be done by considering the number of nodes of the used trees and the dimension of the symbol vectors placed at the nodes of the trees as shown in Eq. (9).

$$D = N_{nodes}(L,A) \cdot N_{sym} \tag{9}$$

With above computations, TreeDE can be initialized as follows:

Compute A by Eq. (8)
Compute D by Eq. (9)
$G \leftarrow 0$
$X_0 \leftarrow \emptyset$
for $i = 1$ **to** NP
    $x_{i,0} \sim U(x_L,x_U)^D$
    $X_0 \leftarrow X_0 \cup \{x_{i,0}\}$
end for i
$x_{best,0} \leftarrow \arg\min_{x_i \in X_0} F(x_i)$

Iteration. The iteration in TreeDE works exactly as in DE as described in section 2. This fact allows for the simple extension of any existing DE implementation to become the TreeDE method. For this, the following main differences between TreeDE and DE should be considered:

- The DE vectors encode static trees as an implicit data structure.
- The lower and upper bounds are not vectors anymore. TreeDE uses the same bounds for all vector components.
- The dimension is no longer a user-defined parameter, but is computed according to the tree structure used.
- A new user-defined parameter is introduced (number of tree levels L).
- The fitness function F interprets the trees mapped in the vectors $x_{i,G}$. For this, it uses the equations of the sections 3.2 and 3.3 to traverse the trees.

4   Experiments

The TreeDE algorithm was tested on two typical problems namely Symbolic Regression and Artificial Ant to evaluate its capabilities. Basically, the standard parameters of DE are used: $x_L = 0$, $x_U = 20$, $F = 0.5$, $\lambda = 0.5$, $CR = 0.1$ and $N_{iter} = 1500$. Merely, the number of levels L is determined by a systematic parameter exploration from $L = 1$ to $L = 10$. The best fitting numbers of levels obtained and used are $L = 6$ for Symbolic

Regression and $L = 5$ for the Artificial Ant problem. Opposed to the suggestion given in [9], NP is not set according to the $5 \cdot D \leq NP \leq 10 \cdot D$ rule. The reason is that for the given benchmark problems with their levels L we have dimensions of $D = 315$ for Symbolic Regression and $D = 1452$ for Artificial Ant leading to huge population sizes. Thus, after a systematic experimentation, these parameters are set to $NP = 20$ for Symbolic Regression and $NP = 30$ for Artificial Ant.

4.1 Symbolic Regression

The task of symbolic regression is to find a mathematical expression fitting a given set of points. Four typical benchmark functions are used: $f_1 = x^3 + x^2 + x$, $f_2 = x^4 + x^3 + x^2 + x$, $f_3 = x^5 + x^4 + x^3 + x^2 + x$ and $f_4 = x^5 - 2 \cdot x^3 + x$. For all benchmark functions f_n 20 points $(x(k), y(k))$ are used with $x(k)$ randomly chosen out of $[-1, 1]$ and $y(k) = f_n(x(k))$.

The set of symbols used is $S = ($ (PLUS,2) , (MINUS,2) , (MUL,2) , (DIV,2) , (X,0) $)$ with $N_{sym} = 5$ and $I_{term} = 5$. The meaning is straight-forward: PLUS $:= a + b$, MINUS $:= a - b$, MUL $:= a \cdot b$, and DIV $:= a/b$, whereby DIV returns a 0 if $b = 0$. The terminal symbol X returns the $x(k)$ value of a point for which a given tree is interpreted.

The used fitness function F is the sum of deviations over all points, i.e., $F(x_{i,G}) = \sum_k |I(x_{i,G}, x(k)) - y(k)|$ with $x_{i,G}$ being a vector (tree) and I the interpreter of trees according to an input value x.

All results as presented in Table 1 are the average over 100 independent runs. A perfect hit means a mathematical expression fitting all points exactly with no deviation (i.e., $F(x_{i,G}) = 0$). The best results and schemes are set in boldface.

All schemes for f_1 and most schemes for f_2 solve these problems reliably indicated by the (near to) 100% perfect hits. As expectable, this rate of perfect hits decreases for the more difficult problems f_3 and f_4, whereby the majority of the schemes for f_3 still have a perfect hit rate over 90%.

An advantage of TreeDE is the few number of needed evaluations compared, e.g., to Genetic Programming or AntTAG. According to [2] for solving f_2 in GP a population size of 500 with 51 iterations, i.e., 25500 evaluations are needed. In AntTAG [1] for solving f_2 6240 evaluations are needed to produce 92% perfect hits and an average fitness of 0.35 (sd: 0.12). For Grammatical Differential Evolution (GDE) an average fitness of $\frac{1}{6} \approx 0.17$ for the best variant is reported in [3]. The best TreeDE variant needs 3000 evaluations in average for producing perfect solutions (0.0 (sd: 0.0) and 100% perfect hits). Even all other variants produce better solutions and perfect hit rates for f_2 than AntTAG and GDE.

What is noticeable is the fact that always at least one of the schemes 'DE/best/1', 'DE/best/2' and 'DE/rand-to-best' is amongst the best or the winner for each function f_n. Symbolic regression problems seem to benefit from the involvement of the global best solution x_{best}. On the other hand, the scheme 'DE/current-to-rand' is either the worst or one of the worse schemes for each function f_n. Also 'DE/current-to-best' produces similar bad performance, whereby in most cases better than 'DE/current-to-rand' what fits to the observation above. It seems that using the current individual x_i as basis for the mutant vector is not the best idea for symbolic regression. The standard scheme 'DE/rand/1' shows only an average performance.

Table 1. Results of TreeDE for the symbolic regression benchmark problems averaged over 100 independent runs. The column 'Avg. Fitness' is the best fitness value reached on average, 's.d.' the appropriate standard deviation, 'Perfect hits (%)' the percentage of perfect solutions ($F(x_{i,G}) = 0$), 'Avg. Gen.' the average number of needed generations for the best solution of a run and 'Avg. Num. Eval.' the number of needed evaluations on average. The best results as well as the winning schemes are typed in boldface.

f_n	DE Scheme	Avg. Fitness	s.d.	Perfect hits (%)	Avg. Gen.	Avg. Num. Eval.
f_1	DE/rand/1	**0.0**	0.0	**100**	52.38	1040
f_1	DE/rand/2	**0.0**	0.0	**100**	64.41	1280
f_1	DE/best/1	**0.0**	0.0	**100**	42.01	840
f_1	DE/best/2	**0.0**	0.0	**100**	41.55	820
f_1	**DE/rand-to-best**	**0.0**	0.0	**100**	**37.68**	740
f_1	DE/current-to-rand	**0.0**	0.0	**100**	67.61	1340
f_1	DE/current-to-best	**0.0**	0.0	**100**	61.32	1220
f_2	DE/rand/1	0.008732	0.086878	99	312.57	6240
f_2	**DE/rand/2**	**0.0**	0.0	**100**	**150.27**	3000
f_2	DE/best/1	**0.0**	0.0	**100**	160.03	3200
f_2	DE/best/2	**0.0**	0.0	**100**	179.23	3580
f_2	DE/rand-to-best	**0.0**	0.0	**100**	170.39	3400
f_2	DE/current-to-rand	0.006692	0.066584	99	312.40	6240
f_2	DE/current-to-best	**0.0**	0.0	**100**	208.06	4160
f_3	DE/rand/1	0.075021	0.308772	94	418.99	8360
f_3	DE/rand/2	0.084723	0.304513	92	531.52	10620
f_3	DE/best/1	0.100675	0.353763	92	401.65	8020
f_3	**DE/best/2**	**0.027460**	0.197065	**98**	422.95	8440
f_3	DE/rand-to-best	0.177271	0.492523	88	505.65	10100
f_3	DE/current-to-rand	0.390757	0.714007	75	661.21	13220
f_3	DE/current-to-best	0.471636	0.779316	70	620.88	12400
f_4	DE/rand/1	0.090039	0.178106	77	732.00	14640
f_4	DE/rand/2	0.095929	0.201207	79	666.06	13320
f_4	**DE/best/1**	**0.036440**	0.090098	84	**542.42**	10840
f_4	DE/best/2	0.165355	0.268143	68	730.52	14600
f_4	**DE/rand-to-best**	**0.038280**	0.118876	**90**	580.69	11600
f_4	DE/current-to-rand	0.143639	0.227404	66	707.55	14140
f_4	DE/current-to-best	0.072713	0.157832	79	570.42	11400

4.2 Artificial Ant

As described in [2], the Ant problem is to find a control program which enables an artificial ant to follow a path of food. Here, the well-known *Santa Fe* trail is used as path of 89 food items placed on a 32x32 grid.

The set of symbols used is $S = ($ (IF-FOOD-AHEAD,2), (PROGN2,2), (PROGN3,3), (MOVE,0), (RIGHT,0), (LEFT,0)) with $N_{sym} = 6$ and $I_{term} = 4$. The meaning is as follows: IF-FOOD-AHEAD processes the 1st child-node (*then* part) if a pice of food is in front of the ant, otherwise it processes the 2nd child-node (*else* part), PROGN2 contains two commands to be executed in sequence, PROGN3 contains three commands to be executed in sequence, MOVE moves the ant one step and collects the food (if any), RIGHT turns the ant to the right (90°) and LEFT turns the ant to the left (90°). The tree composed of these symbols is repetitively executed while the operations (MOVE,RIGHT,LEFT) are counted. The number of allowed operations is restricted to 400 and 600.

The used fitness function F is the maximum number of food items minus the sum of collected food items within the allowed number of operations, i.e., $F(x_{i,G}) = 89 - I(x_{i,G})$

Table 2. Results of TreeDE for the Ant problem averaged over 100 independent runs. The column '# steps' is the number of allowed steps, 'Avg. Fitness' is the best fitness value reached on average, 's.d.' the appropriate standard deviation, 'Perfect hits (%)' the percentage of perfect solutions ($F(x_{i,G}) = 0$), 'Avg. Gen.' the average number of needed generations for the best solution of a run and 'Avg. Num. Eval.' the number of needed evaluations on average. The best results as well as the winning schemes are typed in boldface.

# steps	DE Scheme	Avg. Fitness	s.d.	Perfect hits (%)	Avg. Gen.	Avg. Num. Eval.
400	DE/rand/1	17.020000	6.902145	2	899.32	26970
400	DE/rand/2	17.330000	6.777986	1	920.09	27600
400	**DE/best/1**	**16.750000**	7.695940	2	865.89	25950
400	**DE/best/2**	17.300000	7.381734	**3**	**815.23**	24450
400	DE/rand-to-best	19.050000	5.879413	0	852.84	25560
400	DE/current-to-rand	21.970000	7.812112	1	828.80	24840
400	DE/current-to-best	20.830000	8.574445	1	834.30	25020
600	DE/rand/1	1.660000	5.316427	58	810.64	24300
600	**DE/rand/2**	**1.020000**	2.572858	**69**	764.91	22920
600	DE/best/1	1.960000	3.783966	43	**707.06**	21210
600	DE/best/2	1.140000	3.644228	66	751.87	22530
600	DE/rand-to-best	1.620000	3.913515	61	798.76	23940
600	DE/current-to-rand	5.180000	9.213447	45	918.22	27540
600	DE/current-to-best	5.420000	7.691788	25	798.49	23940

with $x_{i,G}$ being a vector (tree) and I the interpreter of the ant returning the number of collected food items.

As presented in Table 2, the difficulty of the Ant problem depends on the number of allowed operations. With 600 allowed operations, 69% of the runs of the best scheme produce perfect solutions collecting all food items. And with 1.02 missing food items on average, all runs are close to a near optimal solution while needing 22920 evaluations on average. GP needs 52224 evaluations while leaving 13.7 (sd:5.59) food items on the grid[1]. With 400 allowed operations the results are pretty poor. Only 2 - 3% of the runs produce ants collecting all food items. The two best schemes leave approximately 17 food items on the grid needing appr. 25000 evaluation steps. GP misses 19.4 (sd:13.78) food items but produces a perfect hit rate of 20% while needing 49254 evaluations on average. Interestingly, GP has a better perfect hit rate for the more difficult version of the problem.

5 Conclusions

This paper introduced a DE-based method for tree discovery. Symbol-trees are represented by full trees whose nodes represent discrete symbols as vectors. These trees can be mapped to vectors for the DE approach.

DE is a fast optimizer w.r.t. the number of evaluation steps needed. TreeDE 'inherits' this property and (at least for the used benchmark problems it) needs significantly less evaluation steps than other methods. Nevertheless, the obtained results are comparable or better than those obtained by the other methods used in the experiments.

The TreeDE method as presented in this paper is a first step and there are a lot of questions left open: What is the influence of the lower and upper bound on the

---

[1] GP results were obtained by ECJ 18: http://cs.gmu.edu/ eclab/projects/ecj

optimization step? Since $\mathcal{T}_{internal}(\sigma)$, $\mathcal{T}_{leaf}(\sigma)$ and $\mathcal{T}_{arity}(\sigma)$ determine a symbol by considering the maximum, the actual values of the vector components of σ are not of interest, only their mutual relationship. Can these bounds be set to constant values to 'get rid' of them? How about other transforming functions to transform a symbol vector σ to a symbol? Can they improve the results? The standard parameters of DE are used. Can they be optimized by a heuristic to improve the results? As in DE crossover is realized on vector component level or, so to speak, on sub-symbol level. Could crossovering of whole symbols (or even tree branches) be a better strategy?

Furthermore, the dependencies of the parameters should be analysed to derive functions or rules of thumb to determine them for a given problem. This is particularly interesting for the number of tree levels L.

References

1. Abbass, H.A., Hoai, N.X., McKay, R.I.: AntTAG: A new method to compose computer programs using colonies of ants. In: IEEE Congress on Evolutionary Computation (2002)
2. Koza, J.R.: Genetic Programming: On the Programming of Computers by Natural Selection. MIT Press, Cambridge (1992)
3. O'Neill, M., Brabazon, A.: Grammatical Differential Evolution. In: Proc. International Conference on Artificial Intelligence. CSEA Press, Las Vegas (2006)
4. Page, J., Poli, R., Langdon, W.B.: Smooth Uniform Crossover with Smooth Point Mutation in Genetic Programming: A Preliminary Study. In: Langdon, W.B., Fogarty, T.C., Nordin, P., Poli, R. (eds.) EuroGP 1999. LNCS, vol. 1598, pp. 39–48. Springer, Heidelberg (1999)
5. Poli, R., McPhee, N.F.: General Schema theory for genetic programming with subtree-swapping crossover: Part I. Evolutionary Computation 11(1), 53–66 (2003)
6. Shan, Y., Abbass, H., McKay, R.I., Essam, D.: AntTAG: a further study. In: Proc. 6th Australia-Japan Joint Workshop on Intelligent and Evolutionary Systems, Canberra, Australia (2002)
7. Storn, R., Price, K.: Differential Evolution - A Simple and Efficient Adaptive Scheme for Global Optimization Over Continuous Spaces, Univ. California, Berkeley, ICSI, Technical Report TR-95-012 (March 1995), ftp://ftp.icsi.berkeley.edu/pub/techreports/1995/tr-95-012.pdf
8. Storn, R.: On the Usage of Differential Evolution for Function Optimization. In: 1996 Biennial Conference of the North American Fuzzy Information Processing Society (NAFIPS 1996), Berkeley, pp. 519–523. IEEE, USA (1996)
9. Storn, R., Price, K.: Differential Evolution - A Simple and Efficient Heuristic for global Optimization over Continuous Spaces. Journal of Global Optimization 11(4), 341–359 (1997)

A Rigorous Evaluation of Crossover and Mutation in Genetic Programming

David R. White and Simon Poulding

Dept. of Computer Science, University of York,
Heslington, York, YO10 5DD, UK
{drw,smp}@cs.york.ac.uk

Abstract. The role of crossover and mutation in Genetic Programming (GP) has been the subject of much debate since the emergence of the field. In this paper, we contribute new empirical evidence to this argument using a rigorous and principled experimental method applied to six problems common in the GP literature. The approach tunes the algorithm parameters to enable a fair and objective comparison of two different GP algorithms, the first using a combination of crossover and reproduction, and secondly using a combination of mutation and reproduction. We find that crossover does not significantly outperform mutation on most of the problems examined. In addition, we demonstrate that the use of a straightforward Design of Experiments methodology is effective at tuning GP algorithm parameters.

1 Introduction

The role of crossover and mutation in Genetic Programming (GP) has long been the subject of debate in the GP community, particularly whether GP's use of crossover results in a more effective algorithm than approaches that do not, such as local search. Several papers comparing the effectiveness of the two variation operators have been published in the past [1,2,3]. In this paper, we carry out a principled empirical study, comparing the exclusive use of mutation against the exclusive use of crossover. Thus, we do not aim to provide the optimal combination of the two, but hope that our experiments may provide some illuminating evidence to this debate. The application of rigorous experimental method allows us to make a fair comparison between the two, by applying an equivalent amount of effort to the tuning of parameters for both cases across six example problems that are standard in the literature.

It has been stated that both the evolutionary computation and wider heuristic search communities are sometimes less rigorous in their experimental method and analysis than their counterparts in other experimental disciplines, such as the natural sciences. Effective and efficient Design of Experiments (DoE) [4] methods that are applied as a matter of course in those other disciplines have yet to be fully adopted, and often inferior trial-and-error methods are used instead. In this paper, we will demonstrate the ease of application of DoE methods in the hope that they may be more widely employed in subsequent research. The

L. Vanneschi et al. (Eds.): EuroGP 2009, LNCS 5481, pp. 220–231, 2009.

adoption of a DoE methodology enables us to make quantified statements about our findings with a strong degree of confidence.

Johnson [5] commented on the weaknesses of experimental method in previous research and outlined a series of practical guidelines to practitioners in heuristic search. We hope to follow Johnson's excellent guidance, and in particular we are promoting the repeatability of our experiments by adopting as simple a methodology as possible, by selecting tools widely used within the community, and by providing both raw results data, code and scripts via the web [6].

2 Previous Work

2.1 Crossover and Mutation

Luke and Spector [1] published a comparison of crossover and mutation over four problems, which was subsequently revised [2] with some improved statistical analysis. The original paper examined both an exclusive choice between crossover and mutation, and secondly a mixed approach or "blend", whereas the second paper examined only the exclusive choice. We hope to add to their results. Their work across the two papers included the equivalent of 36,000 runs of a typical GP system, and one way we can improve upon their work is by taking advantage of Moore's Law: our results are based on millions of runs. We have also increased the number of parameters evaluated from 4 to 10.

Luke and Spector found little difference in performance between exclusive use of crossover or mutation, and often there was no statistically significant difference. In both papers they note that the situation is more complex than it may appear, and that their results are dependent on parameter settings.

Angeline [3] compared crossover to macromutation, arguing that the crossover operator may in fact function as a mutation operator of sorts. He borrowed the phrase *headless chicken crossover* from the Genetic Algorithms community, where crossover is made with a randomly generated tree rather than a second parent. Angeline compared standard crossover to two variants of headless chicken crossover and found that there was little difference in performance across 3 problems, whilst maintaining fixed settings for the remaining parameters.

Koza includes a section on a simple comparison of genetic programming with and without crossover, using mutation in both cases, in his first book [7]. Very little detail of the comparison is provided and it does not appear to employ statistical methods, but within the instance studied, Koza found crossover was a strongly beneficial addition to the algorithm.

These prior comparisons only considered the importance of parameter settings on the comparison to a limited extent and thus their conclusions cannot be generalised. Whilst t-tests were employed, more sophisticated statistical techniques were not used. Our work addresses this issue, through systematic parameter tuning and large-scale experimentation along with extensive analysis.

2.2   Previous Applications of Rigorous Experimental Method

Feldt et al. [8] investigated the significance of various GP parameters within classification problems by applying fractional factorial designs. A large number of parameters, 17, are considered, many of them specific to the example problems. Their work applied the initial stages of DoE screening methods, to determine which parameters are significant.

Coy et. al [9] also applied two-level factorial designs to optimise parameters, although their work does not involve Genetic Programming. Having run a fractional factorial on a small number of problems, they hillclimb a linear approximation of the response surface to optimise the parameters, and finally average the parameter settings across problems. As described below, our pilot study suggests that such response surface techniques may not be consistently effective for GP algorithm parameters. Generalising from one problem to another is also difficult, and for this reason we apply optimisation separately to each individual problem. Interestingly, they also conclude their paper with a comparison between algorithms, though whether this comparison involves parameter tuning for the other algorithms involved is unclear.

3   Hypotheses

The empirical work is driven by the two hypotheses below. Three GP algorithms are used for the empirical investigation described in this paper, and for clarity within the hypotheses we denote these as follows:

$\mathbf{A_c}$ a GP algorithm that employs two genetic operators: crossover and reproduction. We will tune the parameters of the algorithm, and identify the tuned algorithm using the notation $\mathbf{A_c^*}$.

$\mathbf{A_m}$ a GP algorithm that employs two genetic operators: mutation and reproduction. We identify the tuned algorithm as $\mathbf{A_m^*}$.

$\mathbf{A_d}$ a GP algorithm that employs two genetic operators: crossover and reproduction. This algorithm is not tuned during our investigation; instead the algorithm parameters are set to values that are established within the GP community as 'defaults'.

Hypothesis 1: Crossover and Mutation. There is a significant difference between the performance of algorithms A_c^* and A_m^* for a given problem instance.

In other words, after tuning each algorithm, the performance of the algorithm employing crossover as a genetic operator is significantly different from the performance of algorithm employing mutation. We use this hypothesis to determine the relative effectiveness of crossover and mutation as genetic operators for each of the problems.

Hypothesis 2: Parameter Tuning. There is a significant difference between the performance of A_c^* and A_d for a given problem instance.

In other words, tuning the crossover algorithm results in a significantly different performance from the algorithm using default values. (The 'default' values for a mutation-only GP algorithm are not clearly defined, so we do not make an equivalent comparison for A_m.) We use this hypothesis to demonstrate the effectiveness of the methodology we apply to tune the algorithms.

4 Experimental Design

4.1 Problem Instances

In order to test our hypotheses, it was necessary to select a set of problems. We chose a problem set from the examples distributed with version 16 of the popular ECJ system [10], listed in Table 1. This set includes at least one problem from each category of the examples provided with ECJ. One regression problem includes Ephemeral Random Constants (ERCs), one does not. Note that we have not considered examples using Automatically Defined Functions (ADFs).

Given this selection of problem instances, it was a natural choice to use ECJ as the system to run the algorithm trials described in this paper.

Table 1. ECJ Problems Selected as Problem Instances

Number	Problem
1	Symbolic Regression of $x^4 + x^3 + x^2 + x$ with no ERCs
2	Symbolic Regression of $x^5 - 2x^3 + x$ with ERCs
3	Two Box Problem
4	Santa Fe Ant Trail
5	Boolean 11 Multiplexer
6	Lawnmower

4.2 Response Measure

In comparing the performance of different parameter settings we must select a performance measure. In this work we have used the *fitness of the best individual in the final population* as our response measure. For some of the problems, ECJ will terminate a run as soon as an "ideal" individual is found.

4.3 Parameter Selection

Using ECJ's parameter file system, we selected 9 independent parameters for algorithm A_c and 10 for A_m, as listed in Table 2. A_m has an extra parameter, x_{10}, controlling the expression grown to replace a selected subtree.

Additional ECJ parameters are derived from the set of 10 independent parameters and are listed in the bottom half of Table 2. For example, the probability of reproduction is derived as $1 - x_8$ where x_8 is the probability of crossover or mutation. Similarly, we maintain the product of the population size and number of generations at (approximately) 52500 to ensure the same number of fitness evaluations are made by all algorithms.

Note that tournament selection was chosen as the selection mechanism, hence roulette-wheel, rank selection and other alternatives were not considered. Koza's ramped half-and-half method was used for initialisation and its parameters were treated as experimental parameters, that is, factors to be examined within our experimentation. The justification for these decisions was simply that the number of possible alternatives was in effect infinite.

ECJ provides a series of default parameter values that are the same for all problems. These defaults are based on the work of Koza [7,11], and no attempt has been made by the ECJ team to optimise these parameters for the individual problems concerned. These defaults are used to provide the parameter values for the default algorithm, A_d, and are listed in the right-hand column of Table 2.

Table 2. Algorithm Parameters and their Ranges

	Description	Range Low	Range High	ECJ Default
x_1	Grow prob. initialisation	0.1	0.5	0.5
x_2	Max. depth initial tree	4	10	6
x_3	Min. depth initial tree	$0.34 \times x_2$	$0.75 \times x_2$	2
x_4	Prob. of selecting root	0	0.5	0
x_5	Prob. of selecting terminal	$0.1 \times (1 - x_4)$	$0.5 \times (1 - x_4)$	0.1
x_6	Max. depth child tree	5	19	17
x_7	Population size	30	1500	1024
x_8	Prob. of crossover/mutation	0.1	1.0	0.9
x_9	Tournament selection size	2	9	7
x_{10}	Min. depth mutated subtree	3	9	5
Dependent Parameters				
x_{11}	Max. depth mutated subtree	x_{10}		5
x_{12}	Max. number of generations	$52500/x_7$		51
x_{13}	Prob. of selecting non-terminal	$1 - (x_4 + x_5)$		0.9
x_{14}	Prob. of reproduction	$1 - x_8$		0.1

4.4   Strategy

The performance of the crossover and mutation algorithms, A_c and A_m, depends on the values of the 10 independent parameters identified above. At one choice of parameter values, A_c may perform better than A_m, and the reverse may be true at a different parameter setting. In this way, an arbitrary choice of parameter settings could be a source of significant bias when comparing the two algorithms.

Therefore, a fair and principled comparison of the two algorithms must establish parameter values in an objective and unbiased manner. A sensible choice (which is also consistent with how the algorithms would be used in practice) is to identify, separately for each algorithm, parameter values at which it operates at its optimum performance and then compare these tuned algorithms.

Locating the *absolute* optimal parameter values for stochastic algorithms is often difficult and can require extensive computing resources for experimentation, especially when there are a large number of parameters as in this case. A compromise is to locate *approximations* to the optimal parameter values using a simpler method that requires computing power of a more realistic scale. If an equivalent amount of effort—in terms of both the computing power used and the data analysis performed—is spent in applying this method to each of the algorithms, it is reasonable to expect the approximations to be similarly close to the absolute optimum for each algorithm, and so the comparisons to be fair. This is the approach we take for the experimental work in this paper.

The experimental strategy therefore consists of two phases:

Parameter Tuning. Identify approximations to the optimal parameter values for each algorithm. We tune the algorithms separately for each problem instance, rather than assess a single set of parameter values that give the best performance when averaged over all problem instances.

Performance Comparison. At the optimal parameter values located in the first phase, compare the performance of both the crossover and mutation algorithms in order to test Hypothesis 1. In addition, we compare the performance of the crossover algorithm at its optimal parameter values with the algorithm using default ECJ parameters to test Hypothesis 2.

These phases are described in the next two sections. Each section contains a description of the experimental method for that phase and analysis of the results.

5   Parameter Tuning

5.1   Experimental Method

A possible method of tuning the algorithm parameters is to use Response Surface Methodology (RSM) [4,12]. This is an iterative process that starts in a small region of parameter space and uses experimental results to move the region of interest towards the optimal parameters. The traditional application of RSM is the improvement of industrial processes and it has recently been applied to optimising the parameters of wireless protocols [13].

We performed a pilot study that tested RSM, as well as a more straightforward application of Design of Experiments methodology, as techniques for tuning the parameters of the algorithms A_c and A_m. Contrary to our expectations, the DoE method located parameter values that were closer to the optimum than those found using RSM. (We speculate that the relatively high dimensionality of the parameter space presents a challenge to RSM, but more rigorous experimentation is necessary before we are able to comment on the relative efficacy of each method.) In addition, the RSM required more, potentially subjective, choices from the experimenter than did DoE. We therefore proceeded with the DoE method for the parameter tuning phase described in this section.

Our DoE methodology approximates the performance, y, of the algorithm to the parameter settings, x_i, using a second-order linear model:

$$y = \beta_0 + \sum_i \beta_i x_i + \sum_i \sum_{j>i} \beta_{ij} x_i x_j + \sum_i \beta_{ii} x_i^2 + \epsilon \qquad (1)$$

This model is almost certainly a simplification of the actual relationship between the algorithm's performance and its parameters, especially over the large ranges we allow for the parameter values. However, if we keep in mind the goal of locating *approximations* to the optimal parameter values using an objective methodology, this simplification may be appropriate. The results of the pilot study provided evidence that this is indeed the case.

The β coefficients in the equation define the relationship between algorithm parameters and performance. The objective of the experiments described in this section is to estimate the value of these coefficients in order to fit the model to the actual performance of the algorithm.

The algorithm performance is also affected by the sequence of random numbers that control operations such as initialisation, crossover, mutation, reproduction and selection. We account for this affect using the 'noise' term, ϵ: it quantifies the difference between the predicted mean performance of the algorithm and the observed performance of a single run of the algorithm with a specific random number seed.

5.2   Experimental Design

In order to provide data that is used to estimate the β coefficients in the model, a series of experimental trials are run at different parameter values. The choice of these parameter values is specified by an experimental design. For second-order linear models such as equation (1), a Central Composite Design (CCD) is often chosen since good coefficient estimates can be obtained with relatively few experimental trials [4]. A further pilot study compared the use of CCD with a three-level full factorial design (FD), and suggested the latter to be more effective at locating good approximations to the optimum algorithm parameters, even when a large number of the repetitions of the CCD were used to reduce the effect of the stochastic noise, i.e. the changes in algorithm performance caused by the random seeds. (We speculate that CCD is an efficient and effective design when a second-order linear model is a good approximation of the actual response, but when the model is a necessary simplification as it is here, a three-level FD is more effective since it explores many more points in the parameter space.)

The three-level FD considers three values of each parameter: the points at each end of the range shown in Table 2 and a point in the middle of the range. The ranges are sufficiently large to incorporate most reasonable values for each parameter, but exclude extreme parameter values (e.g. a crossover probability of 0, or a tournament selection size of 1) that might significantly alter the nature of the algorithm and therefore the shape of the response surface in such regions. The full design consists of all possible combinations of the three values for each parameter, resulting in $3^9 = 19{,}683$ design points for A_c, and $3^{10} = 59{,}049$ design points for A_m.

There is redundancy inherent in the three-level factorial design: it contains many more design points than is necessary to fit a second-order model. This allowed us to use a relatively small number of repetitions—two, using a difference random seed in each case—to accommodate the stochastic noise and therefore improve the accuracy of model fitting.

5.3   Analysis

As described in Section 4.2, our response measure is the fitness of the best individual in the final population. We use the 'adjusted fitness' value returned by ECJ for this measure since it has convenient mathematical properties. It is a value in the range $(0, 1]$, with a value of 1 indicating the ideal solution was found by the algorithm.

The model of equation (1) is fitted to the response values using standard linear regression, which estimates the values of the β coefficients.

As a first step in this process, a Box-Cox transform [4] is applied to the adjusted fitness, y:

$$y' = \begin{cases} \frac{y^\lambda - 1}{y} & \text{if } \lambda \neq 0 \\ \ln(y) & \text{if } \lambda = 0 \end{cases} \qquad (2)$$

The value of λ is chosen from the range $[-2, 2]$ so that the fit of the transformed responses, y', to the second-order linear model of equation (1), is maximised. The analysis of the linear model makes particular assumptions, particularly as to the distribution of the random variable ϵ, and this transform enables the responses to satisfy the assumptions as closely as possible.

To locate the optimal parameter values for the algorithm, we apply a deterministic optimisation technique, quadratic programming, to the fitted model. This technique locates the parameter values that give the largest response from the model, and therefore the best performance from the algorithm. We constrain potential parameter values to the ranges listed in Table 2.

Table 3. Parameter Values for Tuned Crossover Algorithm, A_c^*

	Description	Prob 1	Prob 2	Prob 3	Prob 4	Prob 5	Prob 6
x_1	Grow prob. initialisation	0.9	0.1	0.1	0.9	0.9	0.3973
x_2	Max. depth initial tree	4	8	8	4	4	4
x_3	Min. depth initial tree	3	5	4	2	1	3
x_4	Prob. of selecting root	0	0	0	0	0	0
x_5	Prob. of selecting terminal	0.2665	0.4723	0.4612	0.4299	0.5	0.1
x_6	Max. depth child tree	17	19	19	17	16	19
x_7	Population size	1221	1221	1094	1500	1500	143
x_8	Prob. of crossover	1	1	1	0.9861	1	0.9541
x_9	Tournament selection size	2	9	9	9	9	7

Table 4. Parameter Values for Tuned Mutation Algorithm, A_m^*

	Description	Prob 1	Prob 2	Prob 3	Prob 4	Prob 5	Prob 6
x_1	Grow prob. initialisation	0.8364	0.9	0.1	0.9	0.1	0.9
x_2	Max. depth initial tree	4	7	8	4	9	4
x_3	Max. depth initial tree	2	3	3	2	5	1
x_4	Prob. of selecting root	0	0	0	0	0	0
x_5	Prob. of selecting terminal	0.5	0.5	0.5	0.5	0.5	0.5
x_6	Max. depth child tree	18	18	19	14	19	19
x_7	Population size	1458	1193	1010	1419	808	30
x_8	Prob. of mutation	0.9386	0.816	1	0.8488	0.8133	1
x_9	Tournament selection size	9	9	9	8	9	9
x_{10}	Min. depth mutated subtree	3	3	9	3	9	9

5.4 Results

The tuned parameter values are shown in Table 3 for algorithm A_c, and Table 4 for algorithm A_m.

6 Performance Comparison

6.1 Experimental Method

The second phase of experimentation considers the tuned algorithms, A_c^* and A_m^*, and, in addition, the default algorithm A_d. To compare algorithm performance, each combination of algorithm (A_c^*, A_m^* and A_d) and problem (1 to 6) was run 500 times and the response measured. Each algorithm run used a different random seed. The sample size of 500 is an estimate of an appropriate number of trials to demonstrate a statistically significant difference between the algorithms in the presence of stochastic noise.

6.2 Analysis

The hypotheses of Section 3 relate to a *significant* difference between the performance of the algorithms. Our analysis considers two criteria of significance:

Statistical Significance. A statistical analysis of the difference in observed algorithm performance must provide evidence that the observed difference is unlikely to be a chance result.

Scientific Significance. The difference in observed algorithm performance— the effect size—must be sufficiently large in comparison to the stochastic noise (arising from the choice of random seeds) to be scientifically important.

Given a sufficiently large sample size it would be possible to demonstrate a *statistically* significant difference in algorithm performance, even if the observed difference were extremely small. Such a result would unlikely be *scientifically* important, and our second criterion guards against this situation. In particular, it moderates against the possibility that our choice of a sample size of 500 for the comparisons is too large.

We apply *non-parametric* statistical tests to analyse both types of significance. These types of tests make few, if any, assumptions about the distribution from which the observed responses are sampled. Equivalent *parametric* tests, such as a *t*-test, make specific assumptions about the distribution. While such parametric tests can be very effective, even small deviations from these assumptions can invalidate the test results [14]. By using non-parametric tests, we therefore avoid the need to verify our samples satisfy the assumptions and also ensure that the test results are robust to any undetected deviations from the test assumptions.

To analyse *statistical significance*, we apply the Mann-Whitney-Wilcoxon or rank-sum test [15]. The null hypothesis for this test is that the responses of the two algorithms have identical distributions with equal medians; the alternative hypothesis is that the distributions are different. We use a 5% significance level: a *p*-value of $< 5\%$ rejects the test's null hypothesis and indicates a statistically significant difference in algorithm performance.

To analyse *scientific significance*, we calculate the Vargha-Delaney A statistic as a measure of effect size compared to stochastic noise [16]. This statistic is independent of the sample size and has a range of $[0, 1]$: a value of 0.5 indicates no difference in algorithm performance; values between 0.5 and 1.0 indicate increasingly large effect sizes where the first algorithm has the better performance; values between 0.5 and 0.0 indicate increasingly large effect sizes where the second algorithm has the better performance. The choice of what constitutes a significant effect size can depend on context. In this case, we apply the guidelines of Vargha and Delaney [16] that an A statistic greater than 0.64, or less than 0.36, indicates a 'medium' or 'large' effect size. We consider any comparison demonstrating 'medium' or 'large' effect sizes as scientifically significant.

6.3 Results

The results of the comparison between the A_c^* and A_m^* are presented in Table 5, and between the A_c^* and A_d in Table 6. Highlighted in bold are *p*-values that show a statistically significant difference in algorithm performance at the 5% significance level. The Vargha-Delaney A statistic shows which algorithm has the better performance: values greater than 0.5 indicate that A_c^* performs better, while values less than 0.5 indicate that A_m^* and A_d perform better. Values of this statistic highlighted in bold indicate a medium or large effect size which we consider to be scientifically significant.

Table 5. Comparison of Tuned Crossover Algorithm, A_c^*, to Tuned Mutation Algorithm, A_m^*

Statistic	Prob 1	Prob 2	Prob 3	Prob 4	Prob 5	Prob 6
rank-sum *p*-value	$\mathbf{< 10^{-5}}$	0.631	**0.0266**	0.0653	$\mathbf{< 10^{-5}}$	$\mathbf{< 10^{-5}}$
Vargha-Delaney A statistic	**0.757**	0.491	0.541	0.533	**0.939**	0.588

Table 6. Comparison of Tuned Crossover Algorithm, A_c^*, to Default Algorithm, A_d

Statistic	Prob 1	Prob 2	Prob 3	Prob 4	Prob 5	Prob 6
rank-sum p-value	$< 10^{-5}$	0.8662	$< 10^{-5}$	$< 10^{-5}$	$< 10^{-5}$	1.00
Vargha-Delaney A statistic	**0.696**	0.503	0.612	**0.739**	**0.825**	0.500

The results in Table 5 show that Hypothesis 1 holds for two of the six problems: for Problems 1 and 5, the tuned crossover algorithm, A_c^*, demonstrates a superior performance compared to the tuned mutation algorithm, A_m^*, that is both statistically and scientifically significant. Using a similar analysis, the results in Table 6 show that Hypothesis 2 holds for three of the six problems: for Problems 1, 4, and 5, the tuned crossover algorithm, A_c^*, demonstrates a superior performance compared to the default algorithm, A_d, that is both statistically and scientifically significant.

7 Conclusions and Further Work

The use of crossover has a significant effect in improving the performance of only two problems out of six; it does not significantly outperform mutation on the remaining four problems (as discussed above, we require a significant result to demonstrate both statistical and scientific significance). This is consistent with the conclusions of previous work. Moreover, since it is the outcome of rigorous experimental methodology and robust data analysis, we are able to be very confident in the validity of the result.

It is worth noting that A_m is a restricted form of mutation that will only use fixed-length subtrees, and variable-sized mutation may prove even more competitive with A_c. This lends credence to the idea that crossover is often only a macromutation operator, and a population-based paradigm using crossover may not be most effective at solving the majority of these problems.

The next step in the investigation of the role of crossover and mutation is to compare the performance of algorithms that use both genetic operators, and to explore the performance of variations of the mutation algorithm.

The results also demonstrate the effectiveness of using a Design of Experiments (DoE) methodology based on a factorial design in tuning GP algorithm parameters: for three of the six problems this technique produced tuned parameter values that were significantly better that the default values. As described above, we were surprised that a pilot study showed that this straightforward DoE approach resulted in more accurately tuned parameters than a more sophisticated technique, Response Surface Methodology, and a more efficient experimental design, the Central Composite Design.

Further work will investigate the observed superiority of the DoE methodology in tuning algorithm parameters and perform an exploration of the response surfaces of GP algorithms in order to identify a reason for this superiority.

Acknowledgements

This work is supported by an EPSRC grant (EP/D050618/1), SEBASE: Software Engineering By Automated SEarch.

References

1. Luke, S., Spector, L.: A comparison of crossover and mutation in genetic programming. In: Genetic Programming 1997: Proceedings of the Second Annual Conference, pp. 240–248. Morgan Kaufmann, San Francisco (1997)
2. Luke, S., Spector, L.: A revised comparison of crossover and mutation in genetic programming. In: Genetic Programming 1998: Proceedings of the Third Annual Conference, pp. 208–213. Morgan Kaufmann, San Francisco (1998)
3. Angeline, P.J.: Comparing subtree crossover with macromutation. In: Angeline, P.J., McDonnell, J.R., Reynolds, R.G., Eberhart, R. (eds.) EP 1997. LNCS, vol. 1213, pp. 101–112. Springer, Heidelberg (1997)
4. Montgomery, D.C.: Design and Analysis of Experiments, 6th edn. John Wiley & Sons, Inc., Chichester (2005)
5. Johnson, D.S.: A theoretician's guide to the experimental analysis of algorithms. In: Goldwasser, M.H., Johnson, D.S., McGeoch, C.C. (eds.) Data Structures, Near Neighbor Searches, and Methodology: Fifth and Sixth DIMACS Implementation Challenges, pp. 215–250. American Mathematical Society, Providence (2002)
6. Online Experiment Source Code and Scripts,
   http://www.cs.york.ac.uk/~drw/papers/eurogp2009/
7. Koza, J.R.: Genetic Programming: On the Programming of Computers by Means of Natural Selection. MIT Press, Cambridge (1992)
8. Feldt, R., Nordin, P.: Using factorial experiments to evaluate the effect of genetic programming parameters. In: Poli, R., Banzhaf, W., Langdon, W.B., Miller, J., Nordin, P., Fogarty, T.C. (eds.) EuroGP 2000. LNCS, vol. 1802, pp. 271–282. Springer, Heidelberg (2000)
9. Coy, S.P., Golden, B.L., Runger, G.C., Wasil, E.A.: Using experimental design to find effective parameter settings for heuristics. Journal of Heuristics 7(1), 77–97 (2001)
10. ECJ (2008), http://cs.gmu.edu/~eclab/projects/ecj/
11. Koza, J.R.: Genetic Programming II: Automatic Discovery of Reusable Programs. MIT Press, Cambridge (1994)
12. Myers, R.H., Montgomery, D.C.: Response surface methodology: process and product optimization using designed experiments. Wiley Series in Probability and Statistics. John Wiley & Sons, Inc., Chichester (2005)
13. Vadde, K.K., Syrotiuk, V.R., Montgomery, D.C.: Optimizing protocol interaction using response surface methodology. IEEE Trans. Mob. Comput. 5(6), 627–639 (2006)
14. Leech, N.L., Onwuegbuzie, A.J.: A call for greater use of nonparametric statistics. Technical report, US Dept. Education, Educational Resources Information Center (2002)
15. Wilcoxon, F.: Individual comparisions by ranking methods. Biometrics Bulletin 1(6), 80–83 (1945)
16. Vargha, A., Delaney, H.: A critique and improvement of the CL common language effect size statistics of McGraw and Wong. J. Educational and Behavioral Statistics 25(2), 101–132 (2000)

On Crossover Success Rate in Genetic Programming with Offspring Selection*

Gabriel Kronberger, Stephan Winkler, Michael Affenzeller, and Stefan Wagner

Heuristic and Evolutionary Algorithms Laboratory
School of Informatics, Communications and Media - Hagenberg
Upper Austria University of Applied Sciences
Softwarepark 11, A-4232 Hagenberg, Austria
{gkronber,swinkler,maffenze,swagner}@heuristiclab.com

Abstract. A lot of progress towards a theoretic description of genetic programming in form of schema theorems has been made, but the internal dynamics and success factors of genetic programming are still not fully understood. In particular, the effects of different crossover operators in combination with offspring selection are still largely unknown. This contribution sheds light on the ability of well-known GP crossover operators to create better offspring (success rate) when applied to benchmark problems. We conclude that standard (sub-tree swapping) crossover is a good default choice in combination with offspring selection, and that GP with offspring selection and random selection of crossover operators does not improve the performance of the algorithm in terms of best solution quality or efficiency.

1 Motivation

Since the very first experiments with genetic programming a lot of effort has been put into the definition of a theoretic foundation for GP in order to gain a better understanding of its internal dynamics. A lot of progress [7,14,15,17] towards the definition of schema theorems for variable length genetic programming and sub-tree swapping crossover as well as homologous crossover operators [16] has been made. But an overall understanding of the internal dynamics and the success factors of genetic programming is still missing. The effects of mixed or variable arity function sets or different mutation operators in combination with more advanced selection schemes are still not well understood. In particular, the effects of different crossover operators on the tree size and solution quality in combination with offspring selection are largely unknown.

Offspring selection [1] is a generic selection concept for evolutionary algorithms that aims to reduce the effect of premature convergence often observed with traditional selection operators by preservation of important allele [2]. The main

---

* The work described in this paper was done within HEUREKA!, the Josef Ressel centre for heuristic optimization sponsored by the Austrian Research Promotion Agency (FFG).

L. Vanneschi et al. (Eds.): EuroGP 2009, LNCS 5481, pp. 232–243, 2009.

difference to the usual definition of evolutionary algorithms is that after parent selection, recombination and optional mutation, offspring selection filters the newly generated solutions. Only solutions that have a better quality than the best parent are added to the next generation of the population. Winkler et al. used GP with offspring selection to produce accurate classifiers for medical data sets [18].

The recently described hereditary selection concept [9,10] also uses a similar offspring selection scheme in combination with parent selection that is biased to select solutions with few common ancestors. In this research we aim to shed light on the effects of typical genetic programming crossover operators regarding their ability to create improved solutions in the context of offspring selection. We apply GP with offspring selection on four benchmark problems: symbolic regression (Poly-10), time series prediction (Mackey-Glass), classification (Wisconsin diagnostic breast cancer) and the even-parity (4-bit) problem.

In the first part of this paper the focus lies on the achieved solution quality of the different crossover operators on all four benchmarks. In the second part we analyze if it is possible to gain an advantage either in form of better solution quality or in terms of efficiency with a mixture of crossover operators in one GP run.

2   Configuration of Experiments

We ran experiments with five crossover operators: standard (sub-tree swapping) [4] [17], one-point [7], uniform [13], size-fair and homologous size-fair [5] for each of the four benchmark problems: symbolic regression (Poly-10), time series prediction (Mackey-Glass), classification (Wisconsin diagnostic breast cancer) and the even-parity (4-bit) problem.

In the first set of experiments each crossover operator was used separately leading to a total of 20 different genetic programming configurations. The crossover operator, the problem specific evaluation operator and the function set were exchanged for each experiment. All other parameters of the algorithm were the same over all experiments. The random initial population was generated with probabilistic tree creation (PTC2) [8] and uniform distribution of tree sizes in the interval $[3; 50]$. The maximal height and size of trees was set to 8 and 50 respectively to prevent uncontrolled bloating of trees which would blur the results. A single-point mutation operator was used to manipulate 15% of the solution candidates by exchanging either a function symbol (50%) or a terminal symbol (50%). See Table 1 for a summary of all GP parameters.

To analyze the results, the quality of the best solution, the average tree size in the whole population as well as the offspring selection pressure were logged at each generation step together with the number of solutions that have been evaluated so far. Each run was stopped when the maximal offspring selection pressure or the maximal number of solution evaluations had been reached.

The offspring selection pressure of a population is defined as the ratio of the number of solution evaluations that were necessary to fill the population to the population size [1]. A high offspring selection pressure means that the crossover operator is unable to produce better children. A constant low offspring selection pressure over the whole run means that the crossover operator can easily generate better children, but the overall progress in solution quality is too slow to increase the offspring selection pressure.

2.1   Symbolic Regression – Poly-10

The Poly-10 symbolic regression benchmark problem uses ten input variables $x_1, \ldots, x_{10}$. The function for the target variable y is defined as $y = x_1 x_2 + x_3 x_4 + x_5 x_6 + x_1 x_7 x_9 + x_3 x_6 x_{10}$ [6,12]. For our experiments 100 training samples were generated randomly by sampling the values for the input variables uniformly in the range $[-1, 1[$. The usual function set of $+,-,*,\%$ (protected division) and the terminal set of $x_1 \ldots, x_{10}$ without constants was used. The mean squared errors function (MSE) over all 100 training samples was used as fitness function.

2.2   Time Series Prediction – Mackey-Glass

The Mackey-Glass $(\tau = 17)$[1] chaotic time series is an artificial benchmark data set sometimes used as a representative time series for medical or financial data sets [6]. We used the first 928 samples as training set, the terminal set for the prediction of $x(t)$ consisted of past observations $x_{128}, x_{64}, x_{32}, x_{16}, x_8, x_4, x_2, x_1$ and integer constants in the interval $[1; 127]$. The function set and the fitness function (MSE) were the same as in the experiments for Poly-10.

2.3   Classification – Wisconsin Diagnostic Breast Cancer

The Wisconsin diagnostic breast cancer data set from the UCI Machine Learning Repository [3] is a well known data set for binary classification. Only a part (400 samples) of the whole data set was used and the values of the target variable were transformed to values 2 and 4 respectively. Before each genetic programming run the whole data set was shuffled, thus the training set was different for each run.

Again the mean squared errors function for the whole training set was used as fitness function; the validation and test data sets were not used because we are not concerned about overfitting here. In contrast to the previous experiments a rather large function set was used that included functions with different arities and types (see table 1). The terminal set consisted of all ten input variables and real-valued constants in the interval $[-20; 20]$.

2.4   4-Bit Even-Parity Problem

The even-parity problem is a hard problem for genetic programming without special enhancements [11], so in this experiment only a rather small instance with four binary input variables was used. The goal is to find an expression for the even-parity function composed from the basic logic functions AND, OR, NAND, NOR, and the four input variables $D1, D2, D3$, and $D4$. The fitness function is the number of incorrect outputs for all possible assignments of $D1 \ldots D4$.

---

[1] Data set available from: http://neural.cs.nthu.edu.tw/jang/benchmark/

Table 1. General parameters for all experiments and specific parameters for each benchmark problem

General parameters for all experiments	Population size	1000
	Initialization	PTC2 (uniform [3..50])
	Parent selection	fitness-proportional (50%), random (50%)
		strict offspring selection, 1-Elitism
	Mutation rate	15% single point (50% functions, 50% terminals)
	Max. tree size / height	50 / 8
Poly-10	Function set	ADD, SUB, MUL, DIV (protected)
	Terminal set	$x_1 \ldots x_{10}$
	Fitness function	Mean squared errors
	Max. evaluations	1.000.000
Mackey-Glass	Function set	ADD, SUB, MUL, DIV (protected)
	Terminal set	$x_{128}, x_{64}, \ldots, x_2, x_1$, constants: 1..127
	Fitness function	Mean squared errors
	Max. evaluations	5.000.000
Wisconsin	Function set	ADD, MUL, SUB, DIV (protected),
		LOG, EXP, SIGNUM, SIN, COS, TAN,
		IF-THEN-ELSE, LESS-THAN, GREATER-THAN,
		EQUAL, NOT, AND, OR, XOR
	Terminal set	$x_1, \ldots, x_{10}$, constants: $[-20..20]$
	Fitness function	Mean squared errors
	Max. evaluations	2.000.000
Even-4	Function set	AND, OR, NAND, NOR
	Terminal set	$D1, D2, D3, D4$
	Fitness function	Number of incorrect outputs
	Max. evaluations	2.000.000

3   Success Rates of Crossover Operators

Figure 1 shows the quality progress (MSE, note log scale), average tree size, and offspring selection pressure for each of the five crossover operators over time (number of evaluated solutions). Twenty independent repetitions of GP with offspring selection were executed for the Poly-10 benchmark problem. The first row shows the best solution quality, the second row shows average tree size over the whole population and the third row shows offspring selection pressure.

The test using standard crossover were the most successful (16 runs found the best solution), size-fair and homologous size-fair show almost the same behavior finding the optimal solution in 4 and 6 runs. The tests with onepoint and uniform crossover were the least successful. Regarding the development of the average tree size only the runs with standard crossover show a different behavior. It can be easily seen that the average tree size grows a lot faster with standard crossover than with the other crossover operators until it reaches a point where the maximal tree size has a limiting effect. For the development of offspring selection pressure only the runs with uniform crossover show a different pattern. Offspring selection pressure is rising very fast in the first generations. This is an indicator for a rapid reduction of genetic diversity that leads to premature convergence and sub-optimal solution quality at the end of the run.

Figure 2 shows the results of 20 independent GP runs for the Mackey-Glass problem. Again the standard crossover operator shows good performance in terms of solution quality, size-fair and homologous crossover again show almost

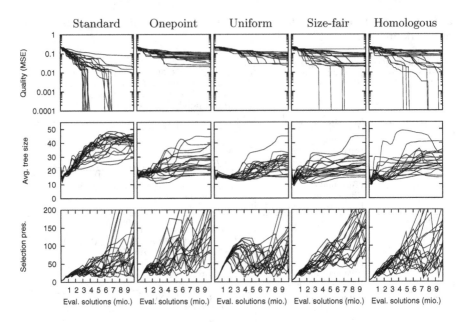

Fig. 1. Best solution quality (MSE, note log scale), average tree size, and offspring selection pressure for 20 runs with each crossover operator for the Poly-10 problem

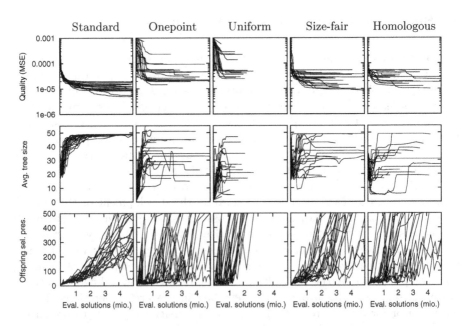

Fig. 2. Best solution quality (MSE, note log scale), average tree size, and offspring selection pressure (note different scale) for 20 runs with each crossover operator for the Mackey-Glass problem

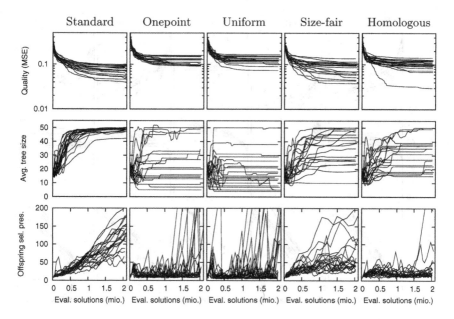

Fig. 3. Best solution quality (MSE, note log scale), average tree size, and offspring selection pressure for 20 runs with each crossover operator for the Wisconsin classification problem

the same behavior and onepoint and uniform are the least effective operators. All runs with standard crossover quickly converge to the upper tree size limit. For all other crossover operators there are a few runs where the solution shape is frozen quickly in the beginning. This has a detrimental effect on solution quality, but this can be easily prevented by adding mutation operators that change the program shapes. The offspring selection pressure charts show that in a few runs with onepoint and uniform crossover the offspring selection pressure rises quickly in the beginning. This coincides with the frozen program shapes because it is more difficult to create better solutions once the tree shape is limited. The runs with standard crossover show almost linear development of offspring selection pressure over time.

Figure 3 shows the results of 20 independent GP runs for the Wisconsin classification problem. The solution quality charts are similar to the charts for Poly-10 and Mackey-Glass except for one very good run with homologous crossover. Both size-fair and homologous crossover reached better solutions than standard crossover but the variance in between runs was also larger. Again onepoint and uniform crossover seem to be least effective. An interesting result is that offspring selection pressure remained low for all homologous crossover operators (Onepoint, Uniform, Size-fair, Homologous) in contrast to the previous results. In the tests with standard crossover offspring selection pressure grew steadily over time, but in the tests with other crossover operators offspring selection pressure stayed at a rather low level. This means that there was a continuous

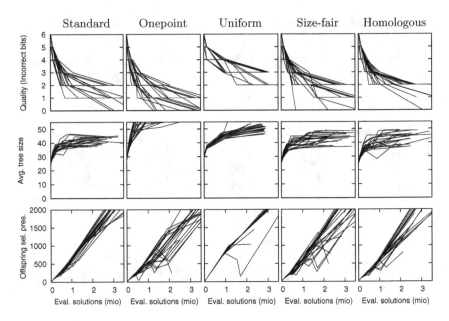

Fig. 4. Best solution quality (number of incorrect outputs), average tree size, and offspring selection pressure for 20 runs with each crossover operator for the even-parity problem

but slow quality improvement over the whole run. The flat offspring selection pressure curves could be caused by either the extended function set or the structure of the data set. Further investigations are necessary to fully explain this observation.

Figure 4 shows the results of 20 independent GP runs for the even-parity problem. This is a rather hard problem with integer quality values, thus the maximal offspring selection pressure was increased to 2000 to allow more evaluations per generation. The limit is necessary to stop the algorithm when it is impossible to create better offspring; without this limit the algorithm would enter an infinite loop. As a result the number of points in the charts is rather low even though two million solutions were evaluated in each run. The optimal solution was never found with uniform crossover. All crossover operators except for the uniform crossover have very similar charts for solution quality, average tree size, and offspring selection pressure.

4   Mixed-Crossover Success Rates

In the second set of experiments a mixed-crossover configuration of genetic programming was applied to the same benchmark problems. Each configuration was changed so that the crossover operator is selected randomly from all five available operators for each crossover event. Even though each crossover operator

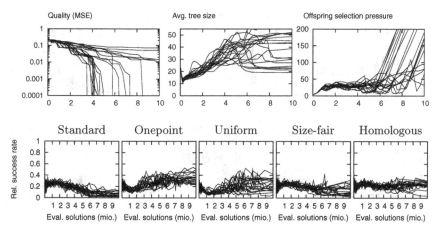

Fig. 5. Best solution quality (MSE; note log scale), average tree size, offspring selection pressure, and success rates of crossover operators for the Poly-10 problem with uniform selection of crossover operators

has the same chance to be selected, each one has a different chance of producing successful children. In combination with offspring selection the relative success rates of each crossover operator can be easily calculated by logging the originating operator for each individual in the population. Except for the changes regarding the crossover operator all other algorithmic settings are exactly the same as in the first set of experiments.

The charts given in the first row of Figure 5 show the development of the best solution quality, average tree size and offspring selection pressure of 20 independent runs of GP for the Poly-10 problem. The charts in the second row show the relative success rate for each crossover operator over time (evaluated solutions). The relative success rate of an operator is the number of solution candidates in the population produced by that operator. In comparison with the results of the single-crossover experiments shown in Figure 1 the solution quality chart is very similar to the chart for standard crossover.

The plots of relative success rates of each crossover operator show that standard, size-fair and homologous crossover couldn't create successful solutions as easily as onepoint and uniform crossover in the later stages of the GP. This is a noteworthy result because onepoint and uniform crossover on their own were not very effective (see Figure 1) but in combination with other crossover operators they could still improve solution quality in the later stages of the GP runs.

Figure 6 shows the results of 20 runs with mixed-crossover for the Mackey-Glass problem. The solution quality chart is similar to the chart of the single-crossover experiment (see Figure 2) with standard crossover. The mixed-crossover configuration has the effect that offspring selection pressure grows more slowly in the beginning but rapidly in the later stages of the GP runs.

The charts of the relative success rates of crossover operators show that standard crossover, as well as size-fair and homologous crossover, couldn't create better solutions after the beginning stages of the GP runs, while the relative

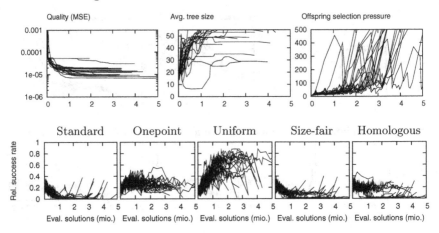

Fig. 6. Best solution quality (MSE, note log scale), average tree size, offspring selection pressure (note different scale), and relative success rates of crossover operators for the Mackey-Glass problem with uniform selection of crossover operators

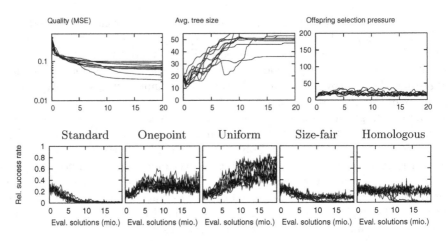

Fig. 7. Best solution quality (MSE, note log scale), average tree size, offspring selection pressure, and relative success rates of crossover operators for the Wisconsin classification problem with uniform selection of crossover operators

success rate of onepoint, and especially uniform crossover, grew to a high level very quickly. Thus, the progress in later stages of GP runs with mixed-crossover was influenced strongly by onepoint and uniform crossover. In comparison to Figure 2 the number of runs with frozen program shapes was reduced by the influence of sub-tree swapping crossovers (standard and size-fair).

Figure 7 shows the results of 20 independent runs with mixed-crossover for the Wisconsin diagnostic breast cancer problem. In terms of solution quality there is not a big difference to the single-crossover experiments (see Figure 3). Frozen program shapes were again reduced by the sub-tree swapping crossover

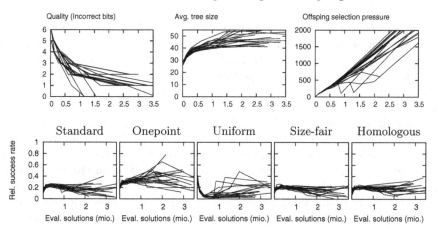

Fig. 8. Best solution quality, average tree size, offspring selection pressure, and relative success rates of crossover operators for the even-parity problem with uniform selection of crossover operators

operators. Offspring selection pressure remained at a rather low level over the whole time. This coincides with the single-crossover results where only the standard crossover had a rising offspring selection pressure curve.

The relative success rates are similar to the results for the Poly-10 and Mackey-Glass problems. In Figure 7 it is very clear that standard crossover was only effective in the early stages of the GP runs while onepoint and uniform crossover could easily create better solutions also in the later stages of the runs. Size-fair and homologous crossover seem to stand in between, in some of the runs they were still effective in the later stages while in other runs they were not effective at all.

Finally, Figure 8 shows the results of 20 independent runs with mixed-crossover for the even-parity problem. In comparison with the results for the single-crossover experiments in Figure 4 solution quality is on par (successful runs: 4) with the experiments using onepoint and size-fair crossover. The relative success rates of crossover operators do not follow a clear pattern. All operators had the same difficulty to produce successful children for this rather hard problem.

5   Conclusion

Based on the analysis of the results in our first experiment set it can be concluded that standard (sub-tree swapping) crossover is a good default choice as long as there are countermeasures to prevent uncontrolled bloating behavior. Standard crossover shows overall good performance except for the even-parity problem that is a special case because it is rather hard to solve with all crossover operators analyzed in this work. The results also show that onepoint and uniform crossover operators do not perform very well on their own. They also have the tendency to quickly freeze the tree shape and thus should be combined with mutation operators which manipulate tree shape.

The aim of the second set of experiments was to find out whether using all five crossover operators in one GP run has a beneficial effect either in terms of achievable solution quality or efficiency. From the results it can be concluded that there is no gain in terms of solution quality or efficiency. For all benchmark problems the mixed-crossover configuration reached a similar solution quality as the best single-crossover configuration.

The results also show that it is difficult for standard crossover to create better solutions after the early stages of mixed-crossover runs. In contrast to this, one-point and uniform crossover can create better solutions even in the later stages. Unfortunately this does not seem to have a strong effect on the best solution quality. Further investigations are needed to find out if this can be used to improve overall solution quality for instance by dynamically adapting the selection probability of each crossover operator guided by offspring selection pressure.

References

1. Affenzeller, M., Wagner, S.: Offspring selection: A new self-adaptive selection scheme for genetic algorithms. In: Adaptive and Natural Computing Algorithms. Springer Computer Series, pp. 218–221. Springer, Heidelberg (2005)
2. Affenzeller, M., Winkler, S.M., Wagner, S.: Effective allele preservation by offspring selection: An empirical study for the TSP. International Journal of Simulation and Process Modelling (2009) (accepted to appear)
3. Asuncion, A., Newman, D.J.: UCI machine learning repository (2007)
4. Koza, J.R.: Genetic Programming. MIT Press, Cambridge (1992)
5. Langdon, W.B.: Size fair and homologous tree genetic programming crossovers. Genetic Programming and Evolvable Machines 1(1/2), 95–119 (2000)
6. Langdon, W.B., Banzhaf, W.: Repeated patterns in genetic programming. Natural Computing (2008) (Published online: May 26, 2007)
7. Langdon, W.B., Poli, R.: Foundations of Genetic Programming. Springer, Heidelberg (2002)
8. Luke, S.: Two fast tree-creation algorithms for genetic programming. IEEE Trans. Evolutionary Computation 4(3), 274–283 (2000)
9. Murphy, G., Ryan, C.: Exploiting the path of least resistance in evolution. In: GECCO 2008: Proceedings of the 10th annual conference on Genetic and evolutionary computation, Atlanta, GA, USA, pp. 1251–1258. ACM, New York (2008)
10. Murphy, G., Ryan, C.: A simple powerful constraint for genetic programming. In: O'Neill, M., Vanneschi, L., Gustafson, S., Esparcia Alcázar, A.I., De Falco, I., Della Cioppa, A., Tarantino, E. (eds.) EuroGP 2008. LNCS, vol. 4971, pp. 146–157. Springer, Heidelberg (2008)
11. Page, J., Poli, R., Langdon, W.B.: Smooth uniform crossover with smooth point mutation in genetic programming: A preliminary study. In: Langdon, W.B., Fogarty, T.C., Nordin, P., Poli, R. (eds.) EuroGP 1999. LNCS, vol. 1598, pp. 39–49. Springer, Heidelberg (1999)
12. Poli, R.: A simple but theoretically-motivated method to control bloat in genetic programming. In: Ryan, C., Soule, T., Keijzer, M., Tsang, E.P.K., Poli, R., Costa, E. (eds.) EuroGP 2003. LNCS, vol. 2610, pp. 204–217. Springer, Heidelberg (2003)

13. Poli, R., Langdon, W.B.: On the search properties of different crossover operators in genetic programming. In: Genetic Programming 1998: Proceedings of the Third Annual Conference, University of Wisconsin, Madison, Wisconsin, USA, pp. 293–301. Morgan Kaufmann, San Francisco (1998)
14. Poli, R., McPhee, N.F.: General schema theory for genetic programming with subtree-swapping crossover: part I. Evol. Comput. 11(1), 53–66 (2003)
15. Poli, R., McPhee, N.F.: General schema theory for genetic programming with subtree-swapping crossover: part II. Evol. Comput. 11(2), 169–206 (2003)
16. Poli, R., McPhee, N.F., Rowe, J.E.: Exact schema theory and markov chain models for genetic programming and variable-length genetic algorithms with homologous crossover. Genetic Programming and Evolvable Machines 5(1), 31–70 (2004)
17. Poli, R., Rowe, J.E., Stephens, C.R., Wright, A.H.: Allele diffusion in linear genetic programming and variable-length genetic algorithms with subtree crossover. In: Foster, J.A., Lutton, E., Miller, J., Ryan, C., Tettamanzi, A.G.B. (eds.) EuroGP 2002. LNCS, vol. 2278, pp. 212–227. Springer, Heidelberg (2002)
18. Winkler, S.M., Affenzeller, M., Wagner, S.: Using enhanced genetic programming techniques for evolving classifiers in the context of medical diagnosis. In: Genetic Programming and Evolvable Machines (2009) (Online First)

An Experimental Study on Fitness Distributions of Tree Shapes in GP with One-Point Crossover

César Estébanez, Ricardo Aler, José M. Valls, and Pablo Alonso

Universidad Carlos III de Madrid
Avda. de la Universidad, 30, 28911, Leganés (Madrid). Spain
{cesteban,aler,jvalls}@inf.uc3m.es, pablo.alopez@alumnos.uc3m.es

Abstract. In Genetic Programming (GP), One-Point Crossover is an alternative to the destructive properties and poor performance of Standard Crossover. One-Point Crossover acts in two phases, first making the population converge to a common tree shape, then looking for the best individual within that shape. So, we understand that One-Point Crossover is making an implicit evolution of tree shapes. We want to know if making this evolution explicit could lead to any improvement in the search power of GP. But we first need to define how this evolution could be performed. In this work we made an exhaustive study of fitness distributions of tree shapes for 6 different GP problems. We were able to identify common properties on distributions, and we propose a method to explicitly evaluate tree shapes. Based on this method, in the future, we want to implement a new genetic operator and a novel representation system for GP.

1 Background and Motivation

Genetic Programming (GP) is a very successful stochastic search technique, widely used by researchers to solve complex problems. Thirty six human competitive results have been documented in the last years, and there are twenty three instances where GP has duplicated the functionality of a previously patented invention, infringed a previously issued patent, or created a patentable new invention [1,2].

However there are many controversial issues related to classical GP. Among the most important ones is the destructive effects of Standard Crossover, which have been widely documented in GP literature [3,4,5,6]. Another major issue is the phenomena known as Code Bloat [3,5,6,7,8,9]. Bloat appears when individuals in the GP population tends to rapidly grow over the generations, producing very large and slow trees. It is a very serious problem because it consumes a lot of resources without a proportional improvement of the average fitness. In [10], Langdon and Poli explain some possible theories of why Code Bloat occurs, and it happens that most of them suggest that bloat is a consequence of destructive effects of Standard Crossover.

In [11] Standard Crossover is compared with two new crossover operators, namely: One-Point Crossover and Uniform Crossover. They conclude that Standard Crossover is a biased and local search operator, which cannot explore the

L. Vanneschi et al. (Eds.): EuroGP 2009, LNCS 5481, pp. 244–255, 2009.
© Springer-Verlag Berlin Heidelberg 2009

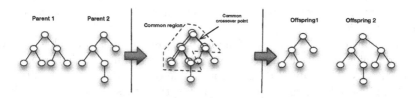

Fig. 1. One-Point Crossover operator

search space quickly, that can only reach certain areas of the search space, and that can get stuck in local maxima. Some documented results show that Standard Crossover has a limited performance advantage compared to mutation based operators [12,13]. On the other hand, One-Point Crossover shows some interesting properties. It works this way: When the two parents are selected, they are aligned. Both trees are traversed to identify the parts that have the same shape. All the links in the common region are stored. The traverse is stopped in a branch as soon as an arity mismatch between corresponding nodes in the two trees is present. Finally, a crossover point is randomly selected among the stored links, and the resulting subtrees are swapped. Figure 1 shows an example of One-Point Crossover.

One-Point Crossover has some interesting features, as we already mentioned. First, it is immune to Bloat [10]. Second, if the initial population is correctly generated, then the individuals of the earlier generations should have very small common regions. This means that crossover points will probably be selected very near to the root of the parents, while in standard crossover every link of the tree has same probability. The consequence is that the amount of genetic material exchanged is greater when using one-point crossover. After a number of generations, (in absence of mutation) one-point crossover makes the population evolve to a common tree shape. From this moment, GP behaves like a Genetic Algorithm (GA), selecting common crossover points in fixed-length-and-shape GP individuals. In this phase, GP intensifies exploitation of the region of the search space bounded by the selected tree shape.

1.1   Evolution of Tree Shapes

From the point of view of this work the most important fact about One-Point Crossover is that, in absence of mutation, it makes the population to converge to a common tree shape. So, One-Point Crossover uses the first generations of the GP run for select one tree shape, and then it makes the whole population to adopt that shape. Based on this fact, we claim that One-Point Crossover is performing an implicit evolution of tree shapes. This makes perfect sense for us: A tree shape can be seen as a region of the search space, namely the region containing all the solutions which are represented by a tree with that shape. Using this point of view, one can imagine One-Point Crossover performing a search in two phases: First, it makes a global search of tree shapes, trying to identify promising regions of the search space; Then, it finally focusses on a specific region, considered to be interesting, and it looks for the best solution inside that region.

If this point of view is correct, One-Point Crossover must be following a criterion to prefer some tree shapes to others. But it has not explicit information on tree shapes. The only information available during the run is the fitness of the GP individuals on the population. Being the fitness function the only information available, it seems to be clear that tree shapes are selected using the fitness of the GP individuals that have that shape, and by the fitness of those individuals containing the building blocks needed to construct the shape[1]. It is very important to remark that those fitness measures are considered one by one, instead of looking at the global picture of all the available matches of the considered tree shape.

We believe that this could be not the best way of evaluate a shape. First, because it is unfair: depending on how the initial population samples the search space of all the possible tree shapes, some shapes will have more matches in the population than others. That means more probabilities of sampling an better-than-average-fitness match, and therefore more opportunities to proliferate. Second, because this method can be deceptive: Big tree shapes can contain billions of different matches. If we have an better-than-average-fitness match in the population, the shape of that match will proliferate, even if the population also contains other matches of that shape with very bad fitnesses. Out intuition says that it will be much more accurate to have a general look on all the available matches and evaluate the shape using a joint fitness measure.

The objective of this work is to find an explicit method to evaluate tree shapes, and for that, we made an experimental study on the fitness distributions of GP tree shapes. First, we need to define what it is a good or a bad tree shape, then, we want to know how to distinguish between them. Finally, we want to obtain a fitness function that can explicitly evaluate a tree shape, that has a reasonable computational complexity, and that can be reused on a great variety of different GP problems. If we can find such a function, then we could improve the search capabilities of One-Point Crossover, making the convergence criterion become clear, fair, and rigorous. In the next section we explain the methodology we followed to study the fitness distributions, and we present the exhaustive experiments we carried out. Then, in Section 3 we define the fitness criterion we will follow to measure tree shapes. We formulate that definition using the data extracted from the experiments. In Section 4 we show how computationally-hungry can be the explicit fitness function we proposed. Finally in Section 5, we explain the conclusions we obtained from this work, and we discuss some

---

[1] Depending on the selection methods and the genetic operators used, the propagation rate of tree shapes from one generation to the next will vary. The GP Schema Theorems proposed by Poli and Langdon in [10] describe the dynamics of tree shapes on GP + One-Point Crossover. In this work it is enough to consider that tree shape can appear in a population because of a verbatim copy (reproduction, elitism, etc.), or because of the combination of building blocks performed by One-Point Crossover.

remaining issues about the future possibilities of this new, explicit fitness function for tree shapes.

2 Fitness Distributions of Matches

The goal of this research is to answer three questions:

- How can we tell whether a tree shape is good or bad?
- Can we estimate that information in a general, accurate, rigorous way?
- Can we make this estimation efficiently?

In order to give answers, we coded six different GP problems in ProGen[2]. Two of them are classical GP problems of two different classes: Symbolic Regression and Boolean 6-Multiplexer. The third problem is a real application of Genetic Programming called GP-Hash [14]. The last three problems are applications of GP to classification, using three UCI datasets: Pima Indians Diabetes [15], SpectHeart [16], and Haberman' Survival [17]. In Table 1 we briefly show the configuration of the six GP problems.

Table 1. Parameters of the 6 GP problems

	Functions	Terminals	Fitness (minimize)	Init Method
Regression	+,−, *, /	X	Prediction Error	Grow, depth 3-6
6-Multiplexer	AND, OR, NOT, IF	A0, A1, D0, D1, D2, D3, D4	Prediction Error	Grow, depth 3-9
GP-Hash	>>, AND, OR, NOT, XOR, MULT, SUM	a0, hval	Avalanche Effect	Grow, depth 2-6
Diabetes	+, −, *, %, * − 1, <, >, =, AND, OR, NOT, IF-ELSE	Attributes of the dataset	Classification Error	Grow, depth 3-5
Haberman	+, −, *, %, * − 1, <, >, =, AND, OR, NOT, IF-ELSE	Attributes of the dataset	Classification Error	Grow, depth 3-5
SPECT Heart	+, −, *, %, * − 1, <, >, =, AND, OR, NOT, IF-ELSE	Attributes of the dataset	Classification Error	Grow, depth 3-5

For each problem, we randomly generated one hundred tree shapes. Shapes were internally represented in ProGen as GP schemata of order 0 (GP trees with don't-care symbols "=" in all their nodes, see [10] and Figure 3). For each shape, we exhaustively generated all its matches and measured their fitness. That allows us to construct the complete fitness distribution of each tree shape for each problem. We show some of those distributions in Figure 2 for illustrative purposes. It

---

[2] ProGen is an Open Source GP framework designed at Universidad Carlos III de Madrid and entirely coded in Java.

took us around one month of intense computation in a 8-Core Intel Xeon with 8Gb of RAM to produce the output files (more than 13Gb of data), and we could only explore relatively small tree shapes. In last column of Table 1 we show the initialization method used in each problem (size of distribution highly depends not only on the tree shape size but also on the size of the function set, so the shape size have to be readjusted for each problem). We had to keep the initial depth at small levels, and we always used grow method. In the future we want to invest more resources to repeat these experiments on bigger shapes (currently, we are working on a clusterized version of ProGen that should drastically reduce the execution times). Anyway, even with relatively small initial depths, most of the tree shapes we evaluated had millions of matches. Furthermore, according to Poli and Langdon [4], the proportion of individuals of a given fitness is generally independent of program size, so we expect a similar behavior on bigger shapes.

Studying in depth the distributions, we can conclude:

- Fitnesses of matches do not follow a normal distribution or any other known distribution (we used Shapiro-Wilkis tests on those problems with gaussian-look distributions).
- The distributions are highly problem dependent.
- Almost all the studied distributions are extremely leptokurtic, with a very large proportion of matches grouping very close to the mode.
- The mode usually coincides with the worst possible fitness.

All those conclusions were expected: As each tree shape encodes a large number of possible solutions, it is reasonable that most of them lack of any sense (worst possible fitness), and only a very few are good solutions that have a decent fitness. We also expected the problem dependency: Obviously, the way fitnesses distribute depends a lot on the definition of the fitness function. We can identify two different types of fitness functions in our problems: First, we have two fitness functions (Regression and GP-Hash) which have no upper bound (it can get as big as infinite). Those distributions have huge tails that we filtered before processing them. The other fitness functions do have an upper bound, and the worst fitness is not that bound, but the middle point between that point and 0: In a binary classification problem, having an error of 100% is as hard as having 0% (just invert the output and you have a perfect classifier).

3   A Fitness Function for Tree Shapes

Fitness distributions of shapes do not follow any known distribution, so we have to find any other statistical information for evaluate tree shapes. We extract a set of statistics from the distribution of each tree shape of each of the 6 GP problems: Minimum, first quartile, median, mean, third quartile, and maximum. From our point of view, the most interesting one is the first quartile: Minimum value is interesting, because it represents the best solution you can find inside that tree shape. But tree shapes are regions of the search space that conglomerate millions of possible solutions. An outstanding minimum lost in a huge population

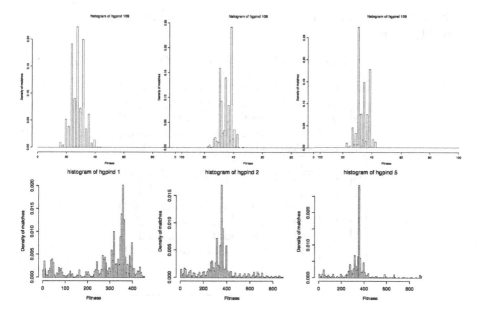

Fig. 2. Some illustrative histograms of fitness distributions of two different problems. First row: Boolean 6-Multiplexer; Second row: Symbolic Regression.

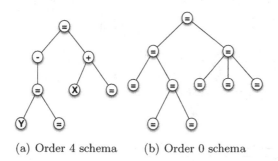

(a) Order 4 schema   (b) Order 0 schema

Fig. 3. Two GP schemata of different orders

of billions of bad solutions is not helpful at all. After finding a good shape, we want to be able to find the good solutions it contains, and that will be much easier in a region with a large proportion of good solutions. This proportion could be measured by the first quartile. A low first quartile guarantees that you will find solutions as good as or better than it with a probability of 25%.

The problem with first quartile is that it is not easy to calculate from small samples, so it could lose its utility when we cannot use complete distributions any more. Mean is very easy to estimate, even with unknown underlying distributions. So, for each GP problem, we generated a table with a row for each tree shape, and a column for each statistic, and we sorted the rows by the mean column. Table 2 shows the head and the tail of the table for the Regression

Table 2. Head and tail of the statistics table of the GP-Hash problem. Rows are ordered by mean.

	Min	FirstQu	Median	Mean	ThirdQu	Max
23	0	52,87	217	176	279,1	318,4
85	0	93,91	274,2	223,1	322,6	353,4
64	5,339	109	279,1	230,9	317,7	346,5
79	0	190,6	288,9	244,7	331,8	353,4
3	0	133	298,2	246,2	351,5	366,3
47	0	155,9	307,9	247,6	346,5	366,3
...	...	...	...	...	...	...
98	0	269,4	351,5	348,4	419,1	804,4
27	0	307,9	356,4	349,5	366,3	842,2
83	0	279,1	351,5	350,7	366,3	1063
1	0	279,1	351,5	351,5	400,2	822,7
31	0	307,9	356,4	353,2	366,3	982
13	0	279,1	356,4	355	385,3	1063

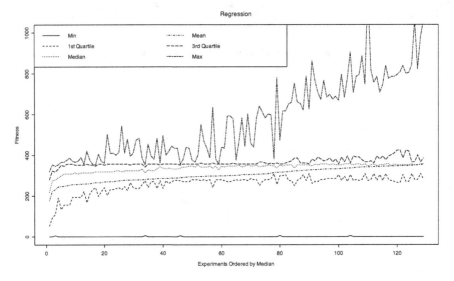

Fig. 4. Graphical representation of the statistics table of the GP-Hash problem when using the mean for ordering the shapes

problem[3]. In the table, the shapes on the top rows have a lower first quartile than those in the tail. In Figure 4, we can see a graphical representation of the whole table (Figure 5 shows similar representations of 3 other problems). All the statistics show a growing tendency when ordering by the mean. That means:

---

[3] Due to space limitations we cannot show all the tables here. We choose the Regression problem because it exemplifies very well the explanation. We have the same problem in Figure 5: we can only show 3 graphical representations in order to keep the figures legible.

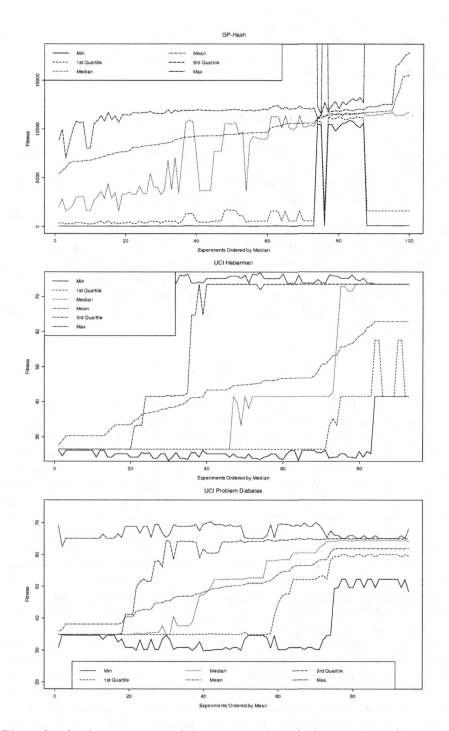

Fig. 5. Graphical representation of the statistics tables of other 3 problems (GP-Hash, Haberman, and Diabetes). Always ordering by mean.

the closer we get to the minimum mean value, the better the tree shapes are (according to our criteria). Of course, the tendency is not perfect: the growths of some statistics are irregular, but they are constant considering the big picture. So the mean could be a good estimator.

4   Efficiency and Computability

The problem now is how to estimate the mean fitness of a tree shape using an affordable amount of CPU power. In our experiments we have the complete distributions, but to obtain them we needed a huge computational effort. If we want to find a practical application of the explicit evolution of shapes, we need to be sure that we can make reliable and efficient estimations out of small random samples.

In a hypothetic implementation of our method, we have a population of tree shapes and we need to obtain the mean fitness of each one. We can extract a random sample from each shape, and estimate the mean. But we need to decide the size of the sample, trying to achieve a balance between estimation accuracy and CPU consumption. With our complete fitness distributions, we can use Chebyshev Inequality:

$$P\left[\frac{|X - \mu|}{\sigma} < k\right] > 1 - 1/k^2 \tag{1}$$

If we fix a value for k, then we can calculate which is the minimum size n which guarantees that the error between our estimation and the population mean is no greater than $k\sigma$ with a probability $p \geq 1 - 1/k^2$. But the problem is that if we do not have the complete distribution (and we will not in a real situation), we need to estimate the values of μ and σ, which means that we need to use a pilot sample, which means more CPU time. Instead, we will try to experimentally find a sample size that could be at least approximately correct for every problem. We tried with four different sample sizes: $n = 100$, $n = 30$, $n = 15$ and $n = 10$. For every tree shape of every problem we take 100 samples of size n. We perform a Wilcoxon/Mann-Whitney contrast (95% confidence level), comparing the mean of each sample with the mean of the complete distribution. If p value of a sample is greater than 0.05, then we have no reasons to say that there are significant differences between the average of the sample and the real mean. We call that an *accepted sample*. Using the information of the 100 tests, we can compute a proportion of *accepted samples* for a given shape. Finally, combining the proportions of each tree shape, we obtain the average proportion of *accepted samples* of size n for a given problem. We show this results on Table 3. We can see that even for the smallest sample size $n = 10$, we can accurately estimate the mean fitness of a distribution around 95% of the times (94.97% in the Regression problem, above 95% in all the other problems). This is an important result, because given that most of the fitness distributions we studied have thousands or millions of matches, it is impressive to be able to make precise

Table 3. Proportion of *accepted samples* for each sample size (100, 30, 15, 10) and for each problem

	$n = 100$	$n = 30$	$n = 15$	$n = 10$
Regression	95.48%	95.34%	95.06%	94.97%
6-Multiplexer	96.69%	95.79%	95.74%	95.13%
GP-Hash	97.77%	96.87%	96.56%	96.28%
Diabetes	96.74%	96.28%	96.02%	95.76%
SpecHeart	95.73%	95.72%	95.71%	95.38%
Haberman	97.04%	96.47%	95.72%	95.62%

estimations using only 10 samples. We believe that this is possible because of the intense grouping around the mode that we observed in the fitness distributions.

5   Conclusions

One-Point Crossover is an alternative to the Standard Crossover operator. It makes the GP population to converge to a common tree shape. The criterion that One-Point Crossover uses to guide the convergence is the fitness of the matches of that tree shape in the population (and also the fitness of individuals that contains the building blocks of that shape). This could be understand as an implicit evolution of tree shapes.

Our hypothesis is that an explicit evolution of tree shapes could improve the results obtained by One-Point Crossover. The goal of this work was to find an alternative way to explicitly measure how good a tree shape is. Our experiments show that just by sampling the distributions of fitness and extracting the mean value, we can obtain a reasonably good criterion to prefer some tree shapes rather than others. This sampling, to be optimum, only needs a small number of samples to give reliable estimations of the mean.

Obviously, the final justification of this work is to implement this explicit evaluation method and see if it improves the GP search capabilities as we expect. The idea is to make One-Point Crossover work in two different phases: in the first phase, it performs a search on partitions of the search space (explicit evolution of tree shapes), evaluating how promising the partitions are. This evaluation is made considering the mean, and relying in the experimental results obtained in this work: tree shapes with a good mean fitness, have a high concentration of better-than-average fitness solutions. Then, in the second phase, we already have a winner shape, and we can focus on the region delimited by it. This new method is analogous to One-Point Crossover, the only difference being that we expect the explicit evolution of shapes to make the first phase more controlled and rigorous. Our long term objective is to check if that really translates into an improvement of the search power of GP.

Acknowledgements

This work has been funded by the Spanish Ministry of Education and Science and FEDER under contract TIN2005-08818-C04 (the OPLINK project) and by

Comunidad de Madrid under contract 2008/00035/001 (Técnicas de Aprendizaje Automático Aplicadas al Interfaz Cerebro-Ordenador).

References

1. Koza, J.R., Andre, D., Bennett III, F.H., Keane, M.: Genetic Programming 3: Darwinian Invention and Problem Solving. Morgan Kaufmann, San Francisco (1999)
2. Koza, J.R., Keane, M.A., Streeter, M.J., Mydlowec, W., Yu, J., Lanza, G.: Genetic Programming IV: Routine Human-Competitive Machine Intelligence. Kluwer Academic Publishers, Dordrecht (2003)
3. Nordin, P., Banzhaf, W.: Complexity compression and evolution. In: Eshelman, L. (ed.) Genetic Algorithms: Proceedings of the Sixth International Conference (ICGA 1995), pp. 310–317. Morgan Kaufmann, San Francisco (1995)
4. Langdon, W.B., Soule, T., Poli, R., Foster, J.A.: The evolution of size and shape. In: Spector, L., Langdon, W.B., O'Reilly, U.-M., Angeline, P.J. (eds.) Advances in Genetic Programming 3, ch. 8, pp. 163–190. MIT Press, Cambridge (1999)
5. McPhee, N.F., Miller, J.D.: Accurate replication in genetic programming. In: Eshelman, L. (ed.) Genetic Algorithms: Proceedings of the Sixth International Conference (ICGA 1995), Pittsburgh, PA, USA, pp. 303–309. Morgan Kaufmann, San Francisco (1995)
6. Blickle, T., Thiele, L.: Genetic programming and redundancy. In: Hopf, J. (ed.) Genetic Algorithms within the Framework of Evolutionary Computation (Workshop at KI 1994, Saarbrücken), pp. 33–38. Im Stadtwald, Building 44, D-66123 Saarbrücken, Germany. Max-Planck-Institut für Informatik, MPI-I-94-241 (1994)
7. Koza, J.R.: Genetic Programming: On the Programming of Computers by Means of Natural Selection. MIT Press, Cambridge (1992)
8. Soule, T.: Code Growth in Genetic Programming. Ph.D thesis, University of Idaho, Moscow, Idaho, USA (May 15, 1998)
9. Bhattacharya, M., Nath, B.: Genetic programming: A review of some concerns. In: Alexandrov, V.N., Dongarra, J., Juliano, B.A., Renner, R.S., Tan, C.J.K. (eds.) ICCS-ComputSci 2001. LNCS, vol. 2074, pp. 1031–1040. Springer, Heidelberg (2001)
10. Langdon, W.B., Poli, R.: Foundations of Genetic Programming. Springer, Heidelberg (2002)
11. Poli, R., Langdon, W.B.: On the search properties of different crossover operators in genetic programming. In: Koza, J.R., Banzhaf, W., Chellapilla, K., Deb, K., Dorigo, M., Fogel, D.B., Garzon, M.H., Goldberg, D.E., Iba, H., Riolo, R. (eds.) Genetic Programming 1998: Proceedings of the Third Annual Conference, University of Wisconsin, Madison, Wisconsin, USA, pp. 293–301. Morgan Kaufmann, San Francisco (1998)
12. Luke, S., Spector, L.: A comparison of crossover and mutation in genetic programming. In: Koza, J.R., Deb, K., Dorigo, M., Fogel, D.B., Garzon, M., Iba, H., Riolo, R.L. (eds.) Genetic Programming 1997: Proceedings of the Second Annual Conference, Stanford University, CA, USA, pp. 240–248. Morgan Kaufmann, San Francisco (1997)
13. Angeline, P.J.: Subtree crossover: Building block engine or macromutation? In: Koza, J.R., Deb, K., Dorigo, M., Fogel, D.B., Garzon, M., Iba, H., Riolo, R.L. (eds.) Genetic Programming 1997: Proceedings of the Second Annual Conference, Stanford University, CA, USA, pp. 9–17. Morgan Kaufmann, San Francisco (1997)

14. Estebanez, C., Hernandez-Castro, J.C., Ribagorda, A., Isasi, P.: Finding state-of-the-art non-cryptographic hashes with genetic programming. In: Runarsson, T.P., Beyer, H.-G., Burke, E., Merelo-Guervos, J.J., Whitley, L.D., Yao, X. (eds.) PPSN 2006. LNCS, vol. 4193, pp. 818–827. Springer, Heidelberg (2006)
15. Smith, J.W., Everhart, J.E., Dickson, W.C., Knowler, W.C., Johannes, R.S.: Using the ADAP learning algorithm to forecast the onset of diabetes mellitus. In: Greenes, R.A. (ed.) Proceedings of the Symposium on Computer Applications in Medical Care, pp. 261–265. IEEE Computer Society Press, Los Alamitos (1988)
16. Kurgan, L.A., Cios, K.J., Tadeusiewicz, R., Ogiela, M.R., Goodenday, L.S.: Knowledge discovery approach to automated cardiac SPECT diagnosis. Artificial Intelligence in Medicine 23(2), 149–169 (2001)
17. Haberman, S.J.: Generalized residuals for log-linear models. In: Proceedings of the 9th International Biometrics Conference, Boston, pp. 104–122 (1976)

Behavioural Diversity and Filtering in GP Navigation Problems

David Jackson

Dept. of Computer Science, University of Liverpool
Liverpool L69 3BX, United Kingdom
djackson@liverpool.ac.uk

Abstract. Promoting and maintaining diversity in a population is considered an important element of evolutionary computing systems, and genetic programpming (GP) is no exception. Diversity metrics in GP are usually based on structural program characteristics, but even when based on behaviour they almost always relate to fitness. We deviate from this in two ways: firstly, by considering an alternative view of diversity based on the actual activity performed during execution, irrespective of fitness; and secondly, by examining the effects of applying associated diversity-enhancing algorithms to the initial population only. Used together with an extension to this approach that provides for additional filtering of candidate population members, the techniques offer significant performance improvements when applied to the Santa Fe artificial ant problem and a maze navigation problem.

1 Introduction

In genetic programming (GP), the evolutionary process is often characterised by a loss of diversity over time [1,2], with the population settling towards a mixture of just a few high-ranking individuals. This may make it impossible for the process to escape from local optima in the fitness landscape, thereby preventing it from discovering solutions. It is therefore generally accepted that it is important to take steps to instil and subsequently preserve a degree of diversity in GP populations.

One of the problems with this is that there are a variety of interpretations of what is meant by diversity, and hence how to promote it. Overviews of diversity measures can be found in [3] and [4], while Burke et al [5] give a more extensive analysis of these measures and of how they relate to fitness.

The most common usage of the term diversity as it applies to GP is concerned with differences in the structure of individual program trees. Almost all GP implementations make a basic attempt to foster some kind of structural or genotypic diversity, at least in the initial population, through their adherence to Koza's suggestions [6] for creating a range of tree sizes and shapes, and perhaps also by the removal of duplicates (see the next Section). More sophisticated structural diversity metrics may be based on edit distance [7], where the similarity between two individuals is measured in terms of the number of edit operations required to turn one into the other.

L. Vanneschi et al. (Eds.): EuroGP 2009, LNCS 5481, pp. 256–267, 2009.

A difficulty with comparing individuals in this way is that program trees which are apparently very different in appearance may in fact compute identical functions. Seeing beyond these surface differences requires the use of graph isomorphism techniques, but these are computationally expensive and become even more so as program trees grow larger over time. A simpler, less costly alternative is to check for pseudo-isomorphism [4], in which the possibility of true isomorphism is assessed based on characteristics such as tree depth and the numbers of terminals and functions present. However, the accuracy of this assessment may be subject to the presence of introns in the code; Wyns et al [8] describe an attempt to improve on this situation through the use of program simplification techniques to remove redundant code.

In contrast, behavioural or phenotypic diversity techniques work by considering the functionality of individuals, i.e. the execution of program trees rather than their appearance. Most usually, behavioural diversity is viewed as corresponding to metrics applied to the fitness values obtained on evaluating each member of the population [9]. One way of doing this is to view the spread of fitness values as an indicator of entropy, or disorder, in the population [8, 10]. Other approaches consider sets or lists of fitness values and use them in combination with genotypic measures [11, 12]. For certain types of problem it may be possible to achieve the effect of behavioural diversity without invoking the fitness function, via the use of semantic sampling schemes [13].

In this paper we also consider the effects of promoting behavioural diversity, but the research differs from previous work in two important ways. Firstly, the concept of 'behaviour' which we employ here does not equate to fitness; indeed, in ensuring diversity the notion of fitness does not even play a part. Nor does behaviour correspond to any outputs produced by programs: for the problems under study here – the well-known Santa Fe ant trail problem and a maze navigation exercise – there are no program outputs as such. Rather, we are concerned with monitoring and recording the activity of programs as they execute, and then using this as the basis for behavioural diversity.

The second difference from many other approaches is that we are not concerned with preserving diversity throughout the lifetime of a run. Instead, we concentrate solely on the role that behavioural diversity can play in the initial population. Studies suggest that the constitution of the population at generation zero can have a significant impact on the dynamics of the remainder of a run [14]. By restricting ourselves in this way, our mechanisms for exploring behavioural diversity are more akin to the simple Koza-style approaches based on genotype diversity, and we will therefore use that as a basis for comparison. Sections 2 and 3 describe these structural and behavioural approaches respectively. In Section 4 we explain how our interpretation of behaviour opens the door to further tailoring of the initial population – a process which we term behavioural *filtering*. Finally, in Section 5, we draw some conclusions and offer pointers to further work.

2 Structural Diversity

In 'vanilla' Koza-style GP, the usual approach to creating an initial population is the so-called 'ramped half-and-half' technique. In this, the population is partitioned into a series of tree depths ranging from 2 to some upper value (usually 6). At each of these depths d, half the program trees are created using the so-called 'full' method and half using the 'grow' method. A tree created using the 'full' method has the property that every path from root to terminal node has the same depth d. Creating such a tree involves choosing nodes from the problem function set at each tree level until the maximum depth is reached. In the 'grow' method, on the other hand, nodes at each tree level lower than the maximum are selected randomly from the function and terminal sets. This means that some path lengths may reach the upper value d, but others may be shorter.

The ramped half-and-half approach is claimed to give a good diversity in the structure of program fragments which can then be combined and integrated to produce more complex and hopefully fitter programs. What it does not do, however, is to ensure that each member of the initial population is unique in its structure. In his first book [6], Koza therefore recommends that the initialisation code in a GP system should also strive to ensure such uniqueness.

We can assess the effectiveness of this advice in the context of two navigation-style problems. The first is the Santa Fe artificial ant problem [6], which is commonly used in assessing the effectiveness of GP algorithms and is known to be difficult to solve [15]. The problem consists of evolving an algorithm to guide an artificial ant through a 32x32 matrix in such a way that it discovers as many food pellets as possible. The 89 pellets are laid in a path which frequently changes direction and which contains gaps of varying sizes. The ant can move directly ahead, move left, move right, and determine if food is present in the square immediately ahead of it. These faculties are all encoded in the function and terminal sets of the problem, along with Lisp-like PROGN connectives for joining actions in sequence. To prevent exhaustive or interminable searching, the ant is allowed a maximum of 600 steps in which to complete the task.

The second test problem we have used is that of navigating a maze. Although less well-known than the ant problem, it has been used as the subject for research on introns in several studies [16, 17, 18]. The maze is shown in Fig. 1, with the initial position and orientation of the agent to be guided through it indicated by the arrow. The agent can turn left or right, move forward or backward, and test whether there is a wall ahead or not. A no-op terminal does nothing except to expend an instruction cycle. Decision making is via an if-then-else function, whilst iteration is achieved via a while function. Program fitness is measured in terms of how close the agent gets to the exit: zero fitness indicates escape from the maze. Navigation continues until the maze is successfully completed, or an upper bound of 1000 instruction cycles is reached.

Parameters for both problems as implemented in our GP system are shown in Table 1, from which it will be seen that the ramped half-and-half approach is used for initialisation. Once a population has been created in this way, it is a simple matter to count how many structurally unique individuals it contains. Experimentation reveals that, for the ant problem, our population of 500 programs contains an average of 370 unique members, while for the maze problem it is a little higher, at 390 unique individuals. In other words, between 22% and 26% of the members are duplicated elsewhere in the population, and on the face of it, this seems a good argument for

Fig. 1. Pre-defined maze used in the maze navigation problem

Table 1. GP parameters for the navigation probems

	Santa Fe ant trail	**Maze navigation**
Objective	To evolve a program that guides an ant along a trail of food particles	To evolve a program that guides an agent out of a maze
Terminal set	left, right, move	left, right, forward, back, no-op, wall-ahead, no-wall-ahead
Function set	if-food-ahead, progn2, progn3	if-then-else, while, progn2
Initial population	Ramped half-and-half	
Evolutionary process	Steady-state; 5-candidate tournament selection	
Fitness cases	One: the Santa Fe trail	One: a pre-defined maze
Fitness	No. of food pellets (0-89) not found by the ant	Closest distance to exit (0-18)
Restrictions	Programs timed-out after 600 steps (left, right or move)	Programs timed out after 1000 instructions
Other parameters	Pop size=500; Gens=51; prob. crossover=0.9; no mutation; prob. internal node used as crossover point=0.9	

disallowing duplication during the creation process. However, these figures are misleading: although over 100 programs are not unique, that does not mean they are all copies of the same single individual. In one typical run of the ant problem, for example, there were 378 unique individuals. The other 122 could be divided up into sets of duplicates as shown in Figure 2.

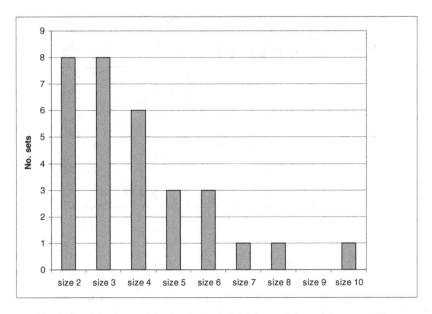

Fig. 2. Sets of structural duplicates in the initial population of the ant problem

From this it can be seen that most of the individuals that are duplicated are copied only a small number of times, e.g. there are 8 sets of identical twins, and another 8 sets of triplets. Only one program appears as many as 10 times in the population for this run, this code being the following: (IF_FOOD_AHEAD MOVE RIGHT).

A similar picture occurs for the maze problem, with the maximum set size always being in single figures. This analysis suggests that the effect of eliminating duplicates during population initialisation may not be as great as desired. To make comparisons, we can look at the number of solutions obtained over 100 runs using both prevention and non-prevention of duplicates, and also the computational effort metric as defined by Koza [6], which computes the minimum number of individuals that need to be processed to give a 99% chance of finding a solution. These figures are presented in Table 2. As can be seen, eliminating duplicates gives a minor improvement for the maze problem, but has a negative effect on performance for the ant problem. This limited experimentation suggests that the case for preventing structural duplication is not proven.

Table 2. Performance figures when allowing and disallowing structural duplication

Approach	Ant problem		Maze problem	
	Success rate (%)	Comp. Effort	Success rate (%)	Comp. Effort
Structural duplicates allowed	13	196,000	14	616,000
Structural duplicates prevented	9	508,500	18	580,000

3 Behavioural Diversity

In using the term behavioural diversity, we differ from other researchers who use it to refer to differences in the fitness of individuals. Rather, we regard behaviour as referring to the computational effects of the algorithm. In the case of our navigation problems, this means the actual observed activity of individual ants or agents. There are various ways in which such activity can be recorded. We have chosen to focus only on the path of an individual as it moves from square to square. In particular, we are not concerned with recording any left or right turns that are executed while remaining on a square, nor with any decision making via execution of statements such as IF_FOOD_AHEAD or WALL_AHEAD.

To record the path histories, we associate with each individual in the population a vector of (row, column) coordinates. Each time the executing program moves to a different square, the new coordinate is added to the end of the vector. Since a program times-out after a fixed number of steps (600 for the ant problem, 1000 for the maze), we know that the vector cannot be longer than that number of elements per individual, and so memory occupancy is not a huge issue.

If we create a population using the standard ramped half-and-half approach, then any structurally identical programs will obviously produce the same path histories when executed. But what about a population in which structural clones are disallowed, as recommended by Koza?

If we compare path histories produced by execution of the members of the initial population, we find that, for the ant problem, an average of only 290 out of the 500 programs behave uniquely, i.e. the behaviour of over 40% of the population is precisely duplicated elsewhere, even though the programs concerned must be structurally dissimilar. A consequence of being a behavioural clone is of course that fitness must

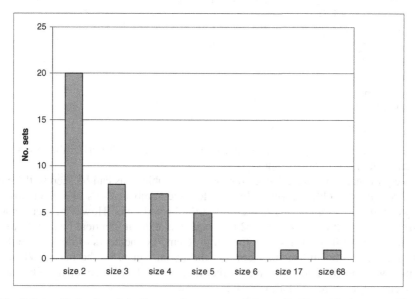

Fig. 3. Sets of behavioural duplicates when structural clones disallowed in the ant problem

also be identical. As before, we can break down the behavioural copies into sets; this is done for a typical run in Figure 3, in which there are 214 individuals which are functionally non-unique.

As with the structural clones, most of the sets are comparatively small. There are, for example, as many as 20 sets of behavioural twins. However, for this run, there is also one set of size 17 and another of size 68. Further investigation reveals that the larger set consists entirely of programs for which the ant does not move off its initial square, and so has a zero length path history. Examples of programs in this set are the following:

```
(PROGN2 RIGHT LEFT)

(IF_FOOD_AHEAD LEFT RIGHT)

(PROGN2 (PROGN3 LEFT RIGHT RIGHT) (IF_FOOD_AHEAD RIGHT LEFT))
```

and even:

```
(PROGN3
    (IF_FOOD_AHEAD
        (IF_FOOD_AHEAD
            LEFT
            (PROGN2 (PROGN2 MOVE MOVE)
                    (IF_FOOD_AHEAD LEFT RIGHT)
            )
        )
        (IF_FOOD_AHEAD
          ( PROGN3
            (IF_FOOD_AHEAD RIGHT LEFT)
            LEFT
            (PROGN2 RIGHT RIGHT)
          )
          RIGHT
        )
    )
    RIGHT
    RIGHT
)
```

These programs are all very different in appearance, yet all perform identically on execution and have minimal fitness.

A very different picture emerges for the maze problem, as can be seen in the chart of one run of the problem given in Fig. 4. Here the number of sets of a given size are always small; for example, there are only 3 duplicated pairs. However, there is also one set of size 402; i.e. there are 402 programs that exhibit identical behaviour to the other members of that set. As with the ant problem, the members of this set are those programs that do not move the agent off the initial square of the maze. When put together with the other sets, this means that there are only 17 individuals with unique behaviour in the entire population.

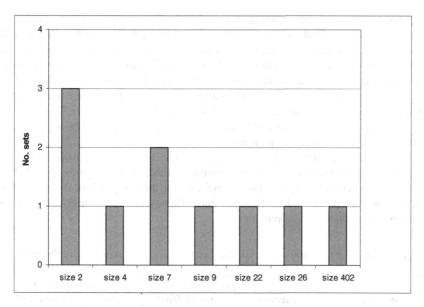

Fig. 4. Sets of behavioural duplicates when structural clones disallowed in the maze problem

A natural question is what happens when we prevent behaviourally identical individuals being created in the initial population, and for this we can define the following pseudo-code algorithm:

```
function duplicated(prog)
    for each member of population created prior to prog
        if path history of member matches path history of prog
            return TRUE
        endif
    endfor
    return FALSE
endfunction

function initialise_population
    for each prog in the population
        attempts = 0
        depth = assigned depth of this program tree
        do
            create prog tree using ramped half-and-half
                                        at current depth
            attempts = attempts + 1
            if (attempts >= MAX_ATTEMPTS and depth <= MAX_DEPTH)
                attempts = 0
                depth = depth + 1
            endif
            test_fitness(prog)
        while (duplicated(prog))
    endfor
endfunction
```

As described earlier, the ramped half-and-half method partitions the population into a sequence of initial depths, usually from 2 up to 6. For each individual, the algorithm above creates a program tree with the assigned depth, then immediately tests its fitness so that a path history can be generated. This history is then compared with the paths of previously created members to check for duplicate behaviour. If the new program is a duplicate, the tree creation code is executed again and another check made. If a pre-defined number of attempts using the assigned tree depth is ever exceeded, that depth is incremented to give the algorithm a greater chance of finding a program that differs from other members.

Applying this algorithm to our initial population and then executing it over 100 runs gives us the performance statistics shown in Table 3. The figures from Table 2 have been repeated here for ease of comparison. For the ant problem there is a slight improvement, but it is not hugely significant. However, for the maze problem the improvement is much more apparent.

Table 3. Performance comparison – structural versus behavioural diversity in initial population

Approach	Ant problem		Maze problem	
	Success rate (%)	Comp. Effort	Success rate (%)	Comp. Effort
Structural duplicates allowed	13	196,000	14	616,000
Structural duplicates prevented	9	508,500	18	580,000
Behavioural duplicates prevented	18	187,000	51	37,500

4   Behavioural Filtering

Once a mechanism has been established for recording the behaviour of individuals, the door is open to making additional decisions about programs which are allowed to become members of the initial population. We see this as a form of filtering or screening procedure for initial candidature, based on phenotypic behaviour.

In the ant problem, any ant which is successful must pick up all 89 pellets, and so it must make at least that number of moves in the matrix. In fact it has to make quite a few more moves because of the need to negotiate the gaps in the trail. However, a close look at the path histories of our initial population reveals that there are a number of individuals that have much shorter path histories. Without even examining their fitness we know for sure that they cannot possibly be solutions to the problem. On average, the number of programs with path lengths less than 80 is 35 out of 500. A simple example of such a program is (IF_FOOD_AHEAD MOVE LEFT) which keeps the ant moving ahead while there is food in front of it, but as soon as the ant is surrounded by empty cells it will just keep performing left turns.

A bigger problem is the category of programs which enter loops. Ants which have any hope of discovering the full trail of food should not keep revisiting the same square. Having path histories available to us means that we can look for such repeated landings on the same cell of the matrix. If we define a looping program to be one which hits the same square more than 3 times, we find (perhaps surprisingly) that an average of 440 out of the 500 individuals (i.e. 88%) are looping programs. Again, a simple example is the program (PROGN3 MOVE LEFT RIGHT) which just keeps moving along the top line of the grid, wrapping round to the first column when it reaches the rightmost edge because of the grid's toroidal form.

Initial populations in the maze problem suffer similar afflictions. For this problem we might expect even reasonably fit programs to hit a given square more than once, and so we will define a looping program to be one which visits a square more than 5 times (rather than the limit of 3 used in the ant problem). In doing so, we find that an average of 190 programs get stuck in loops. If we also define a short path history to be anything less than 18 steps (since the horizontal distance to the exit is 18 squares), then we find that almost every non-looping program is a short-path program. In fact, it is almost impossible to generate an initial population that consists entirely of programs which neither loop nor generate short paths.

Suppose, then, that in addition to disallowing behavioural duplicates, we also impose the following additional filtering. For both problems we will rule out programs that visit the same square more than a predefined maximum of times (3 for the ant problem, 5 for the maze). Additionally, for the ant problem, we will rule out individuals with short path lengths (i.e. less than 80). Table 4 shows the results; again, earlier results are repeated here for ease of comparison. It can be seen that this additional filtering of the initial population has a markedly positive effect on performance for both problems.

Table 4. Effects on performance using additional filtering of initial population

Approach	Ant problem		Maze problem	
	Success rate (%)	Comp. Effort	Success rate (%)	Comp. Effort
Structural duplicates allowed	13	196,000	14	616,000
Structural duplicates prevented	9	508,500	18	580,000
Behavioural duplicates prevented	18	187,000	51	37,500
Behavioural duplicates prevented, plus additional filtering	53	40,000	67	5,500

5 Conclusions

In this paper we have taken a somewhat different view of behaviour from that usually adopted when considering diversity in genetic programming. Rather than equating behaviour with fitness, we have instead chosen to compare the actions performed by individuals during execution. With this interpretation in mind we have shown that, even when structural duplicates are removed in the initial populations of the Santa Fe and maze traversal problems, a significant amount of behavioural duplication usually remains. Elimination of these behavioural clones, together with further filtering to remove programs which loop or remain stuck on a square, has been seen to offer significant performance advantages.

The other chief difference between the work reported here and existing research on diversity is that we focus purely on the initial population. One of the advantages of this is that it reduces computation costs, since it is only at this first stage of a run that we need discard newly-created programs that do not fit our criteria for entering the population. However, one of our plans for further work is to examine the effects of promoting diversity and other forms of filtering throughout the lifetime of a run – for example, when evaluating offspring created during crossover.

It has already been mentioned that the memory costs of the approach are also low, since only moves from square to square are recorded and there is a pre-defined upper limit to the number of such moves. These path records have been used as our basis for phenotypic diversity, but our notion of what makes two individuals different is simplistic: paths need differ by only one square in order for the individuals to be considered non-similar. It would be interesting to apply alternative definitions of diversity that are based on more extensive path divergences.

Finally, a more obvious extension to the research here is to apply it to other problem domains. The navigation problems used here have their own peculiarities that make them an interesting subject for behavioural analysis, but there seems no reason why the approach could not be introduced for other problems.

References

1. McPhee, N.F., Hopper, N.J.: Analysis of Genetic Diversity through Program History. In: Banzhaf, W., et al. (eds.) Proc. Genetic and Evolutionary Computation Conf., Florida, USA, pp. 1112–1120 (1999)
2. Daida, J.M., Ward, D.J., Hilss, A.M., Long, S.L., Hodges, M.R., Kriesel, J.T.: Visualizing the Loss of Diversity in Genetic Programming. In: Proc. IEEE Congress on Evolutionary Computation, Portland, Oregon, USA, pp. 1225–1232 (2004)
3. Hien, N.T., Hoai, N.X.: A Brief Overview of Population Diversity Measures in Genetic Programming. In: Pham, T.L., et al. (eds.) Proc. 3rd Asian-Pacific Workshop on Genetic Programming, Hanoi, Vietnam, pp. 128–139 (2006)
4. Burke, E., Gustafson, S., Kendall, G., Krasnogor, N.: Advanced Population Diversity Measures in Genetic Programming. In: Guervós, J.J.M., Adamidis, P.A., Beyer, H.-G., Fernández-Villacañas, J.-L., Schwefel, H.-P. (eds.) PPSN 2002. LNCS, vol. 2439, pp. 341–350. Springer, Heidelberg (2002)
5. Burke, E., Gustafson, S., Kendall, G.: Diversity in Genetic Programming: An Analysis of Measures and Correlation with Fitness. IEEE Transactions on Evolutionary Computation 8(1), 47–62 (2004)
6. Koza, J.R.: Genetic Programming: On the Programming of Computers by Means of Natural Selection. MIT Press, Cambridge (1992)
7. de Jong, E.D., Watson, R.A., Pollack, J.B.: Reducing Bloat and Promoting Diversity using Multi-Objective Methods. In: Spector, L., et al. (eds.) Proc. Genetic Evolutionary Computation Conf., San Francisco, CA, USA, pp. 11–18 (2001)

8. Wyns, B., de Bruyne, P., Boullart, L.: Characterizing Diversity in Genetic Programming. In: Collet, P., Tomassini, M., Ebner, M., Gustafson, S., Ekárt, A. (eds.) EuroGP 2006. LNCS, vol. 3905, pp. 250–259. Springer, Heidelberg (2006)

9. Rosca, J.P.: Genetic Programming Exploratory Power and the Discovery of Functions. In: McDonnell, J.R., et al. (eds.) Proc. 4th Conf. Evolutionary Programming, San Diego, CA, USA, pp. 719–736 (1995)

10. Rosca, J.P.: Entropy-Driven Adaptive Representation. In: Rosca, J.P. (ed.) Proc. Workshop on Genetic Programming: From Theory to Real-World Applications, Tahoe City, CA, USA, pp. 23–32 (1995)

11. D'haeseleer, P., Bluming, J.: Effects of Locality in Individual and Population Evolution. In: Kinnear, K.E., et al. (eds.) Advances in Genetic Programming, ch. 8, pp. 177–198. MIT Press, Cambridge (1994)

12. Ryan, C.: Pygmies and Civil Servants. In: Kinnear, K.E., et al. (eds.) Advances in Genetic Programming, ch. 11, pp. 243–263. MIT Press, Cambridge (1994)

13. Looks, M.: On the Behavioural Diversity of Random Programs. In: Thierens, D., et al. (eds.) Proc. Genetic and Evolutionary Computing Conf. (GECCO 2007), London, England, UK, pp. 1636–1642 (2007)

14. Daida, J.M.: Towards Identifying Populations that Increase the Likelihood of Success in Genetic Programming. In: Beyer, H.-G., et al. (eds.) Proc. Genetic and Evolutionary Computing Conf. (GECCO 2005), Washington DC, USA, pp. 1627–1634 (2005)

15. Langdon, W.B., Poli, R.: Why Ants are Hard. In: Koza, J.R., et al. (eds.) Genetic Programming 1998: Proceedings of the Third Annual Conference, pp. 193–201. Morgan Kaufman, San Francisco (1998)

16. Soule, T.: Code Growth in Genetic Programming. Ph.D Thesis, University of Idaho (1998)

17. Langdon, W.B., Soule, T., Poli, R., Foster, J.A.: The Evolution of Size and Shape. In: Spector, L., et al. (eds.) Advances in Genetic Programming, vol. 3, pp. 163–190. MIT Press, Cambridge (1999)

18. Jackson, D.: Dormant Program Nodes and the Efficiency of Genetic Programming. In: Beyer, H.-G., et al. (eds.) Proc. Genetic and Evolutionary Computing Conf. (GECCO 2005), Washington DC, USA, pp. 1745–1751 (2005)

A Real-Time Evolutionary
Object Recognition System

Marc Ebner

Eberhard-Karls-Universität Tübingen
Wilhelm-Schickard-Institut für Informatik
Abt. Rechnerarchitektur, Sand 1, 72076 Tübingen
marc.ebner@wsii.uni-tuebingen.de
http://www.ra.cs.uni-tuebingen.de/mitarb/ebner/welcome.html

Abstract. We have created a real-time evolutionary object recognition system. Genetic Programming is used to automatically search the space of possible computer vision programs guided through user interaction. The user selects the object to be extracted with the mouse pointer and follows it over multiple frames of a video sequence. Several different alternative algorithms are evaluated in the background for each input image. Real-time performance is achieved through the use of the GPU for image processing operations.

1 Motivation

Current vision systems are usually not adaptive to their environment. Algorithms which have been developed in the laboratory often break when the vision system is moved to a different environment. The only component of a vision system which is currently adaptive is the automatic white balance of the camera. Recently, Ebner [1] proposed building an adaptive on-line evolutionary visual system. Such a system would adapt itself to the current environment. Evolutionary algorithms would be used to search for an algorithm which would always perform optimally for the given environmental conditions.

Suppose we are given some algorithm which extracts or detects a person in an image. The algorithm would perform flawlessly provided that there is enough light available to illuminate the person in the image. During sunset, the algorithm may have to be modified to extract or detect the same person. At night when there is likely going to be a lot of noise in the image data, additional modifications have to be made to the algorithm to perform the same task it did during daylight.

Ebner [1] suggested to run multiple, slightly modified algorithms in the background in addition to the main algorithm. The main algorithm would take the input image, compute an output and present this output to the user. The system would also evaluate the modified algorithms using the same input image. The algorithm with the best performance would become the main algorithm for the next input image. Clearly, such a system would require enormous computational powers.

A computer vision algorithm usually applies several image processing operators to an input image. Such algorithms often struggle to maintain real-time

L. Vanneschi et al. (Eds.): EuroGP 2009, LNCS 5481, pp. 268–279, 2009.

performance. How will it then be possible to run multiple algorithms in parallel? The rise of powerful graphics processing units provides a solution to this problem.

Below, we will show how we created an experimental real-time evolutionary vision system by exploiting the power of the graphics processing unit. The paper is structured as follows. We first provide a brief review of related research in the field of evolutionary computer vision in Section 2. In Section 3, we describe how our real-time evolutionary vision system works. Details on how this system is mapped to the graphics hardware is presented in Section 4. The system is evaluated on several image sequences in Section 5. Conclusions are provided in Section 6.

2   Evolutionary Computer Vision

In evolutionary computer vision, evolutionary algorithms are often used to search for a solution which is not immediately apparent to those skilled in the art or to improve upon an existing evolution. Early work on evolutionary computer vision started in the early 1990s. Lohmann, a pioneer in the field, advocated the idea and showed how an Evolution Strategy may be used to find an algorithm which computes the Euler number of an image [2]. In relatively early work, evolutionary algorithms were mostly used to evolve low-level operators, e.g. edge detectors [3], feature detectors [4], or interest point detectors [5]. They were also used for target recognition [6].

However, early on, it was clear that in principle these techniques could be used to create fully adaptive operators which would be optimal or near optimal for the task at hand [7]. It was also clear that Genetic Programming would be particularly useful for image analysis [8]. Johnson et al. were successful at using genetic programming to evolve visual routines [9]. Today, evolutionary computer vision has become a very active research area. Current work ranges from the evolution of low-level detectors [10], to object recognition [11,12] or camera calibration [13]. A taxonomic tutorial and introduction into the field is given by Cagnoni [14].

Due to the enormous computational requirements, experiments in evolutionary computer vision are usually performed off-line (see Mussi and Cagnoni [15] for a notable exception). Once an appropriate algorithm has been evolved, it may of course be used in real time. Genetic algorithms [16] or Evolution Strategies [17] are mostly used to improve already existing algorithms, i.e. they are used for parameter optimization while Genetic Programming [18,19] is used to evolve an algorithm from scratch. Since we are interested in the most general scheme for the evolution of algorithms, we will be using Genetic Programming to search the space of possible solutions.

3   A Real-Time Evolutionary Object Recognition System

In searching for an optimal computer vision algorithm, one has to answer the question in what order and with which parameters should the known operators from the literature be applied to the input image to achieve the desired output.

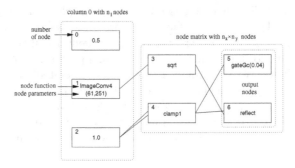

Fig. 1. Cartesian Genetic Programming representation of an individual

Fig. 2. (a) one byte is used to specify an image offset. The offset is determined by moving a vector clockwise and radially outward as the byte value increases. (b) image at scale 0 (c) image at scale 4.

The instructions of the evolved computer program correspond to the operators from the computer vision literature. Possible representations which could be used to evolve image processing algorithms include tree-based GP, linear GP or Cartesian GP. For our experiments, we will be using Cartesian Genetic Programming [20] because this GP variant is readily mapped to the graphics hardware as we will describe below.

Similar to the Cartesian GP paradigm, we will be working with a $(n_x \times n_y)$ matrix of image processing operators as shown in Figure 1. In addition to this matrix, we will be using a vector of n_1 input images (column 0) which will serve as input to the operators stored in the matrix. The output computed by one operator located at position (x, y) in the matrix can be computed by combining the output computed by any of the cells located at column $x - 1$. The processed image data is finally available in column n_x. We denote this representation as a $n_1 + n_x \times n_y$ representation. Many algorithms known from the literature can be fit into this scheme.

Table 3 shows the list of operators which are available for column 0. The first three operators, 0.0, 0.5 and 1.0 simply return a gray image where all colors are set to the color vec3(0.0,0.0,0.0), vec3(0.5,0.5,0.5), vec3(1.0,1.0,1.0) respectively. The operator Image outputs the input image. Argument 1 is used to offset the image by mapping the one byte argument to an offset vector as shown in

Figure 2(a). A byte value of 0 is mapped to no offset. The vector moves clockwise and radially outward as the byte value increases. The maximum distance from the center is 10 pixels. The second byte argument is used to set the appropriate scale (see Figure 2(b) and (c)). The range $[0, 255]$ is mapped to the scale $[0, 4]$. A scale of 0 is just the original input image. A scale of 1 denotes a down-sampled version of the original image where a 2×2 area of pixels is averaged for every image pixel.

The operators DX and DY compute the derivative in the x- and y-direction respectively. The operator Lap computes the Laplacian and the operator Grad computes the gradient magnitude. For all these operators, the two byte arguments again denote offset and scale of the input image which is used for the computation. The operator ImageLogDX outputs the logarithm of the derivative in the x-direction. This is a so called color constant descriptor [21] which only depends on the reflectance on the patch but not on the illuminant. Such descriptors are particularly useful for object detection when the lighting conditions change.

The operator ImageIdw, can be used to compute a gray scale image from the input image. The output of this operator $o_i = w_r c_r + w_g c_g + w_b c_b$ with $i \in \{r, g, b\}$ is computed using RGB weights $w_i \in \{-1, 0, 1, 2\}$ The RGB weights are stored in the second byte argument. The first byte argument is used to store the offset and the scale of the image as before. Similarly, the operator ImageChrom can be used to compute chromaticities. For this operator, the output is computed using $o_i = w_i c_i / (c_r + c_g + c_b)$.

The operators ImageGC1, ImageGC4, ImageGC16 provide the input image convolved with a filter matrix which is stored in the two byte arguments. A 3×3 filter matrix is used with two bits per weight $w_i \in \{-1, 0, 1, 2\}$ with $i \in \{1, ..., 8\}$. The center element is not included. Let N be the neighborhood of the current pixel, then the output is computed as

$$o_i = \sum_{(x,y) \in N} w_i c_i. \tag{1}$$

The operator ImageGC1 places the neighboring pixels directly around the current pixel. The operator ImageGC4 increases the distance of the neighboring pixels from the current pixel radially outward by a distance of 4. The operator ImageGC16 increases this distance to 16 pixels. An additional convolution operator ImageGCd is provided which uses a variable sized distance of the pixels from the current pixel. The first byte argument is used to specify the weights $w_i \in \{-0.5, 0.5\}$ with $i \in \{1, ..., 8\}$. The second byte argument is used to specify the distance of the pixels within the range $[0, 64]$.

The operator ImageSeg is used to segment or discretize the image into a discrete number of regions. The first byte argument specifies the number of allowed pixel values and the second byte argument specifies the scale of the image which is used for the computation.

A $n_x \times n_y$ matrix of operators is used to combine the image data which has been made available in column 0. The list of operators which can be used in this $n_x \times n_y$ matrix is shown in Table 4. Many of these operators are taken directly

Function of Operator	Name	Args	Arg 1	Arg 2
zero	0.0	0	unused	unused
half	0.5	0	unused	unused
one	1.0	0	unused	unused
identity	Image	2	offset	scale
horizontal derivative	DX	2	offset	scale
vertical derivative	DY	2	offset	scale
Laplacian	Lap	2	offset	scale
gradient	Grad	2	offset	scale
color constant desc.	ImageLogDX	2	offset	scale
gray scale image	ImageIdw	2	offset/scale	RGB weights
chromaticities	ImageChrom	2	offset/scale	RGB weights
convolution (dist. 1)	ImageGC1	2	conv. weights	conv. weights
convolution (dist. 4)	ImageGC4	2	conv. weights	conv. weights
convolution (dist. 16)	ImageGC16	2	conv. weights	conv. weights
convolution	ImageGCd	2	conv. weights	distance
segmentation	ImageSeg	2	levels	scale

Fig. 3. Operators available for column 0

from the specification of the OpenGL Shading Language (OpenGLSL) [22]. The data is fed through the matrix from left to right. First, column 1 is evaluated, then column 2 and so on. Each node of the matrix either takes one or two arguments. The first byte argument specifies from which node in the previous column the input value v1 is read out. A modulo operation is used to map the byte argument to the range $[0, n_x]$. Similarly, the second byte argument specifies the input node used for v2. Some operations only take a single argument. In this case, the second byte argument is ignored. Some operators use the second byte argument as a constant c2. This constant is computed using c2=b/255.

The functions performed by the operators are shown in the last column of Table 4 using the syntax of the OpenGL shading language. Functions include standard operations such as addition, subtraction, multiplication, division as well as specialized functions for image processing which extract one of the channels or uses a color channel as a gate function. Other functions, which were included in the function set, are functions which are mostly used for computer graphics applications. Those functions were included simply because they are readily available from the OpenGLSL function set.

The overall output of the detector can be computed once the last column has been evaluated. To compute the overall output, we average the output of all operators in column n_x. In other words, we output the average behavior of the detectors in the last column. Because of this, modifications to a single element towards the right hand side of the matrix have a relatively small effect on the overall output.

Obviously, it would also be possible to use a different method to compute the overall output. For instance, we could also multiply the output of the detectors located in the last column. In this case, all detectors would have to have a high

function of operator	name	args	computed function
pass through	id	1	v1
absolute value	abs	1	abs(v1)
scalar product	dot	1	vec3(dot(v1,v1))
square root	sqrt	1	sqrt(v1)
normalize	norm	1	normalize(v1)
clamp	clamp(0,1)	1	clamp(v1,0.0,1.0)
step function	step(0)	1	step(vec3(0),v1)
step function	step(0.5)	1	step(vec3(0.5),v1)
smooth step function	smstep(0,1)	1	smoothstep(vec3(0),vec3(1),v1)
red channel	red	1	vec3(v1.r,0,0)
green channel	green	1	vec3(0,v1.g,0)
blue channel	blue	1	vec3(0,0,v1.b)
channel average	avg	1	vec3((v1.r+v1.g+v1.b)/3)
channel minimum	min	1	vec3(min(min(v1.r,v1.g),v1.b))
channel maximum	max	1	vec3(max(max(v1.r,v1.g),v1.b))
mark minimum comp.	equalMin	1	equal(v1,vec3(min(min(v1.r,v1.g),v1.b)))
mark maximum comp.	equalMax	1	equal(v1,vec3(max(max(v1.r,v1.g),v1.b)))
use red channel as gate	gateR	1	?(v1.r>0) vec3(v1.g):vec3(a.b)
use green channel as gate	gateG	1	?(v1.g>0) vec3(v1.r):vec3(a.b)
use blue channel as gate	gateB	1	?(v1.b>0) vec3(v1.r):vec3(a.g)
use red channel as gate	gateRc	2	?(v1.r>c2) vec3(v1.g):vec3(v1.b)
use green channel as gate	gateGc	2	?(v1.g>c2) vec3(v1.r):vec3(v1.b)
use blue channel as gate	gateBc	2	?(v1.b>c2) vec3(v1.r):vec3(v1.g)
step function	step	2	step(vec3(c2),v1)
addition	+	2	v1+v2
subtraction	-	2	v1-v2
multiplication	*	2	v1*v2
division	/	2	v1/v2
minimum	min	2	min(v1,v2)
maximum	max	2	max(v1,v2)
clamp	clamp0	2	clamp(v1,v2,vec3(1))
clamp	clamp1	2	clamp(v1,vec3(0),v2)
mix	mix	2	mix(v1,v2,0.5)
step	step	2	step(v1,v2)
less than	lessThan	2	vec3(lessThan(v1,v2))
greater than	greaterThan	2	vec3(greaterThan(v1,v2))
dot	dot	2	vec3(dot(v1,v2))
cross	cross	2	cross(v1,v2)
reflect	reflect	2	reflect(v1,v2)
refract	refract	2	refract(v1,v2,0.1)

Fig. 4. Operators available for the nodes of the $n_x \times n_y$ matrix

output for the overall detector to respond at all. However, such an object detector would most likely be highly fragile. All detector outputs have to be non-zero for the overall detector to output anything at all.

Experiments in the field of evolutionary computer vision are usually very expensive with respect to the computational requirements. Each individual of

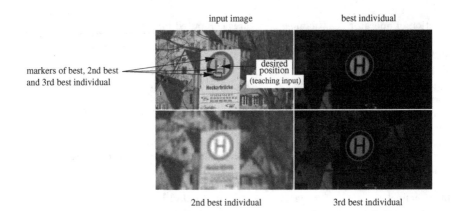

Fig. 5. System overview. The user manually specifies the position of the object which should be extracted in the upper left hand sub-window. The output of the three best individuals is shown using markers (overlayed on the input image). The images which are computed by the three best individuals are also shown.

the population represents a possible algorithmic solution for a given problem. Each solution has to be evaluated on the input image or sets of input images. We will be working with an input stream of images which is continually being processed by the system. Due to clever use of the graphics processing unit, our system is able to evaluate multiple individuals for every input image.

Our task will be to locate certain objects in the input stream. The user is able to tell the system which object should be extracted by moving the mouse pointer over the object and then pressing the left mouse button (see Figure 5). The position of the mouse pointer is then used for computing the fitness of the individuals as long as the button is pressed. The detected object position as well as the processed image of the three best individuals of the population is always output onto the screen. Once the user releases the mouse button, the images are still being processed, however, evolution is halted and the output of the three best individuals is continued to be shown.

Figure 1 shows the entire setup for a sample individual. It is straightforward to map this representation to a linear bit string genome by concatenating all the parameters. The parameters of column 0 are stored first, followed by the parameters for the $n_x \times n_y$ matrix read out from top to bottom and from left to right.

4   GPU Accelerated Image Processing

We have used the graphics processing unit (GPU) to accelerate the image processing operators. Fung et al. [23] noticed very early on that image processing tasks can be mapped to the graphics hardware. Today, graphics hardware is used to accelerate a variety of tasks from the simulation of reaction-diffusion equations to fluid dynamics or image segmentation [24]. Nvidia has developed the Compute Unified

Device Architecture (CUDA) [25]. This architecture allows the programmer to use the GPU as a massively parallel computing device. The CUDA architecture is highly suited for accelerating image processing operations. However, we have chosen to use the OpenGL shading language for our implementation. OpenGLSL has the advantage of providing easy access to scale spaces through mip mapping. Currently, it is not known whether a CUDA implementation would provide a significant speedup over the approach followed here.

Our choice of operators and representation which were described in the previous section were largely fixed by the syntax of the OpenGL shading language. OpenGLSL is usually used to render realistic computer graphics in real-time. It is highly optimized to render triangles and planar polygons. For added realism, textures can be applied to these triangles and polygons. So called pixel shaders can be used to individually program the operation which is performed by the GPU whenever the rasterizer renders a single pixel.

The pixel shaders can be programmed using a C-like language, the OpenGL shading language. Each pixel shader has access to multiple textures. We implemented the above representation by rendering a single rectangle which has the same size as the input image and which consists of four vertices. The rasterizer executes the code of the pixel shader for every pixel of the rectangle. The current input image is provided to the pixel shader as a texture. The operations listed in Table 3 are implemented by reading out the texture. For instance, the following OpenGLSL code reads out the input image with an offset of $(3,4)$ and at a scale of 2.

```
vec3 c=texture2D(texture0,gl_TexCoord[0].st+vec2(3,4),2).rgb;
```

The three component vector c then holds the down-sampled data from the input image.

All high level operations such as edge detection, computation of the Laplacian, convolution or segmentation are implemented through the OpenGLSL. The result of these operations can then be combined through binary operations such as multiplication, computation of maximum values or through gate functions. Obviously, in computer vision one also wants to apply multiple image processing operators in sequence. For instance, one would like to apply a convolution and then an edge detector to the convolved image. However, it is currently not possible to implement such operations using the OpenGLSL with a single pass through the graphics pipeline. One would have to use multiple passes through the graphics pipeline and the image data would have to be exchanged between the GPU and the CPU for each pass. This transfer between GPU and CPU would be a severe bottleneck. In the future, the OpenGLSL may allow write operations to textures. If this were possible, we would be able to compute multiple image processing operations with a single pass through the pipeline. Unfortunately, at present, this is not a possibility.

5   Experiments

Our system works with a parent population of μ individuals. We basically use a $(\mu + \lambda)$ Evolution Strategy to evolve our pixel shader programs. All parents are re-evaluated for each input image because the input image is continuously changing. New parents are selected among the parents and the offspring by sorting them in ascending order according to their fitness and then selecting the best μ parents. Since the representation that we use is highly redundant, the μ best parents will all be identical after several iterations. That's why we sort parents and offspring according to fitness and then select the μ best parents which all have different fitness values. This provides a simple form of diversity maintenance.

In addition, half of the offspring for every iteration are generated from scratch using the random number generator. This also provides a steady influx of alternative algorithms and prevents the system from converging to a single point inside the search space. The other half of the offspring are generated by recombining two parent individuals with a crossover probability of $p_{\text{cross}} = 0.7$.

All offspring are mutated by selecting one of two mutation operations at random. The first mutation operation is a standard GA-like mutation operation with a mutation probability of $p_{\text{mut}} = \frac{2}{l}$ where l is the size of the genome in bits. The second mutation operation selects one byte of the individual at random and increases or decreases that byte at random. This allows for small incremental or decremental changes to arguments which specify offsets or scale. Alternatively, a gray code could also have been used to achieve the same effect.

The task of the individuals is to detect objects in an image sequence which are specified interactively by a human operator by moving the mouse over the object and holding the mouse button pressed as long as the evolved object detector is not good enough. Our training set consists of those input images where the mouse button is pressed and our testing test consists of all other images. No image is used twice during an evolutionary run.

The evolved object detector outputs an entire image. The detected object position is determined by locating the pixel with the highest value. Each pixel with RGB components $[r, g, b]$ is interpreted as a 24-bit number. If several pixels have the value $FFFFFF$, we compute the center of gravity of these pixel positions. Fitness is computed based on the difference (measured in pixels) between the detected object position and the object position which is manually determined by a human operator.

We tested our system on several different image sequences. Each detector receives as input only a single image. Naturally, the shape of an object can vary from one image to the next. Its colors will also vary slightly from one image to the next (this is also the case for a stationary camera and a stationary object due to noise in the sensor). A successful detector must therefore become robust against small distortions or noise in the data.

We were able to evolve detectors which detect ducks in a pond, interest points on a building or traffic signs. The object was usually located with a reasonably small error after relatively short time. Manually writing detectors which perform the same task would have taken considerably more time than was required to evolve the detectors.

input image   output created by best individual

Fig. 6. Output of three evolved object detectors. The located output of the three best evolved individuals are marked in red, green, and blue respectively. In all three cases, we were able to evolve an individual which is able to detect the object or part of the image which should be located.

In some cases, the object color provided a strong cue on where the object is located. However, we were also able to show that it is possible to evolve detectors which were not based on color. Figure 6 shows three different image sequences which were used to test our system. The task for the first sequence was to detect ducks in a pond. The task for the second image sequence was to detect the sign of a bus stop. The task for the third sequence was to detect an interest point located on a building. In many cases, we were surprised on the robustness of the detectors. Quite often, the detectors were able to tolerate medium changes in appearance and/or the scale of the object.

Our system is able to achieve a frame rate of 4.5 Hz while evaluating 23 alternative individuals in the background and also visualizing the results of the three best individuals. This frame rate is achieved with a $2 + 2 \times 2$ representation on an Intel Core 2 CPU running at 2.13GHz and a GeForce 9600GT/PCI/SEE2. In other words, more than one hundred individuals are evaluated per second. The image sequences had a size of 320x240.

6 Conclusions

We have shown how a real-time evolutionary system can be built based on the OpenGL shading language which evaluates multiple alternative algorithms in the background for every input image. Our system is based on a Cartesian Genetic Programming representation. High-level image processing operations are used as input nodes. The output of these operations is then recombined using elementary functions which are available from the OpenGL shading language. At present, it is not possible to apply multiple image processing operations such as edge detection or a convolution in sequence without transferring the output of one operation back to the CPU. Our representation is streamlined to match the architecture of the GPU hardware and thereby fully exploiting the power of the GPU. With future GPU hardware, it could be possible that write operations to texture are also allowed. This would increase the power of the present approach considerably.

References

1. Ebner, M.: An adaptive on-line evolutionary visual system. In: Hart, E., Paechter, B., Willies, J. (eds.) Workshop on Pervasive Adaptation, Venice, Italy. IEEE, Los Alamitos (2008) (in press)
2. Lohmann, R.: Bionische Verfahren zur Entwicklung visueller Systeme. Ph.D thesis, Technische Universität Berlin, Verfahrenstechnik und Energietechnik (1991)
3. Harris, C., Buxton, B.: Evolving edge detectors with genetic programming. In: Koza, J.R., Goldberg, D.E., Fogel, D.B., Riolo, R.L. (eds.) Genetic Programming 1996. Proc. of the 1st Annual Conf., pp. 309–314. The MIT Press, Cambridge (1996)
4. Rizki, M.M., Tamburino, L.A., Zmuda, M.A.: Evolving multi-resolution feature-detectors. In: Fogel, D.B., Atmar, W. (eds.) Proc. of the 2nd American Conf. on Evolutionary Programming, Evolutionary Programming Society, pp. 108–118 (1993)
5. Ebner, M.: On the evolution of interest operators using genetic programming. In: Poli, R., Langdon, W.B., Schoenauer, M., Fogarty, T., Banzhaf, W. (eds.) EuroGP 1998. LNCS, vol. 1391, pp. 6–10. Springer, Heidelberg (1998)
6. Katz, A.J., Thrift, P.R.: Generating image filters for target recognition by genetic learning. IEEE Transactions on Pattern Analysis and Machine Intelligence 16(9), 906–910 (1994)
7. Ebner, M., Zell, A.: Evolving a task specific image operator. In: Poli, R., Voigt, H.-M., Cagnoni, S., Corne, D.W., Smith, G.D., Fogarty, T.C. (eds.) EvoIASP 1999 and EuroEcTel 1999. LNCS, vol. 1596, pp. 74–89. Springer, Heidelberg (1999)
8. Poli, R.: Genetic programming for image analysis. In: Koza, J.R., Goldberg, D.E., Fogel, D.B., Riolo, R.L. (eds.) Genetic Programming 1996, Proc. of the 1st Annual Conf., pp. 363–368. The MIT Press, Cambridge (1996)
9. Johnson, M.P., Maes, P., Darrell, T.: Evolving visual routines. In: Brooks, R.A., Maes, P. (eds.) Artificial Life IV, Proc. of the 4th Int. Workshop on the Synthesis and Simulation of Living Systems, pp. 198–209. The MIT Press, Cambridge (1994)
10. Trujillo, L., Olague, G.: Synthesis of interest point detectors through genetic programming. In: Proc. of the Genetic and Evolutionary Computation Conf., Seattle, WA, pp. 887–894. ACM, New York (2006)

11. Krawiec, K., Bhanu, B.: Visual learning by evolutionary and coevolutionary feature synthesis. IEEE Transactions on Evolutionary Computation 11(5), 635–650 (2007)
12. Treptow, A., Zell, A.: Combining AdaBoost learning and evolutionary search to select features for real-time object detection. In: Proc. of the IEEE Congress on Evolutionary Computation, Portland, OR, vol. 2, pp. 2107–2113. IEEE, Los Alamitos (2004)
13. Heinemann, P., Streichert, F., Sehnke, F., Zell, A.: Automatic calibration of camera to world mapping in robocup using evolutionary algorithms. In: Proc. of the IEEE International Congress on Evolutionary Computation, San Francisco, CA, pp. 1316–1323. IEEE, Los Alamitos (2006)
14. Cagnoni, S.: Evolutionary computer vision: a taxonomic tutorial. In: 8th Int. Conf. on Hybrid Intelligent Systems, pp. 1–6. IEEE Comp. Society, Los Alamitos (2008)
15. Mussi, L., Cagnoni, S.: Artificial creatures for object tracking and segmentation. In: Giacobini, M., Brabazon, A., Cagnoni, S., Di Caro, G.A., Drechsler, R., Ekárt, A., Esparcia-Alcázar, A.I., Farooq, M., Fink, A., McCormack, J., O'Neill, M., Romero, J., Rothlauf, F., Squillero, G., Uyar, A.Ş., Yang, S. (eds.) EvoWorkshops 2008. LNCS, vol. 4974, pp. 255–264. Springer, Heidelberg (2008)
16. Mitchell, M.: An Introduction to Genetic Algorithms. The MIT Press, Cambridge (1996)
17. Rechenberg, I.: Evolutionsstrategie 1994. frommann-holzboog, Stuttgart (1994)
18. Koza, J.R.: Genetic Programming. In: On the Programming of Computers by Means of Natural Selection. The MIT Press, Cambridge (1992)
19. Banzhaf, W., Nordin, P., Keller, R.E., Francone, F.D.: Genetic Programming - An Introduction: On The Automatic Evolution of Computer Programs and Its Applications. Morgan Kaufmann Publishers, San Francisco (1998)
20. Miller, J.F.: An empirical study of the efficiency of learning boolean functions using a Cartesian Genetic Programming approach. In: Banzhaf, W., Daida, J., Eiben, A.E., Garzon, M.H., Honavar, V., Jakiela, M., Smith, R.E. (eds.) Proc. of the Genetic and Evolutionary Computation Conf., pp. 1135–1142. Morgan Kaufmann, San Francisco (1999)
21. Ebner, M.: Color Constancy. John Wiley & Sons, England (2007)
22. Rost, R.J.: OpenGL Shading Language, 2nd edn. Addison-Wesley, Upper Saddle River (2006)
23. Fung, J., Tang, F., Mann, S.: Mediated reality using computer graphics hardware for computer vision. In: Proc. of the 6th Int. Symposium on Wearable Computers, pp. 83–89. ACM, New York (2002)
24. Owens, J.D., Luebke, D., Govindaraju, N., Harris, M., Krüger, J., Lefohn, A.E., Purcell, T.J.: A survey of general-purpose computation on graphics hardware. In: Eurographics 2005, State of the Art Reports, pp. 21–51 (2005)
25. NVIDIA: Compute Unified Device Architecture. Programming Guide V.1.1 (2007)

On the Effectiveness of Evolution Compared to Time-Consuming Full Search of Optimal 6-State Automata

Marcus Komann[1], Patrick Ediger[2], Dietmar Fey[1], and Rolf Hoffmann[2]

[1] Friedrich-Schiller-University Jena, Germany
Chair for Computer Architecture
Ernst-Abbe-Platz 2, D-07743 Jena, Germany
marcus.komann@googlemail.com, fey@uni-jena.de
[2] Technische Universität Darmstadt, Germany
FB Informatik, FG Rechnerarchitektur
Hochschulstr. 10, D-64289 Darmstadt, Germany
{ediger,hoffmann}@ra.informatik.tu-darmstadt.de

Abstract. The Creature's Exploration Problem is defined for an independent agent on regular grids. This agent shall visit all non-blocked cells in the grid autonomously in shortest time. Such a creature is defined by a specific finite state machine. Literature shows that the optimal 6-state automaton has already been found by simulating all possible automata. This paper tries to answer the question if it is possible to find good or optimal automata by using evolution instead of time-consuming full simulation. We show that it is possible to achieve 80% to 90% of the quality of the best automata with evolution in much shorter time.

1   Introduction

Evolution has been in the interest of computer scientists for a long time now. In the first place, it can be used to simulate nature, social behaviour, or even economic processes. But it can not only be used for simulation. It has also proven its strengths in solving different problems from the computer science world. Especially if analytic descriptions of such problems are difficult, all kinds of genetic approaches have helped in the search for optimal or at least near-optimal solutions. But using evolution often comprises a drawback for the using scientist. He often doesn't know about the overall, absolute quality of the results the evolution creates. These results might be good, but it is uncertain if there are better results and how much better those would be. This statement is especially true for analytically incomprehensible problems and/or for large problem spaces that cannot be traversed completely.

One such difficult problem is the *Creature's Exploration Problem* (CEP) where an agent shall visit all non-blocked cells in one or several regular grid structures. A practical application of this problem are for example cleaning robots moving around in several rooms of a university. Another application is the steering of so-called Marching Pixels agents which cooperatively and emergently traverse images in order to retrieve data about objects in the images [1]. The CEP can

L. Vanneschi et al. (Eds.): EuroGP 2009, LNCS 5481, pp. 280–291, 2009.

be compared to the Lawnmower Problem, too [2]. If a CEP agent is provided with a large number of states (which is required for sophisticated grids), the problem space of possible state machines/automata becomes enormous. Halbach and Hoffmann [3] undertook quite some efforts to find optimal automata for the CEP with six and less states by simulating all possible automata. Details are presented in Sect. 2.1.

In this paper, we want to investigate how well evolution behaves concerning the solution of such sophisticated problems by the example of the CEP. By using the research about optimal automata from Halbach [4][3], we are in the position to know absolute best automata for the vast, nonlinear problem space of CEP. We compare them to the best automata we evolved in order to find out more about the quality that can be achieved by evolutionary means.

Quite some work has been done using evolution for multi-objective optimisation (EMO). [5] gives an overview about different aspects of evolution for solving such problems but concludes that each evolution has to be customised for the specific problem it is applied to. One example from this field is the successful application of Cartesian Genetic Programming to real-valued, multimodal, continuous functions by Walker and Miller [6]. Another recent example from EMO are Nadhye's interplanetary trajectory optimisation [7].

The question for effectiveness of computer-simulated evolution is as old as the strategy itself. Surprisingly, literature that compares *calculated* global optima to *evolved* optima for highly nonlinear problem spaces is more difficult to find. [8] used a specific mutation operator on the Travelling Salesperson Problem with 16 cities. He found out that the evolution had to look at 0,000067% of all possible tours during evolution on average in order to find the global optimal tour. Bruce [9] compared Tree-Based and Stack-Based Genetic Programming with automatically defined functions (ADF) for the formerly mentioned Lawnmower Problem which is very similar to the CEP. On a 8 × 12 grid, Tree-Based GP without ADF needed to evaluate 4,539,000 different programs in order to deliver a solution to the problem with 99% probability while Stack-Based GP did not deliver that in 20,000,000 evaluations. With ADFs, they required 16,000 and 130,000 evaluations respectively.

The paper is structured as follows. After the introduction, we define the CEP along with the specification of automata in Sect. 2 before optimal automata for the problem are listed. Sect. 3 describes the details of the evolution of the problem like representation of the individuals and fitness function. The results are shown and discussed in Sect. 4 before we summarise and give an outlook in the last section.

2 The Creature's Exploration Problem and Its Optimal Solutions

2.1 The Agent's Task and Its Behaviour Described by an Automaton

The *Creature's Exploration Problem* (CEP) is defined in [10]: Given is a two-dimensional cellular field (regular grid) consisting of blocked cells and

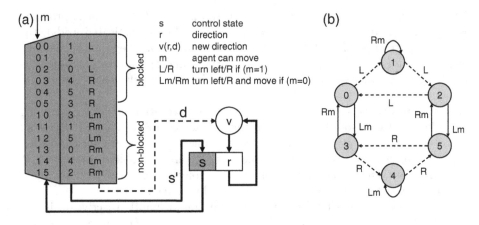

Fig. 1. A state machine (a) models an agent's behaviour. Corresponding 6-state graph (b), dashed line for $m = 0$, solid line for $m = 1$.

non-blocked cells. This grid serves as a fixed environment for one agent which moves around stepwise and has to fulfil a certain task. This task is visiting all non-blocked cells at least once in shortest time. The agent has a moving direction in which it always tries to take a step. The agent can look ahead one cell in that direction. It is only allowed to move onto non-blocked cells and can perform four different actions:

- L (turn Left),
- R (turn Right),
- Lm (turn Left and move) move forward and then turn left.
- Rm (turn Right and move) move forward and then turn right,

Action L or R is performed if the *front cell* (the cell ahead in the moving direction) is blocked. Action Lm or Rm is performed if the front cell is not blocked.

The behaviour of the agent is modeled by a finite state machine (see Fig. 1). The state machine has n states $s \in \{0, 1 \ldots, n - 1\}$, one binary input m (blocked or non-blocked front cell) and one binary output d (turn left or right). A state table can be coded into a string by concatenating the contents line by line. For the example of Fig. 1 it would be 1L2L0L4R5R3R-3Lm1Rm5Lm0Rm4Lm2Rm. The behaviour can also be depicted by a state graph (Fig. 1 (b)). E.g., being in current state $s = 3$ and receiving input $m = 0$ (do not move), this agent will perform the action R and the state transition to the next state $s' = 4$.

The direction $r \in \{North, East, West, South\}$ and new direction calculation $v(r, d)$ with turning direction $d \in \{L, R\}$ shown in Fig. 1 (a) are not part of the automaton itself. They are shown here because they are required for the simulation of the agent if implemented for example in software or on an FPGA where new coordinates of the agent depend on old coordinates and r. The agent has no knowledge about directions. It just moves or stays and turns left or right. Thus, r can be seen neither in the string representation nor in Fig. 1 (b).

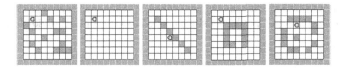

Fig. 2. The initial five grids that were simulated in hardware

The amount of state machines which can be coded is $M = (\#s\#y)^{(\#s\#x)}$ where $n = \#s$ is the number of states, $\#x$ is the number of different input values and $\#y$ is the number of different output actions. In this paper, we focus on state machines with six states. Thus, $12^{12} = 8,916,100,448,256$ possible behaviours can be defined by state machines. Note that not all of the agent behaviours described by state machines are distinct (e. g., permutation of the states leads to equivalent behaviours) or useful (e. g., state graphs which make little use of the inputs or which are weakly connected).

2.2 Optimal 6-State Automata for the CEP

In [3], *all* possible behaviours for six states were enumerated and simulated on a particular set of five initial grids as shown in Fig. 2. This task was aided by specific configurable logic (FPGAs) and needed approximately one day for the enumeration and 29 days for the simulation. 312,948 state machines were then preselected by a minimum level of performance, i. e., the number of visited cells. This preselection was further simulated in software extended by the criterion speed and on an additional set of 21 grids with different sizes, distribution of blocked cells, and start positions of the agent in order to evaluate the robustness of those preselected state machines. The state machines were first ranked by the average percentage of visited cells and second by speed, denoting the number of newly visited non-blocked cells per step in completely visited grids (see Tab. 1) [4]. Only completely visited grids were included in the speed measure because speed was defined as the number of steps agents needed to visit all cells. The speed measure was not defined for incompletely visited grids. In this paper, we are later using these previously found state machines for comparison with evolution in Sect. 4.

Modelling the behaviour with a state machine with a restricted number of states and evaluation by enumerations was also undertaken in the SOS (State-Observation-State) project [11]. There, they were able to learn state machines for different problems, e. g. the Santa-Fe trail. [12] showed that random turning directions improve the results of 2-state automata of this model.

3 Evolving 6-State-Automata for the Creature's Exploration Problem

Evolving solutions for problems on a computer always comes down to two major topics. The first one is the *representation of individuals* by chromosomes, in

Table 1. The ten best 6-state machines ranked by 1st percentage of visited cells and 2nd by speed. Speed is the number of steps needed to visit all cells divided by all empty cells.

Nbr.	String representation	Visited cells	Speed
1:	1L2L0L4R5R3R-3L1R5L0R4L2R	99.92%	0.260
2:	1R2R0R4L5L3L-3R1L5R0L4R2L	99.92%	0.252
3:	1L2L0L4R5R3R-3L4L2R0R1R5L	99.87%	0.268
4:	1R2R0R4L5L3L-3R4R2L0L1L5R	99.29%	0.259
5:	1R2L0R4L5L3L-3R4R5R0L1L2R	99.12%	0.223
6:	0R2R3R4L5L1L-1R5R4R0L2L3L	99.07%	0.136
7:	1L2L3L4L2R0L-2L4L0R3R5R5L4R	98.91%	0.193
8:	1L2L3R4L2R0L-2L4L0R3L5L4R	98.91%	0.169
9:	1R2R3R1L5L1L-1R0L2L4R3L1L	98.63%	0.251
10:	1R2L0L4R5R3R-3L4L5L0R1L2R	97.94%	0.238

biology called the genotype. Individuals are thereby possible solutions of the problem without regard to their actual effectiveness. All individuals together span the problem space. The second major topic is the *fitness* or quality of these individuals. It represents how well an individual copes with the given problem. For the CEP, those topics were handled as follows. Furthermore, we discuss the problem of consistency of the problem space and describe the evolutionary setting used in the tests in Sect. 4.

3.1 Representation of the Individuals

An individual for the CEP is a state machine, resp. an automaton, which is executed by an agent on a regular grid. This agent can also be steered by a small program which we want to evolve. Such a program consists of basic operations. According to the problem definition in Sect. 2.1, we have to provide this functionality for the agent: Turn_Left, Turn_Right, Move_Turn_Left, Move_Turn_Right, If_Obstacle_Ahead, If_Old_State_Was_{One, ..., Six}, and New_State_Is_{One, ..., Six}. All programs are made of a set of these 17 operations.

In order to compare evolved programs to optimal automata, we force the programs to have the structure as presented in Fig. 3. We only describe the procedure for the first state. The other states are treated accordingly. The agent first checks its state. If it is *One*, it looks ahead if the front cell is blocked. If that is true, the agent turns to either side and switches to a new state from *One* to *Six*. If the front cell is not blocked, it moves there and turns to either side before switching to a new state. If the agent's state is not *One*, it checks if its state equals *Two* and executes the same branches as described and so on.

The evolution is allowed to change the operations which are depicted in italics in the figure. Operations in black stay fixed. Evolution is only allowed to

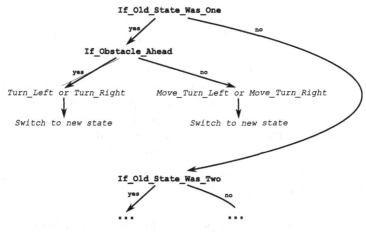

Fig. 3. Program structure

change operations to an operation from that operation's family, for example New_State_Is_One can be changed to New_State_Is_Three but not to Turn_Left or If_Old_State_Was_Four.

As for technical implementation, all basic operations are given a specific number. An individual/chromosome thus consists of $states * operations\ per\ state = 6*6 = 36$ numbers, of which 24 are changeable. The individuals are mutated and recombined with other individuals under the rules described before (details on the operators can be found in Sect. 3.4). By using this structure and evolutionary constraints, we are able to evolve programs for the CEP which are directly comparable to the optimal automata described in Sect. 2.2.

We use small programs here because of two reasons. The first one is that it is easier to explain what an agent really does by presenting operations and their structure instead of strings of numbers. The other reason is that we want to use Genetic Programming for evolving CEP programs with different numbers of states in the future. The programs shall then consist of the same basic operations but shall not be forced to a specific structure or length.

3.2   Fitness Function

In this paper, we want to compare the effectiveness of evolution to the optimal automata presented in Sect. 2.2. Thus, we use the same set of 26 grids as input to our evolved automata. An agent which executes the program/automata represented by an individual simulates the 26 grids, counts the number of pixels it visits, and measures the time it takes it to do this.

The first idea for a fitness function is to simply copy the quality measure presented in Tab. 1. This measure rates automata highest that visit most non-blocked cells. If two automata visit the same number of cells, these are also rated by speed. First tests showed that the probability of two automata in the population with the same number of visited cells was very low. Thus, we cut out the speed term and receive a very simple first fitness function:

$$Fitness_{preliminary}(Individual) = \sum_{i=grid_{first}}^{grid_{last}} cells_{visited}(i). \quad (1)$$

A fitness function not only rates quality but also somehow steers the evolutionary process. The results using this first fitness function were not satisfactory because the convergence was poor and resulting programs did not visit enough cells. But it showed that adding a penalty factor improved the results noticeably. This penalty is applied each time an agent is not able to visit 50% or more non-blocked cells of a grid. Let this be the penalty function:

$$Penalty(grid) = \begin{cases} value \text{ , if } cells_{visited}(grid) < 0.5 * cells_{non-blocked}(grid) \\ 0 \qquad \text{, otherwise} \end{cases}.$$
$$(2)$$

We played around, changing the penalty values and the minimum percentage of cells that had to be visited. It showed that a boundary of 50% yielded good automata. Results for various penalty values can be found in Sect. 4. The final, enhanced fitness that was used for evolution thus was:

$$Fitness(Individual) = \sum_{i=grid_{first}}^{grid_{last}} cells_{visited}(i) + Penalty(i). \quad (3)$$

3.3   Consistency of the Problem Space

A very important factor in evolution is consistency/continuity of the problem space. In other words: Do direct neighbours of individuals have similar fitness? Or does the fitness value change dramatically if individuals are slightly altered? Evolution is also successful if the problem-space is difficult [6][7][13][14]. But spaces with higher consistency provide better results and are more predictable [15].

In order to gain knowledge about the consistency of the vast problem-space of the CEP, we took each optimal automata from Sect. 2.2 and calculated all automata that differ least from them. Therefore, these automata were changed in exactly one position which is either the turning direction or the resulting new state after an agent's step. This equals changing or shifting exactly one of the arrows in Fig. 1 b) and creates one new automaton. It was done for all possible states at each possible position creating 72 direct neighbour automata for every optimal automata from Sect. 2.2. Each of these new automata differs from its original optimal automaton in exactly one position.

These new automata were then simulated and their results were calculated. It showed that the problem space was difficult but not completely irregular. Some of the new automata created very unsuccessful agents which did not have any correlation to the optimal automaton. On the other hand, some of the neighbours indeed represented automata that had fitness values close to the optimal ones. On average, the neighbouring automata had clearly better fitness values than randomly generated automata.

3.4 Evolutionary Parameters and Operators

We defined the population size to be 20 for the tests in Sect. 4. The initial population was randomly chosen. For selection, the roulette wheel operator was used where the fitness of individuals defines the size of their segments on a virtual wheel. The wheel is turned and and stops at a certain segment. This segment's individual is selected to the intermediate population [16].

50% of the intermediate population were then recombined randomly using one-point crossover. In this case, one of the numbers that represent operations is chosen randomly for two parent individuals. Two new individuals are created by taking all operations before the chosen number from the first parent and all operations afterwards from the other parent and vice versa.

10% of the intermediate population were mutated. We chose to limit the mutation by allowing it to change three operations maximally. Normally, mutation is responsible for leaving local optima and is thus allowed to change an individual randomly. In this case, the recombination operator creates individuals that are relatively far away from their parents concerning semantics of their programs and thus fitness. This is why we uncommonly restrict mutation and use it to stay close to the parent individuals in order to do local search.

The remaining 40% of the intermediate population were left unchanged and were reproduced in the new population. The overall best program was saved even if it was not transferred to the new population. The definition of these operators takes care of on one hand not staying in local optima and on the other hand searching near good individuals. It has to be noted that it is possible that the same individual is created and evaluated more than once during evolution due to the design of the operators. Keeping a history makes no sense, especially with growing amount of generations and thus growing history that results in ever larger costs for comparison.

4 Results of the Evolution

We used the approach described in Sect. 3 to evolve good automata for the CEP. Several runs were started which differed in the penalty value of the fitness function. This section describes the final results of the computation.

Tab. 2 shows the results of different evolution runs. The first and the last columns contain the ordinal number of the runs and the penalty term of the fitness function (as described in Sect. 3) of that run respectively. No agent running times can be seen in the table. This is due to the major objective that the number of visited cells is more important than the number of agent steps. No two automata in the table have the same number of visited cells. Thus, agent running time comparison is obsolete.

The second column shows the most important fact, namely the amount of cells that were visited by the agent of the best automata found during that run. The sum of all cells that could possibly be visited was 6,137. It is the sum of all non-blocked cells of the 26 different input grids as described in Sect. 2.2. The best automata in the worst runs were able to visit more than 3/4 of free cells

Table 2. Results of different evolution runs

Run	Cells Visited	Total Generations	Best Found in Generation #	Improve- ments	Penalty
01	4,876/6,137 (79.5%)	8,036	3,820	13	0
02	4,761/6,137 (77.6%)	8,644	4,871	15	0
03	4,779/6,137 (77.9%)	8,569	7,075	17	2,000
04	5,567/6,137 (90.7%)	8,332	613	24	2,000
05	5,620/6,137 (91.6%)	38,103	24,465	28	450
06	5,270/6,137 (85.9%)	37,644	30,450	21	450
07	5,322/6,137 (86.7%)	50,000	11,414	20	800
08	5,088/6,137 (82.9%)	50,000	4,366	9	800
09	5,193/6,137 (84.6%)	50,000	17,229	21	1,500
10	5,147/6,137 (83.9%)	50,000	14,615	25	1,500

Table 3. Statistics of visited cells of evolution and random walks

	All Individuals		Best per 20 Individuals	
	Average	Deviation	Average	Deviation
Evolution	3,938.7/6,137 (64.2%)	319,1	4,459.1/6137 (72.7%)	115.8
Random walk	470.6/6,137 (7.6%)	139.5	2,209/6,137 (36%)	902

while the best ones in the best runs reached over 90%. On average, the best automata found 84.1% of free cells.

For comparison, we created 20 random automata 100,000 times. The results can be seen in Tab. 3. These statistics show that the majority of automata is much worse than the evolved ones. The comparison between the average best automata per evolved generation over all runs and a random set, each consisting of 20 individuals, favours evolution, too.

Tab. 2 also shows that the penalty term should not be discarded. The two runs with penalty term equal zero resulted in two of the three worst automata. A penalty of 450 granted the best results on average although this should not be taken too seriously because the amount of these random samples is too small to imply general conclusions and needs further investigations.

Column number three shows the number of evolution steps that were taken out during that run. Those numbers differ between runs because tests five to ten were planned as long-term tests in order to see if results improve significantly with longer execution times.

The specific evolution step/generation in which the best automaton of a run was found can be seen in column four. The values there indicate that in most cases the best automaton was found relatively late indicating that longer evolution times might bring better results. Only in runs four and eight, the best automaton was already found in less then 10% of the evolution steps. The question is now if those early results hint that the evolution has run into local optima?

In Sect. 3.3, we showed that the problem space is relatively inconsistent. Creating direct or indirect neighbours of automata by recombination or mutation thus often leads to completely different fitness values. Therefore, the fear of local optima can be discarded. Apart from that, the best automata in runs six and ten visit similarly many cells like the other runs indicating that in those cases it was simply "luck" that these automata were found this early.

In column five, the amount of improvements, which occurred during a run, is presented. Apart from number six, every run improved double-digit times. A bad result would have been if "somehow good" automata would have been found in the initial random populations and few or no improvements would have happened. Then, the conclusion would have been that evolution was inappropriate for the CEP. Anyhow, the number of improvements is low in comparison to the total amount of generations. But the fitness values of different improvements (not shown here) differed by at least 10% in most cases and more than 30% for the first ones after the start population. Thus, few but significant improvements were made instead of many sub-1% improvements.

One automaton simulation of the 26 grids took about 0.13 seconds on a 2GHz AMD PC. In the tests, we used a population of 20 individuals per generation. Thus, simulating one generation took about 2.6 seconds and 50,000 generations required 36 hours. During such a run, 1,000,000 automata were simulated.

5   Summary and Outlook

In this paper, we compared different approaches to find solutions for the Creature's Exploration Problem for single autonomous agents. The big question was: Can time-consuming full search of all automata of the CEP be replaced by evolution and how successful would that be?

The results show that the answer depends on some influencing factors. Full search of automata of course is able to find the best automata for specific goal functions. But it requires large computation times to fulfil this goal. Concerning the cost function of "visited cells first", evolution found automata that visited 80% to 90% of possible cells and performed much better than random walks. This is quite a nice result concerning that this evolution simulated maximal one million automata, which is only 0.00001% of all possible automata. The optimal automata were found in more than one month exploiting dedicated hardware while evolution required one and a half day on a 2GHz AMD PC.

If the user seeks the one optimal automaton, evolution is not the tool to use because finding that one best individual in such a large set is very improbable. But if he/she is satisfied with 80+% quality and/or if he/she needs good but not the best automata fast, evolving such automata provides an easy to handle, fast, and robust opportunity.

In the future, we are going to extend the works described in this paper concerning different aspects. At first, it might be useful to change the operators. Especially two- or multiple-point recombination is interesting. The tight mutation operator needs some examination, too. Apart from that, the simple cost function "success over time" might not be the ultimate goal. For some scenarios, it might be more important to have fast agents than to visit one percent more or less cells. Or there might be a lower barrier of minimally visited cells which should be achieved for each grid. Our goal is to use different cost functions where time is more important and other aspects like for example the standard deviation of success come into play. This could be done for example by weighing the visited pixels, the required time, and statistical data and using this new measure as fitness.

Apart from different measures, we also want to investigate if agents become much more powerful if they are given more or less states by extending the length and structure of their programs. And another very interesting topic is the usage of multiple agents. More agents should be more successful and might be able to visit all cells faster. Several questions have to be answered: How expensive are multiple agents? How many agents are needed for 100% visits? Shall the agents communicate or not? How long does evolution take then? And how many states are required for multiple agents to fulfil these goals? All these questions are part of the future work arising from this paper.

References

1. Komann, M., Fey, D.: Realising emergent image preprocessing tasks in cellular-automaton-alike massively parallel hardware. International Journal of Parallel, Emergent and Distributed Systems 22(2), 79–89 (2007)
2. Koza, J.R.: Scalable learning in genetic programming using automatic function definition, pp. 99–117 (1994)
3. Halbach, M.: Algorithmen und Hardwarearchitekturen zur optimierten Aufzählung von Automaten und deren Einsatz bei der Simulation künstlicher Kreaturen. Ph.D thesis, Technische Universität Darmstadt (2008)
4. Halbach, M., Hoffmann, R., Both, L.: Optimal 6-state algorithms for the behavior of several moving creatures. In: El Yacoubi, S., Chopard, B., Bandini, S. (eds.) ACRI 2006. LNCS, vol. 4173, pp. 571–581. Springer, Heidelberg (2006)
5. Konak, A., Coit, D.W., Smith, A.E.: Multi-objective optimization using genetic algorithms: A tutorial. Reliability Engineering & System Safety 91(9), 992 (2006); Special Issue - Genetic Algorithms and Reliability
6. Walker, J.A., Miller, J.F.: Solving real-valued optimisation problems using cartesian genetic programming. In: GECCO 2007: Proceedings of the 9th annual conference on Genetic and evolutionary computation, pp. 1724–1730. ACM, New York (2007)
7. Padhye, N.: Interplanetary trajectory optimization with swing-bys using evolutionary multi-objective optimization. In: GECCO 2008: Proceedings of the 2008 GECCO conference companion on Genetic and evolutionary computation, pp. 1835–1838. ACM, New York (2008)
8. Fogel, D.B.: An evolutionary approach to the traveling salesman problem. Biological Cybernetics 60(2), 139–144 (1988)
9. Bruce, W.S.: The lawnmower problem revisited: Stack-based genetic programming and automatically defined functions. In: Genetic Programming 1997: Proceedings of the Second Annual Conference, pp. 52–57. Morgan Kaufmann, San Francisco (1997)

10. Halbach, M., Heenes, W., Hoffmann, R., Tisje, J.: Optimizing the behavior of a moving creature in software and in hardware. In: Sloot, P.M.A., Chopard, B., Hoekstra, A.G. (eds.) ACRI 2004. LNCS, vol. 3305, pp. 841–850. Springer, Heidelberg (2004)

11. Mesot, B., Sanchez, E., Peña, C.-A., Perez-Uribe, A.: SOS++: Finding smart behaviors using learning and evolution. In: Standish, R., Bedau, M., Abbass, H. (eds.) Artificial Life VIII: The 8th International Conference on Artificial Life, pp. 264–273. MIT Press, Cambridge (2002)

12. Di Stefano, B.N., Lawniczak, A.T.: Autonomous roving object's coverage of its universe. In: CCECE, pp. 1591–1594. IEEE, Los Alamitos (2006)

13. Holland, J.H.: Genetic algorithms - computer programs that "evolve" in ways that resemble natural selection can solve complex problems even their creators do not fully understand. Scientific American 267, 66–72 (1992)

14. Mitchell, M., Hraber, P.T., Crutchfield, J.P.: Revisiting the edge of chaos: evolving cellular automata to perform computations. Technical Report Santa Fe Institute Working Paper 93-03-014 (1993)

15. Storn, R., Price, K.: Differential evolution – a simple and efficient heuristic for global optimization over continuous spaces. Journal of Global Optimization 11(4), 341–359 (1997)

16. Michalewicz, Z.: Genetic algorithms + data structures = evolution programs, 3rd edn. Springer, London (1996)

Semantic Aware Crossover for Genetic Programming: The Case for Real-Valued Function Regression

Quang Uy Nguyen[1], Xuan Hoai Nguyen[2], and Michael O'Neill[1]

[1] Natural Computing Research & Applications Group, University College Dublin, Ireland
[2] School of Computer Science and Engineering, Seoul National University, Korea
quanguyhn@yahoo.com, nxhoai@gmail.com, m.oneill@ucd.ie

Abstract. In this paper, we apply the ideas from [2] to investigate the effect of some semantic based guidance to the crossover operator of GP. We conduct a series of experiments on a family of real-valued symbolic regression problems, examining four different semantic aware crossover operators. One operator considers the semantics of the exchanged subtrees, while the other compares the semantics of the child trees to their parents. Two control operators are adopted which reverse the logic of the semantic equivalence test. The results show that on the family of test problems examined, the (approximate) semantic aware crossover operators can provide performance advantages over the standard subtree crossover adopted in Genetic Programming.

Keywords: crossover, semantic, and genetic programming.

1 Introduction

Since genetic programming was born, it has been seen by some researchers in and out of the field that GP is a potentially powerful method for automated synthesis of computer programs by evolutionary means. The 'program' is usually presented in a language of syntactic formalism such as s-expression trees [7], a linear sequence of instructions, grammars, or graphs [1, 12]. The genetic operators in such GP systems are usually designed to ensure the syntactic closure property, i.e. to produce syntactically valid children from any syntactically valid parent(s). Using such purely syntactical genetic operators, GP evolutionary search is conducted on the syntactical space of programs with the only semantic guidance from the fitness of program measured by the difference of behavior of evolving programs and the target programs (usually on a finite input-output set called fitness cases). Although GP has been shown to be effective in evolving programs for solving different problems using such (finite) behavior-based semantic guidance and pure syntactical genetic operator, this practice is somewhat unusual from real programmers' perspective. Computer programs are not just constrained by syntax but also by semantics. As a normal practice, any change to a program should pay heavy attention to the change in semantics of the program and not just those changes that guarantee to maintain the program syntactical validity. To amend this deficiency in GP resulting from the lack of semantic guidance on genetic

L. Vanneschi et al. (Eds.): EuroGP 2009, LNCS 5481, pp. 292–302, 2009.

operators, recently, Beadle and Johnson have proposed a semantic-based crossover operator for genetic programming [2]. They showed that their semantically driven crossover operator could help GP in achieving better results and less code bloat on some standard Boolean test problems.

In this paper, we extend the ideas from [2] to investigate the effect of some semantic based guidance on the crossover operator in GP on a family of real-valued symbolic regression problems. We also propose a new form of semantic-aware crossover for Genetic Programming, which considers approximations of the semantics of the exchanged subtrees. In contrast to the approach of Beadle and Johnson [2], the new semantic-aware crossover can be applied to both Boolean and continuous problem domains.

The paper is organized as follows. In the next section, we give a review of related work on crossover operators and semantic based operations in GP. Section 3 contains the descriptions of some possible methods for using semantic information to guide the crossover operator in GP. These methods are used in experiments described in section 4 of the paper. The results of the experiments are then given and discussed in section 5. Section 6 concludes the paper and highlights some potential future extension of this work.

2   Related Work

In a series of work, Johnson has advocated for the use of semantic information in the evolutionary process of GP [3, 4, 5, 6]. He proposed a number of possible ways for incorporating program semantics extracted by static analysis techniques into the fitness function of GP. In [2], the authors elaborated on their suggestions by investigating the effect of using semantic information to guide the crossover operation in GP for Boolean domains, the resultant operator is called semantically driven crossover (SDC). Their main idea is to check the semantic equivalence between newly born children as the results of the crossover operator with their parents. The semantic equivalence checking of two Boolean expression trees is done by transforming the trees to reduced ordered binary decision diagrams (ROBDDs). They have the same semantic if and only if they are reduced to the same ROBDD. The semantic equivalence checking is then used to determine which of the individual participating in crossover operation will be copied to the next generation. If the children born as the result of crossover are semantically equivalent with their parents, they are not copied to the next generation, instead their parents are copied. By doing this, the author argued that it helps to increase the semantic diversity of evolving population of programs. Indeed the idea is rather similar to deterministic fitness crowding, one of the common mechanisms for promoting diversity in evolutionary algorithms [8, 9]. In the deterministic fitness crowding method, the children must compete directly with the parents to be copied to the next generation and they are copied only if they are different enough (usually measured by a distance function) from their parents. The important difference between work in [2] and deterministic fitness crowding is the use of semantic information (which seems to be attributed to GP only). Semantic driven crossover was reported useful in [2] in both increasing GP performance but also reducing code bloat on some standard Boolean test problems (even though it seems that reducing bloat was not intentionally the motivation for the introduction of semantic driven crossover).

Prior to Beadle and Johnson [2], there were a number of works on analyzing the effect of crossover and attempts to improve them. However, the main focus of this work has been on code bloat, fitness-destructive effect of crossover [15, 16, 19] and length distributions, and how to improve it in this respect [17, 18].

The work on semantic driven crossover operators in this paper is different from [2] in two ways. Firstly, the domain for testing semantically driven crossovers is real-valued rather than Boolean. For real-valued domains, the idea of checking semantic equivalence by reducing to common ROBDDs is no longer possible. Secondly, the semantic guidance of the crossover operator is not just derived from the whole program tree behavior but also from subtrees. This is inspired by recent work in [10] for calculating subtree semantics. This method is used in [14] as a way for measuring the fitness in GP. However, the subtree semantic calculated in this paper is for real-valued domains but not Boolean domains as in [2, 10].

3 Methodologies

The aim of this study is to extend earlier work [2, 10] to real-valued domains. For such problems it is not easy to compute the semantics or semantic equivalence of two expression trees by reducing them to a common structure as for Boolean domain as in [2]. In fact, the problem of determining semantic equivalence between two real-valued expressions is known to be NP-hard [20]. Similarly, the problem of completely enumerating and comparing fitness of subtree expressions as in [10] is also impossible on real domains. We have to calculate the approximate semantics. In this paper, a simple method for measuring and comparing the semantics of two expressions is used both for individual trees and subtrees and the two methods are compared experimentally. To determine the semantic equivalence of two expressions, we measure them against a random set of points sampled from the domain. If the output of the two trees on the random sample set are close enough (subject to a parameter called semantic sensitivity) then they are designated as semantically equivalent. It can be written in pseudo-code as follows:

```
If Abs(Value_On_Random_Set(P1)-
    Value_On_Random_Set(P2))<∑ then
        Return P1 is semantically equivalent to P2.
```

Where Abs is the absolute function and $\sum$ is a predefined constant called the *semantic sensitivity*. This method is inspired by the simple technique for simplifying s-expression program trees proposed in [11] called equivalence decision simplification (EDS), where complicated subtrees could be replaced by much simpler and template subtrees if they are semantically equivalent. The method of checking semantic equivalence of EDS is similar to the method used here and EDS has shown to be an efficient tool for removing redundant codes for tree-based genetic programming[1] [11]. To test the effect of using semantic information to guide the subtree crossover in genetic programming, we propose 4 possible scenarios based on [2, 10].

In scenario 1, we constrain the crossover in such a way that if two crossover points are chosen the two subtrees under the crossover points are checked for semantic equivalence. If they have equivalent semantics, the operator is forced to be executed on two new crossover points. The pseudo-code for this scenario can be summarized as follows:

```
1.1.Select two parents: P1, P2
1.2.Choose at random crossover points at Subtree1 in P1
    Choose at random crossover points at Subtree2 in P2
   if(Subtree1 is not equivalent with Subtree2{
            Execute crossover
            Add the children to the new population
            Return TRUE
   }
   else{
            Add P1 and P2 to new population
            Return TRUE
   }
```

The motivation for scenario 1 is to encourage GP individual trees to exchange subtrees that have different semantics, which is expected to encourage the change in semantics of the whole trees after each crossover.

In the second scenario, we reverse the semantic bias in the crossover operator compared to the first scenario in that the two subtrees are selected for crossover if and only if they are semantically equivalent. The only change in code is the condition of if statement in 1.2 as follows:

```
   if (Subtree1 is equivalent with Subtree2)
```

The third scenario is similar to the method used in [2]. Here the semantics of newly created children are checked against the semantics of their parents. If their semantics are found to be equivalent with their parents then they are discarded and the parents would be passed to the new generation instead. The objective of such an implementation of crossover, as stated in [2], is to enforce the semantic diversity (i.e., keep generating new individual programs with new behavior). The pseudo-code for this scenario is as follows:

---

[1] The JAVA code of EDS could be freely downloaded from: sc.snu.ac.kr

```
2.1. Select two parents: P1, P2
2.2. If (condition to make crossover is satisfied)
     {
     Make Crossover to have Children C1 and C2
     if (C1 is equivalent with P1 Or P2)
         Add P1 to the new population
     else
         Add C1 to the new population
     if (C2 is equivalent with P1 Or P2)
         Add P2 to the new population
     else
         Add C2 to the new population
     }
     else
         Add P1 and P2 to new population
```

The last and fourth scenario tested in this paper is the reverse of the third, where the children are accepted to go in to the new population if and only if they are semantically equivalent with their parent(s).

4 Experiments

To investigate the possible effects of semantic aware crossovers given in the previous section, they are tested on the symbolic regression problems with target functions in a family of polynomials of increasing degree given in [13] and they are:

$F_1 = X^3 + X^2 + X.$
$F2 = X^4 + X^3 + X^2 + X.$
$F3 = X^5 + X^4 + X^3 + X^2 + X$
$F4 = X^6 + X^5 + X^4 + X^3 + X^2 + X$

The parameters setting for GP in the experiment is as follows:

- Population size: 500.

- Number of generation: 50

- Tournament selection size: 3

- Crossover probability: 0.9

- Mutation probability: 0.1

- Max depth of program tree at the initial generation: 6

- Max depth of program tree at all time: 15

- Non-terminals: +, -, *, / (protected version), sin, cos, exp, log (protected)

- Terminals: X, 1

- Number of sample: 20 random points from [-1...1].

- Hit: when an individual has an absolute error < 0.01 on a fitness case.

- Termination: when a program score 20 hits or running out of generations.

The *semantic sensitivities* (Σ) used in the experiments are: 0.01, 0.02, 0.04, 0.05, 0.06, 0.08, 0.1, 0.5, and 1. For each crossover scenario, each target problem, and semantic sensitivity, 100 runs are performed, which makes the total number of runs 14800.

5 Results and Discussion

The number of successful runs (out of 100 runs) is given in Table 1. It can be seen that the effect of using semantic aware crossover depends on the manner in which it is applied. When semantics is calculated on subtrees (Scenario 1), it can improve the performance of GP in terms of the number of successful runs (e.g., 71 versus 62 on F1, and 36 versus 28 on F2). While semantics as used in a similar way to [2] by comparing the resulting children to their parents only give slight improvements in some cases (Scenario 3). When we reverse the logic of the semantic equivalence test in the Control Scenarios 2 and 4 no improvement in the number of successful runs is achieved when compared with standard GP.

However, the improvement in Scenario 1 also depends on the value *of semantic sensitivity* (Σ).The results suggest that the value of sensitivities from 0.01 to 0.1 are suitable for all four test functions, and the greater sensitivities seem to confer less improvement.

In Table 2 we show mean and standard deviation of best fitness for the 50th generation of each run. The results shown in this table is consistent with the results in Table 1. The semantic aware crossover as used in scenario 1 where the semantics of subtrees to be exchanged are compared, can slightly improve the performance of GP in terms of the mean and standard deviation of best fitness. Whereas in Scenario 3, where the semantics of children are compared to their parents only results in an improvement in some settings of semantic sensitivity. Again, no improvements are observed for the control scenarios 2 and 4 where the logic of the equivalence test is reversed. From these results it would appear that the more difficult problems benefit more in terms of the mean and standard deviation of the best fitness when semantic aware crossover is adopted. This claim can be confirmed when we consider the result of performing t-tests of the results in Scenario 1. These results are shown in Table 3.

Table 1. The number of successful runs over 100 runs

Scenario	Sensitivity	F1	F2	F3	F4
GP		62	28	15	10
Scenario 1	0.01	68	33	22	10
	0.02	**70**	**33**	**22**	**14**
	0.04	**70**	**34**	**20**	**19**
	0.05	**71**	**33**	**19**	**14**
	0.06	**71**	**32**	**20**	**17**
	0.08	**70**	**35**	**20**	**17**
	0.1	**66**	**36**	**17**	**14**
	0.5	62	33	17	12
	1	65	27	12	8
Scenario 2	0.01	41	31	13	7
	0.02	40	18	13	10
	0.04	40	15	14	8
	0.05	44	26	12	7
	0.06	41	17	13	7
	0.08	41	19	15	7
	0.1	42	24	12	10
	0.5	45	18	10	10
	1	55	25	18	5
Scenario 3	0.01	52	24	18	12
	0.02	57	32	16	12
	0.04	52	32	23	8
	0.05	54	33	15	7
	0.06	52	29	16	10
	0.08	50	26	13	11
	0.1	44	26	13	8
	0.5	38	15	15	5
	1	39	17	14	3
Scenario 4	0.01	2	0	0	0
	0.02	3	0	0	0
	0.04	3	0	0	0
	0.05	2	0	0	0
	0.06	2	0	1	0
	0.08	3	1	0	0
	0.1	3	1	1	0
	0.5	7	0	0	1
	1	12	1	0	1

Table 2. Mean and standard deviation of the best fitness at 50 generations

	Sensitivity	F1 Mean (Stdev)	F2 Mean (Stdev)	F3 Mean (Stdev)	F4 Mean (Stdev)
GP		0.128 (0.170)	0.262 (0.241)	0.302 (0.252)	0.397 (0.355)
Scenario 1	0.01	0.135 (0.205)	0.230 (0.214)	0.274 (0.215)	0.335 (0.271)
	0.02	**0.133** **(0.209)**	**0.236** **(0.218)**	**0.275** **(0.216)**	**0.329** **(0.275)**
	0.04	**0.125** **(0.182)**	**0.231** **(0.214)**	**0.268** **(0.212)**	**0.328** **(0.278)**
	0.05	**0.126** **(0.185)**	**0.224** **(0.206)**	**0.266** **(0.213)**	**0.330** **(0.277)**
	0.06	**0.121** **(0.181)**	**0.227** **(0.207)**	**0.265** **(0.215)**	**0.319** **(0.270)**
	0.08	**0.137** **(0.215)**	**0.223** **(0.205)**	**0.279** **(0.218)**	**0.333** **(0.281)**
	0.1	**0.133** **(0.185)**	**0.225** **(0.208)**	**0.274** **(0.215)**	**0.334** **(0.268)**
	0.5	0.120 (0.157)	0.217 (0.203)	0.276 (0.213)	0.336 (0.266)
	1	0.123 (0.174)	0.214 (0.149)	0.287 (0.215)	0.397 (0.292)
Scenario 2	0.01	0.258 (0.357)	0.275 (0.222)	0.325 (0.235)	0.459 (0.421)
	0.02	0.221 (0.318)	0.307 (0.234)	0.330 (0.237)	0.458 (0.348)
	0.04	0.229 (0.321	0.310 (0.238)	0.332 (0.244)	0.435 (0.314)
	0.05	0.253 (0.356)	0.289 (0.216)	0.336 (0.256)	0.439 (0.395)
	0.06	0.241 (0.333)	0.302 (0.237)	0.327 (0.238)	0.443 (0.314)
	0.08	0.239 (0.325)	0.315 (0.256)	0.334 (0.245)	0.434 (0.305)
	0.1	0.250 (0.355)	0.279 (0.197)	0.337 (0.260)	0.436 (0.396)
	0.5	0.236 (0.395)	0.280 (0.237)	0.314 (0.195)	0.377 (0.293)
	1	0.168 (0.205)	0.328 (0.300)	0.294 (0.237)	0.398 (0.262)
Scenario 3	0.01	0.229 (0.373)	0.239 (0.215)	0.321 (0.321)	0.362 (0.305)
	0.02	0.208 (0.326)	0.218 (0.184)	0.229 (0.286)	0.348 (0.292)
	0.04	0.224 (0.334)	0.229 (0.211)	0.306 (0.311)	0.353 (0.241)
	0.05	0.212 (0.312)	0.216 (0.197)	0.303 (0.232)	0.364 (0.307)
	0.06	0.204 (0.333)	0.229 (0.230)	0.280 (0.241)	0.315 (0.218)

Table 2. (*Continued*)

	0.08	0.234 (0.350)	0.248 (0.225)	0.335 (0.302)	0.347 (0.251)
	0.1	0.224 (0.329)	0.252 (0.214)	0.359 (0.302)	0.418 (0.305)
	0.5	0.225 (0.332)	0.297 (0.211)	0.344 (0.310)	0.406 (0.275)
	1	0.216 (0.224)	0.343 (0.247)	0.378 (0.266)	0.460 (0.247)
	0.01	1.144 (0.911)	1.287 (0.959)	1.590 (1.219)	1.641 (1.215)
	0.02	1.153 (0.908)	1.304 (0.964)	1.569 (1.251)	1.662 (1.228)
	0.04	1.122 (0.911)	1.280 (1.002)	1.561 (1.247)	1.650 (1.214)
	0.05	1.204 (0.946)	1.259 (0.967)	1.513 (1.203)	1.658 (1.236)
Scenario 4	0.06	1.127 (0.914)	1.281 (0.997)	1.473 (1.236)	1.640 (1.223)
	0.08	1.121 (0.910)	1.205 (0.999)	1.495 (1.999)	1.604 (1.225)
	0.1	1.144 (0.938)	1.180 (0.980)	1.519 (1.191)	1.650 (1.216)
	0.5	0.994 (0.895)	1.038 (0.834)	1.249 (1.015)	1.575 (1.233)
	1	0.752 (0.787)	0.870 (0.645)	1.177 (1.107)	1.350 (1.277)

Table 3. Mean and standard deviation of best fitness and p-value of t-test

Scenario	Sensi-tivity	f1 Mean Stdev	P-value	f2 Mean Stdev	P-value	f3 Mean Stdev	P-value	f4 Mean Stdev	P-value
GP		0.128 0.170		0.262 0.241		0.302 0.252		0.397 0.355	
	0.01	0.135 0.205	0.647	0.230 0.214	0.351	0.274 0.215	0.405	0.335 0.271	0.165
	0.02	0.133 0.209	0.687	0.236 0.218	0.462	0.275 0.216	0.412	0.329 0.275	0.129
	0.04	0.125 0.182	0.932	0.231 0.214	0.367	0.268 0.212	0.304	0.328 0.278	0.128
	0.05	0.126 0.185	0.894	0.224 0.206	0.282	0.266 0.213	0.274	0.330 0.277	0.137
Scenario 1	0.06	0.121 0.181	0.982	0.227 0.207	0.225	0.265 0.215	0.264	0.319 0.270	**0.080**
	0.08	0.137 0.215	0.604	0.223 0.205	0.255	0.279 0.218	0.495	0.333 0.281	0.157
	0.1	0.133 0.185	0.678	0.225 0.208	0.161	0.274 0.215	0.395	0.334 0.268	0.162
	0.5	0.120 0.157	0.928	0.217 0.203	0.131	0.276 0.213	0.414	0.336 0.266	0.171
	1	0.123 0.174	0.986	0.214 0.149	0.090	0.287 0.215	0.543	0.397 0.292	0.997

6 Conclusions and Future Work

The effect of semantics with crossover in Genetic Programming is investigated in this paper. In this study we focus on the family of real-valued problem domains in form of polynomials (symbolic regression). The investigation is performed on four scenarios. The experimental results show that semantic aware crossover as adopted in Scenario 1, where the semantics of subtrees to be exchanged are analysed, can improve performance of Genetic Programming in both number of successful runs and the mean best fitness on the problems examined. In the alternative Scenario 3, where the semantics of the children resulting from crossover are compared to their parents performance gains can be observed in some cases. Comparing subtree semantic aware crossover to the individual-based form, there is an advantage for the subtree approach on the four symbolic regression instances examined. The two control scenarios (2 and 4) where the logic of the semantic equivalence tests are reversed show no improvement over standard subtree crossover.

There are some interesting areas for future investigation. In contrast to the approach adopted in an earlier study on Boolean problems [2] the semantic aware crossover operators adopted here can be applied to both Boolean and real-valued domains. We also wish to examine the utility of these semantic aware operators to more difficult symbolic regression problems, and more classic benchmark problems from the literature to determine more clearly the generality of these findings.

Acknowledgements

This research was funded under a Government of Ireland Postgraduate Scholarship from the Irish Research Council for Science, Engineering and Technology (IRCSET).

References

1. Banzhaf, W., Nordin, P., Francone, F.D., Keller, R.E.: Genetic Programming: An Introduction - On the Automatic Evolution of Computer Programs and Its Applications. Morgan Kaufmann Publishers, San Francisco (1998)
2. Beadle, L., Johnson, C.G.: Semantically Driven Crossover in Genetic Programming. In: Proceedings of the IEEE World Congress on Computational Intelligence, pp. 111–116. IEEE Press, Los Alamitos (2008)
3. Johnson, C.G.: Deriving Genetic Programming Fitness Properties by Static Analysis. In: Foster, J.A., Lutton, E., Miller, J., Ryan, C., Tettamanzi, A.G.B. (eds.) EuroGP 2002. LNCS, vol. 2278, pp. 299–308. Springer, Heidelberg (2002)
4. Johnson, C.G.: Genetic Programming with Guaranteed Constraints. In: Lofti, A., John, B., Garibaldi, J. (eds.) Recent Advances in Soft Computing. Physica/Springer-Verlag, Heidelberg (2002)
5. Johnson, C.G.: Genetic Programming with Fitness based on Model Checking. In: Ebner, M., O'Neill, M., Ekárt, A., Vanneschi, L., Esparcia-Alcázar, A.I. (eds.) EuroGP 2007. LNCS, vol. 4445, pp. 114–124. Springer, Heidelberg (2007)

6. Johnson, C.G.: What Can Automatic Programming Learn from Theoretical Computer Science? In: Yao, X. (ed.) Proceedings of the UK Workshop on Computational Intelligence, University of Birmingham (2002)
7. Koza, J.R.: Genetic Programming: On the Programming of Computers by Means of Natural Selection. MIT Press, Cambridge (1992)
8. Mahfoud, S.W.: Crowding Preselection Revisited. In: Manner, R., Manderick, B. (eds.) Parallel Problem Solving from Nature, vol. 2, pp. 27–36. Elsevier, Amsterdam (1992)
9. Mahfoud, S.W.: Niching Methods for Genetic Algorithms. Doctoral Dissertation at University of Illinois at Urbana-Champaign (1995)
10. McPhee, N.F., Ohs, B., Hutchison, T.: Semantic Building Blocks in Genetic Programming. In: O'Neill, M., Vanneschi, L., Gustafson, S., Esparcia Alcázar, A.I., De Falco, I., Della Cioppa, A., Tarantino, E. (eds.) EuroGP 2008. LNCS, vol. 4971, pp. 134–145. Springer, Heidelberg (2008)
11. Mori, N., McKay, R.I., Nguyen, X.H., Essam, D.: Equivalent Decision Simplification: A New Method for Simplifying Algebraic Expressions in Genetic Programming. In: Proceedings of 11th Asia-Pacific Workshop on Intelligent and Evolutionary Systems (2007)
12. Poli, R., Langdon, W.B., McPhee, N.F.: A Field Guide to Genetic Programming (2008), http://lulu.com http://www.gp-field-guide.org.uk
13. Nguyen, X.H., McKay, R.I., Essam, D.: Solving the Symbolic Regression Problem with Tree-Adjunct Grammar Guided Genetic Programming: The Comparative Results. In: Proceedings of the 2002 Congress on Evolutionary Computation (CEC 2002), pp. 1326–1331. IEEE Press, Los Alamitos (2002)
14. Krysztof, K., PremysBaw, P.: Potential Fitness for Genetic Programming. In: Proceedings of Genetic and Evolutionary Computation Conference (GECCO 2008), Late-Breaking Papers, pp. 2175–2180. ACM, New York (2008)
15. Langdon, W.B., Poli, R.: Fitness causes bloat: Mutation. In: Koza, J. (ed.) Late Breaking Papers at the GP 1997 Conference, Stanford, CA, USA, July 13-16, pp. 132–140. Stanford Bookstore (1997)
16. Langdon, W.B., Soule, T., Poli, R., Foster, J.A.: The evolution of size and shape. In: Spector, L., Langdon, W.B., O'Reilly, U.-M., Angeline, P.J. (eds.) Advances in Genetic Programming 3, ch. 8, pp. 163–190. MIT Press, Cambridge (1999)
17. Dignum, S., Poli, R.: Crossover, Sampling, Bloat and the Harmful Effects of Size Limits. In: O'Neill, M., Vanneschi, L., Gustafson, S., Esparcia Alcázar, A.I., De Falco, I., Della Cioppa, A., Tarantino, E. (eds.) EuroGP 2008. LNCS, vol. 4971, pp. 158–169. Springer, Heidelberg (2008)
18. Dignum, S., Poli, R.: Operator Equalisation and Bloat Free GP. In: O'Neill, M., Vanneschi, L., Gustafson, S., Esparcia Alcázar, A.I., De Falco, I., Della Cioppa, A., Tarantino, E. (eds.) EuroGP 2008. LNCS, vol. 4971, pp. 110–121. Springer, Heidelberg (2008)
19. Banzhaf, W., Langdon, W.B.: Some considerations on the reason for bloat. In: Genetic Programming and Evolvable Machines, vol. 3, pp. 81–91. Springer, Netherlands (2002)
20. Ghodrat, M.A., Givargis, T., Nicolau, A.: Equivalence Checking of Arithmetic Expressions. In: CASES 2005, San Francisco, California. ACM, New York (2005)

Beneficial Preadaptation in the Evolution of a 2D Agent Control System with Genetic Programming

Lee Graham, Robert Cattral, and Franz Oppacher

School of Computer Science, Carleton University, 1125 Colonel By Drive,
K1S 5B6, Ottawa, Ontario, Canada
lee@stellaralchemy.com, rob@rencode.com,
oppacher@scs.carleton.ca

Abstract. We examine two versions of a genetic programming (GP) system for the evolution of a control system for a simple agent in a simulated 2D physical environment. Each version involves a complex behavior-learning task for the agent. In each case the performance of the GP system with and without initial epoch(s) of preadaptation are contrasted. The preadaptation epochs involve simplification of the learning task, allowing the evolved behavior to develop in stages, with rewards for intermediate steps. Both versions show an increase in mean best-of-run fitness when preadaptation is used.

Keywords: exaptation, preadaptation.

1 Introduction

This paper documents an attempt to make easier a complex behavior-learning task in GP [1] by starting simple and increasing the difficulty through a sequence of exaptation events. *Exaptation* (or co-option) is said to occur when a trait, evolved for some role or for none, is put to new use [2], [3], [4]. The term was proposed in part to counter implications of the term *preadaptation*, which carries with it a connotation of foresight not warranted in biological evolution [5]. When exaptation is put to deliberate use in evolutionary computing, however, the term "preadaptation", in reference to adaptation occurring prior to a predetermined exaptation event, is apt.

In the system described here, each exaptation event punctuates two epochs and allows behavior learned in one epoch to be a starting point in learning the task of the next. The task chosen is the control of an agent in a simulated 2D physical environment. Two versions of the system, called the *2D GP Navigator*, are presented, each with different learning tasks and exaptation events.

2 Version One

The task is to gather 20 target objects and bring them to a depot while avoiding collisions with walls and obstacles. The environment is a rectangular region bounded by walls. Obstacles are fixed polygons with which objects can collide. The depot is a fixed non-physical circle that acts as a drop-off destination during epochs in which the

L. Vanneschi et al. (Eds.): EuroGP 2009, LNCS 5481, pp. 303–314, 2009.
© Springer-Verlag Berlin Heidelberg 2009

learning task involves collecting targets and bringing them to the depot. Targets are small circles that move (in some cases of their own accord, in some cases due to external forces).

2.1   The Learning Tasks

The learning task depends on the epoch. There are six epochs divided into two sequences, one exaptive, the other not. The settings for each epoch are shown in Table 1. Sequence 1 constitutes the exaptive configuration, sequence 2 the non-exaptive. The latter's runs are split into two epochs to keep the total evaluation time equal to that of sequence 1. Both sequences involve 90 generations at an evaluation time of 160 simulation cycles followed by 60 generations at 1000 cycles. All of sequence 2, as well as the final epoch of sequence 1, use the same fitness function.

Tracked Aspects of Performance
Four measures are tracked in the various epochs and used to compute fitness values.

1) Distance from target(s). Whenever targets are present, one is designated as the *current target*. In each cycle, the distance from the navigator to this target is added to the *target distance sum*. The number of such measurements is the *target distance count*. During epochs with multiple targets, in which a target is collected, brought to the depot, and then a new one is sought for collection, dropping off a collected target at the depot results in an update of a second target distance statistic. This is the *multiple target-closeness measure*. It is initialized to 1.0, and, with each new target collected, is multiplied by

$$(S + 1) / (400 * \max(1, N)) \tag{1}$$

where S is the target distance sum, and N is the target distance count. Once the *multiple target-closeness measure* is updated, the values of S and N are reset to zero. This rewards both the collection of many targets and a speedy approach to each target.

During epochs in which targets are being collected, a target is automatically grasped and carried upon contact with the navigator provided that it is not already carrying one. Contact with any target will do; it need not be the designated *current target*.

2) Distance from depot. The distance from the navigator to the depot is tracked in the same way as its distance from the current target. Once a target is carried, the goal is to bring it to the depot. For this reason, distance measurements are only made while a target is being carried. The sums are called the *depot distance sum* and the *depot distance count*. A *depot-closeness measure* is also adjusted in the same manner as the *multiple target-closeness measure*. It is adjusted whenever a target is deposited at the depot. Depositing a target at the depot is done automatically when the navigator comes into contact with the depot, provided it is carrying a target at the time.

3) Maximum net distance traveled. At each cycle, the distance between the navigator and its starting location is computed. The maximum observed distance, called the *maximum distance reached*, is tracked.

4) Number of collisions with walls and obstacles. A check is made during each cycle to determine whether the navigator is in collision with walls or obstacles. If so, the *collision sum* is incremented. In either case the *collision check count* is also incremented. These allow the fitness function to determine the fraction of time spent in collision with objects.

Fitness Functions

There are three fitness functions used in the six epochs.

1) Fitness Function A. Fitness function A is

$$F_A = (T_s / \max(1, T_c)) - (D_m / 4) \tag{2}$$

where T_s is the *target distance sum*, T_c is the *target distance count*, and D_m is the *maximum distance reached*. It is used in epoch 1 of sequence 1. It rewards the navigator both for approaching its target and for moving away from its starting point.

2) Fitness Function B. Fitness function B is

$$P = (0.7 + 0.3[1 - (C_s / max(1, C_c))]), \quad F_B = P * F_A \tag{3}$$

where P is a collision penalty, C_s is the *collision sum* and C_c is the *collision check count*. This is an augmented version of fitness function F_A that takes into account collisions with walls and obstacles. Up to 30% of F_A can be lost due to collisions.

Table 1. Parameters and Settings for Epoch Sequences 1 & 2

	Seq. 1 Epochs				Seq. 2 Epochs	
	1	2	3	4	1	2
Tracked aspects of performance						
Distance from *current target*	*	*	*	*	*	*
Distance from depot				*	*	*
Max net distance traveled	*	*	*			
Collisions (walls & obstacles)		*	*	*	*	*
Fitness function	A	B	B	C	C	C
Settings						
Epoch termination generation	30	60	90	150	90	150
Obstacles present		*	*	*	*	*
***Current target* moves**			*	*	*	*
Targets	1	1	1	20	20	20
Cycles per evaluation	160	160	160	1000	160	1000
Target(s) are collectable				*	*	*

3) Fitness Function C. Fitness function C is

$$F_C = 10^5 P * ([(T_m(T_s + 1)) / 400m(T_c)] + [(P_m(P_s + 1)) / 400m(P_c)]) \tag{4}$$

where T_m is the *multiple target-closeness measure*, P_m is the *depot-closeness measure*, P_s is the *depot distance sum*, P_c is the *depot distance count*, and $m(x) = \max(1, x)$.

The final bracketed expression is not simply $(T_m + P_m)$ as this would not take into account the final target being pursued or carried when evaluation ends. The factor of 10^5 merely makes text output in a console window easier to read. F_C punishes wall and obstacle collisions and rewards the quick collection and delivery of many targets. Experiments are conducted to test the idea that fitness functions F_A and F_B can preadapt the population for F_C.

2.2 The GP System

The GP system settings for version one of the *2D GP Navigator* are listed in Table 2. Function nodes *FA*, *LA*, and *RA* correspond to *front*, *left*, and *right* antenna sensors, respectively. *TDR* and *TDS* are target *direction* and *distance* sensors. *DDR* and *DDS* are depot *direction* and *distance* sensors. *S* returns a *steering* value. *V* returns a *velocity* value. *ERC* is an "ephemeral" random constant. *TAR* returns 1 when the navigator is carrying a target, otherwise 0. *SS* takes a single argument and uses it to set the *steering* value, returning the new value. *SV* does the same for the *velocity* value. +, −, *, and / are bounded arithmetic operators, returning $min(1, max(-1, f(a,b)))$ for arguments *a* and *b*. / increases the absolute value of the divisor to 0.00001 if it is smaller. *SQR* is the sign-preserving *square* function. *SQRT* is the sign-preserving *square-root* function. *IF* evaluates its first argument and, if the result is negative, returns the third, else the second. *NEG* reverses sign. All nodes take inputs in the range [-1.0,+1.0] and return values in the same range.

Table 2. GP System Settings

Setting, Parameter, or Option	Value
Function set	{*FA, LA ,RA, TDR, TDS, DDR, DDS, S, V, ERC, TAR, SS, SV, +, −, *, /, SQR, SQRT, IF, NEG*}
Halting condition	150 generations
Population size	50 (*small due to costly fitness evaluation in terms of processing time*)
Elitism	1 elite individual
Duplicates	Duplicates are permitted
Variation operators	50% crossover, 50% mutation
Selection	Tournaments of size 4
Population initialization	Randomly-generated trees with between 1 and 100 nodes.

Half of all crossovers use normal GP crossover; the rest use a crossover that traverses parents in lock-step pre-order, reverting to recursive copying wherever the two parents differ, with each parent having an equal chance of its version of a subtree being copied to the child. Half of all mutations replace randomly-chosen subtrees with randomly-generated subtrees, with half of the remaining using a node-lowering mutation that picks a node at random from the tree and makes it a child of its randomly-chosen replacement (should it require siblings to satisfy the arity of its new parent node, these are provided as randomly-generated trees of 1 to 8 nodes), and half using a node-raising mutation that picks a node at random and uses it to replace its parent in the tree. All operators limit the resulting tree size to 1000 nodes or less.

The growth method used to generate random trees begins with a single node and a desired tree size chosen at random from a predetermined range. It then repeatedly performs leaf-expansion until the desired tree size is reached or exceeded. If it is exceeded, repeated leaf-raises are performed until the tree is less than or equal to the desired size. This cycle is repeated a maximum of 10 times. The resulting tree is no bigger than the desired size.

2.3 The Navigator

The navigator is a triangle that is moved by having its control system set linear and angular components of its velocity at regular intervals. The GP tree executes once every cycle. A cycle advances the simulation by 0.4 seconds. The physics engine used is the open-source "Chipmunk" engine [6]. Each cycle involves four steps.

1. The navigator executes its GP tree.Components of the triangle's velocity may be set multiple times. Only the final values will be assigned.
2. The physics simulation is advanced by 0.12 seconds.
3. All forces on the navigator's triangle are zeroed.
4. The physics simulation is advanced by 0.28 seconds.

Sensors and Effectors

The navigator possesses seven sensors and two effectors. Three sensors, called the *left antenna sensor*, the *right antenna sensor*, and the *front antenna sensor*, report on the presence of obstacles or walls within a small radius around the tips of three antennas protruding from the left, right, and front vertices of the triangle. Neither the antennas nor the circles are objects that collide with other objects. All objects pass through them unaffected. Each antenna sensor, when queried, returns a value between 0.0 and 1.0. A value of 0.0 indicates that no wall or obstacle is within a distance of r from its tip, where r is the radius of sensitivity. A value of 1.0 indicates that a wall or obstacle is touching the tip. Intermediate values reflect different degrees of proximity.

Two sensors provide information on the position of the current target relative to the navigator. The *target direction sensor* gives a reading of 0.0 if the target is directly ahead of the navigator, with negative and positive values indicating that the target is to the right or left. The *target distance sensor* returns a value equal to

$$(1 - (3500 / (d^2 + 3500)))^2 \qquad (5)$$

where d is the Euclidean distance between the navigator and the target. The value is small when the target is close and large when far away. Two sensors provide information on the position of the depot relative to the navigator. These operate in the same manner as their target sensor counterparts. The navigator has only two effectors. The first is controlled using the *SS* node. This takes a single argument which is then used directly to set the angular velocity of the navigator (the *steering* value). The second is controlled using the *SV* node. This also takes a single argument. The linear velocity of the navigator is then reduced by 20%, and a new velocity vector is added to it, with magnitude proportional to the node's input, directed in the navigator's forward direction. The resulting velocity, if it exceeds a fixed maximum magnitude, is reduced to that maximum.

2.4 Early Termination of Evaluations

To reduce processing time, individuals that do not seem promising are terminated early before evaluation time expires. Checks occur at the one-, two-, and three-quarter marks in an evaluation. At each check point, fitness is compared to that of the individuals in the previous generation as they were at the time of same check point. If fitness is below the median, evaluation ends. These checks only take effect after the third generation in any epoch.

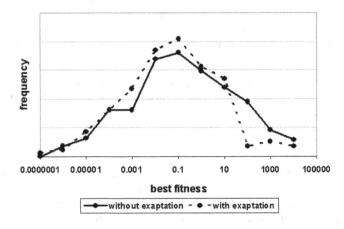

Fig. 1. Distribution showing exaptation producing many fewer of the least fit results

Table 3. Statistics for Best-of-Run Log-Fitness

Configuration	Without exaptation	With exaptation
Mean best log-fitness	-1.24	-1.75
Standard deviation	2.24	2.06
Confidence interval (95%)	±0.26	±0.21
Samples	283	347

2.5 Results

Fitness function F_C returns values that range over many orders of magnitude. The mean of many such values is dominated by the worst values and does not reflect any central tendency in the data. Thus, each fitness value, f, recorded from function F_C is converted to $f = \log_{10}(f)$. Table 3 gives the best-of-run fitness statistics for both configurations. Both show roughly the same variability; neither is more robust or consistent. Sufficient runs were gathered to provide a good degree of statistical power in separating the two means. A Mann-Whitney U test [7] comparing the best-of-run values gives a p value of 0.0049. Fig. 1 shows that the outcome distributions are roughly the same when considering the frequency of highly fit solutions (at left). Differences are in the mid-range and poor outcomes. The exaptive system produces more of the former and fewer of the latter.

3 Version Two

Version two of the 2D navigator reflects an attempt to get around problems that arose in version one, and to amplify the benefit of exaptation. Version one's most significant departure from an ideal demonstration of the potential power of exaptation stems from the ability of the GP system to frequently excel at the learning task even without preadaptation. Ideally, we would like to demonstrate the benefit of preadaptation on a learning task in which the non-exaptive approach fares much more poorly. A number of modifications are made in version two. The two greatest enhancements, in terms of rendering the task more difficult for the non-exaptive GP system, are additions of two crucial new navigator responsibilities. In version one, the navigator's sensors track a new target-depot pair automatically upon deposition of a carried target. Version two moves this burden onto the navigator itself. It must determine on its own when to make this switch. In version one, a target is carried or deposited at the depot by having the navigator touch the appropriate object. Version two requires that the target be pushed to the depot. The environment encountered by the navigator differs from that of version one in the following four ways.

1) Open arena. Version one kept objects within a rectangular region using four walls. These are gone in version two, granting the navigator the ability to travel far in any direction. This makes learning more difficult since the navigator might not remain near the targets with which it is meant to interact.

2) Dispersed targets. The targets are distributed roughly evenly throughout what was originally a bounded rectangular region. Results from version one showed that navigators were able to take advantage of the clustered targets, picking up many without using sensors to locate them.

3) Multiple depots. There are 20 depots, one for each target, also distributed roughly evenly. A target should only be brought to its associated depot. Target-depot pairings are chosen at random each generation. This makes learning more difficult since each target's position relative to its depot will rarely be the same twice in a row.

4) No obstacles. Obstacles are absent in version two.

3.1 The Learning Tasks

There are three epochs divided into two sequences, one exaptive, the other not. The final learning task in each is to maximize the number of targets that are at their depots when evaluation is finished. Settings for each epoch are shown in Table 4. Sequence 1 constitutes an exaptive approach while sequence 2 corresponds to a non-exaptive approach. Both sequences involve the same total amount of simulation time. Sequence 1 begins with a preadapting epoch that rewards intermediate work in bringing targets to their depots, using only a single target-depot pair for training, and short evaluations. The second epoch involves the same fitness function as is used in the sequence 2. It rewards navigators based entirely on the number of targets at their depots when evaluation ends.

Table 4. Parameters and Settings for Epoch Sequences 1 & 2

	Seq. 1 Epochs		Seq. 2 Epochs
	1	2	1
Tracked aspects of performance			
Timing of first attempted target/depot switch	*		
Total navigator movement	*		
Distance of navigator from target	*		
Total target movement	*		
Distance of target from depot	*		
Fitness function	D	E	E
Settings			
Generations	80	71	79
Targets	1	20	20
Depots	1	20	20
Evaluation cycles per individual	300	3000	3000
Evaluation cycles per epoch	1.2×10^6	1.065×10^7	1.185×10^7

Tracked Aspects of Performance

Five performance aspects are measured during evaluation in the preadapting epoch of sequence 1. The fitness function of the final epoch in each sequence involves none of these five. It makes a single measurement – the number of targets at their depots – only after evaluation ends.

1) Timing of First Attempted Target/Depot Switch. Since the preadapting epoch involves only a single target-depot pair, using the SW node to divert the navigator's sensors to a new pair has no effect. Nevertheless, in order to preadapt the population for the more complex task of the second epoch, the use of the SW node is monitored and factored into the fitness function. The *SW* node switches the sensors to a new target-depot pair whenever the last *SW* call in a GP tree receives an argument with a value greater than 0.5. A GP tree may include many *SW* nodes; only the last of these to be executed is heeded. The first time such a switch is made, the evaluation cycle number (between 1 and 300) is recorded for use by the fitness function.

2) Total Navigator Movement. The preadapting epoch divides fitness values into four tiers. The lowest of these rewards navigators for simply moving. The navigator's movement is tracked by computing the Euclidean distance between its current position and its position in the previous cycle. These measurements are summed over all 300 cycles.

3) Distance of Navigator from Target. The second tier rewards not just for moving, but for moving closer to the target. The distance between the navigator and the target is measured before the evaluation begins and after it ends. The difference is used to determine whether the net movement was toward or away from the target.

4) Total Target Movement. The third tier is for navigators that move close enough to the target to bump it. Navigators in this tier are rewarded for simply pushing the target. Like total navigator movement, above, the distance between the target's current position and its position in the previous cycle is summed over all 300 cycles.

5) Distance of Target from Depot. The fourth tier rewards pushing the target closer to the depot. The distance between the two is measured before and after evaluation. The distance of the target from the depot is also recorded at the moment an SW node is first used to switch sensors to a new target-depot pair. This allows the fitness function to reward based on the timing of the switch with respect to the final crucial step in its task – pushing the target to its depot – alongside the usual reward based on the timing relative to the overall evaluation time.

Fitness Functions
Version two uses two fitness functions, F_D and F_E. Only F_D uses the five measures described above. F_E measures only the number of targets at their depots when evaluation ends. Early termination of fitness evaluation is not used.

1) Fitness Function D. Fitness function D's return values are divided into four tiers, each separated by one million points to prevent tier overlap. Tier 1 simply rewards movement. Tier 2 rewards moving closer to the target. Tier 3 rewards pushing the target. Tier 4 rewards pushing the target closer to the depot. Each tier takes into account the timing of the use of the *SW* node. This occurs in an environment with a single target and depot. Thus, the use of the *SW* node can be factored into the fitness function without the severe negative consequences that its misuse can produce when there are many targets and depots. When there are many targets and depots, switching to a new target-depot pair can mean abandoning partial work done on the current target-depot pair. Many randomly-generated individuals make switches often, resulting in little to no work being done in bringing targets to their depots. They move erratically for the duration of an evaluation. By contrast, navigators that never switch cannot expect to bring more than a single target to its depot. To bring 20 targets to their depots, use of the *SW* node must be carefully timed.

Dividing fitness values into non-overlapping tiers, allows the population to learn in stages and at its own pace. Individuals in higher tiers will win out against those in lower tiers but will defeat individuals in the same tier only when outperforming them at the task evaluated in that tier. Pseudo-code for F_D is given below.

```
timing = 1 // Used to penalize early use of SW node
if (navigator switched target-depot pair)
    delta = cycle number of first target-depot switch
    timing = 9 * delta / (total evaluation time)
if (target was pushed closer to depot)                      // TIER 4
    if (SW node was activated at all)
        a = initial target-to-depot distance
```

```
        b = distance when SW was first activated
            boost = 100 - (10 * b / a)
     else boost = 10
     d = initial - final target-to-depot distance
     return( 1000000 - (d * timing * boost) )
if (total target movement > 5.0)                    // TIER 3
     return( 2000000 - (total target movement * timing) )
d = initial - final navigator-to-target distance
if (d > 5.0) return( 3000000 - (d * timing) )       // TIER 2
return( 4000000 - (total navigator movement * timing) )  // TIER 1
```

2) Fitness Function E. Fitness function E is a count of the number of targets at their depots when evaluation ends. This provides *very* impoverished performance feedback to the evolutionary process, and no reward for intermediate work. Its value is

$$F_E = 1 / (T + 1) \tag{6}$$

where T is the number of targets at their depots.

Table 5. GP System Settings

Setting, Parameter, or Option	Value
Function set	{*FA, LA, RA, TDR, TDS, DDR, DDS, S, V, ERC, TAR, SS, SV, +, −, *, /, SQR, SQRT, IF, NEG, SW*}
Halting condition	151 generations (Sequence 1) 79 generations (Sequence 2)

3.2 The GP System

GP system settings are listed in Table 5. Only those that differ from version one are given. The function set is identical to that of version one except for the addition of *SW*. Because there are no walls or obstacles, and because targets are pushed instead of carried, *FA*, *LA*, *RA*, and *TAR* act as constants.

3.3 Results

Since fitness function F_E uses inversion to convert a maximization problem into a minimization problem, all logged fitness values are converted to $T = (1 / f) - 1$, called the *converted fitness*.

Table 6. Statistics for Best-of-Run Converted Fitness

Configuration	Without exaptation	With Exaptation
Mean best converted fitness	2.96	6.10
Standard deviation	2.84	3.62
Confidence interval (95%)	±0.63	±0.98
Number of samples	78	52

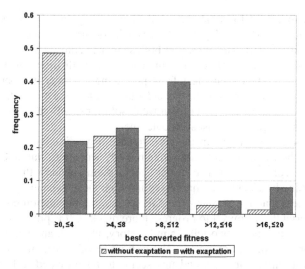

Fig. 2. Distribution showing that exaptation produces many fewer of the least fit results

The best-of-run fitness statistics for each sequence are given in Table 6. A Mann-Whitney U test comparison gives $p \leq 0.0001$. Both the exaptive and non-exaptive configurations have a comparable amount of variation in their outcomes. The overlap in outcome distributions, however, is less extreme than that of version one, showing an even greater benefit for the exaptive approach.

Fig. 2 shows how the outcomes are spread across the full range of 0 to 20. As with version one, the most obvious difference between the exaptive and non-exaptive approaches is in the poorest outcomes, with the exaptive approach being about half as likely as the non-exaptive approach to produce control systems that perform extremely poorly. The balance of outcomes shows the exaptive approach exceeds the non-exaptive approach, to greater or lesser degrees, in all other bins in the graph.

4   Conclusion

The *2D GP Navigator* provides examples of complex task learning with and without prior training on simplified versions of the learning task or rewards for intermediate steps. The exaptive approaches, which add new complexity to the task with each epoch or tier, successfully outperform otherwise identical non-exaptive approaches. In version one the benefit was moderate, whereas the improvement in version two, with its harsher task and impoverished feedback, was more substantial.

Several unexpected aspects of the behavior of the system seem to be evident in the fitness data and in animations of navigators in action [8], [9]. In version one the non-exaptive configuration was able to produce highly fit solutions. At the outset this had seemed unlikely given the complexity of the task. Observations of the behavior of fit individuals, such as [8], indicate that the placement of targets allowed many to be collected without requiring the ability to sense them. Because targets are clustered, the navigator can expect to gather many of them simply by repeatedly traversing a small area. Many target and depot collisions will occur, and many targets will be deposited with little effort. This occurs in both sequence configurations. Nevertheless, the exaptive configuration is more efficient. The outcome distributions in Fig. 1 show that the boosted advantage stems less from producing a greater number of highly fit solutions, and more from producing considerably less of the poorest. The exaptive approach seems less likely to converge on weak solutions, with its epochs of preadaptation acting as a safeguard against such outcomes.

One other noteworthy aspect of the performance in version one is that in the final 60 generations, the exaptive configuration is initially worse off, but quickly overtakes the non-exaptive configuration. The earlier epochs seem to endow the population with beneficial building blocks that are harnessed later on. Despite entering the final epoch with poorer fitness than the non-exaptive GP system, the population possesses a store of genetic material that allows it to progress more quickly. It is preadapted.

The results from version two tell a similar story, but with an even more exaggerated benefit for the exaptive approach. The task was made more difficult than that of version one in a variety of ways and this seems to have helped in highlighting the potential strength of exaptation and learning in stages.

What is perhaps most remarkable about version two is the ability of the non-exaptive system to solve the problem at all. Given the crude and impoverished feedback provided by the fitness function, we expected no progress. Its successful runs, despite being infrequent, are still somewhat surprising. This speaks to the strength and tenacity of the GP problem-solving technique.

References

1. Koza, J.R.: Genetic Programming. MIT Press, Cambridge (1992)
2. Gould, S.J., Vrba, E.: Exaptation: A Missing Term in the Science of Form. Paleobiology 8(1), 4–15 (1982)
3. Andrews, P.W., Gangestad, W., Matthews, D.: Adaptationism – How to Carry Out an Exaptationist Program. Behavioral and Brain Sciences 25(4), 489–553 (2002)
4. Gould, S.J.: The Structure of Evolutionary Theory, ch. 11. The Belknap Press of Harvard University Press, Cambridge (2002)
5. Frazzetta, T.H.: Complex Adaptations in Evolving Populations. Sinauer Associates, Sunderland (1975)
6. Lembcke, S.: Chipmunk Game Dynamics (last accessed, July 2008), http://wiki.slembcke.net/main/published/Chipmunk
7. Mann, H.B., Whitney, D.R.: On a Test of Whether One of Two Random Variables is Stochastically Larger Than the Other. Annals of Mathematical Statistics 18, 50–60 (1947)
8. Graham, L.: Animation of an Evolved 2D Navigator (last accessed, October 2008), http://www.youtube.com/watch?v=GrnHnGwmZ7A
9. Graham, L.: Animation of an Evolved 2D Navigator (last accessed, October 2008), http://www.youtube.com/watch?v=i4f5gT-GV9M

Adaptation, Performance and Vapnik-Chervonenkis Dimension of Straight Line Programs

José L. Montaña[1], César L. Alonso[2], Cruz E. Borges[1], and José L. Crespo[3]

[1] Departamento de Matemáticas, Estadística y Computación,
Universidad de Cantabria, 39005 Santander, Spain
{montanjl,borgesce}@unican.es
[2] Centro de Inteligencia Artificial, Universidad de Oviedo
Campus de Viesques, 33271 Gijón, Spain
calonso@aic.uniovi.es
[3] Departamento de Matemática Aplicada, Estadística y Ciencias de la Computación,
Universidad de Cantabria, 39005 Santander, Spain
luis.crespo@unican.es

Abstract. We discuss here empirical comparison between model selection methods based on Linear Genetic Programming. Two statistical methods are compared: model selection based on Empirical Risk Minimization (ERM) and model selection based on Structural Risk Minimization (SRM). For this purpose we have identified the main components which determine the capacity of some linear structures as classifiers showing an upper bound for the Vapnik-Chervonenkis (VC) dimension of classes of programs representing linear code defined by arithmetic computations and sign tests. This upper bound is used to define a fitness based on VC regularization that performs significantly better than the fitness based on empirical risk.

Keywords: Genetic Programming, Linear Genetic Programming, Vapnik-Chervonenkis dimension.

1 Introduction

Throughout these pages we study some theoretical and empirical properties of a new structure for representing computer programs in the GP paradigm. This data structure –called *straight line program* (slp) in the framework of Symbolic Computation ([1])– was introduced for the first time into the GP setting in [2]. A slp consists of a finite sequence of computational assignments. Each assignment is obtained by applying some functional (selected from a specified) to a set of arguments that can be variables, constants or pre-computed results. The slp structure can describe complex computable functions using less amount of computational resources than GP-trees. The key point for explaining this feature is the ability of slp's for reusing previously computed results during the evaluation process. Another advantage with respect to trees is that the slp structure

L. Vanneschi et al. (Eds.): EuroGP 2009, LNCS 5481, pp. 315–326, 2009.

can describe multivariate functions by selecting a number of assignments as the output set. Hence one single slp has the same representation capacity as a forest of trees (see [2] for a complete presentation of this structure).

Linear Genetic Programming (LGP) is a GP variant that evolves sequences of instructions from an imperative programming language or from a machine language. The structure of the program representation consists of assignments of operations over constants or memory variables called registers, to another registers (see [3] for a complete overview on LGP). The GP approach with slp's can be seen as a particular case of LGP where the data structures representing the programs are lists of computational assignments.

We study the practical performance of *ad-hoc* recombination operators for slp's. We apply the SLP-based GP approach to solve some instances of the symbolic regression problem. Experimentation done over academic examples uses a weak form of structural risk minimization and suggests that the slp structure behaves very well when dealing with bounded length individuals directed to minimize a compromise between empirical risk and non-scalar length (i.e. number of non-linear operations used by the structure). We have calculated an explicit upper bound for the Vapnik-Chervonenkis dimension (VCD) of some particular classes of slp's. This bound constitutes our basic tool in order to perform structural risk minimization of the slp structure.

2   Straight Line Programs: Basic Concepts and Properties

Straight line programs are commonly used for solving problems of algebraic and geometric flavor. An extensive study of the use of slp's in this context can be found in [4]. The formal definition of slp's we provide in this section is taken from [2].

Definition 1. *Let $F = \{f_1, \ldots, f_n\}$ be a set of functions, where each f_i has arity a_i, $1 \leq i \leq n$, and let $T = \{t_1, \ldots, t_m\}$ be a set of terminals. A straight line program (slp) over F and T is a finite sequence of computational instructions $\Gamma = \{I_1, \ldots, I_l\}$, where for each $k \in \{1, \ldots, l\}$,*

$I_k \equiv u_k := f_{j_k}(\alpha_1, \ldots, \alpha_{a_{j_k}}); with \ f_{j_k} \in F,$
$\alpha_i \in T$ *for all i if $k = 1$ and $\alpha_i \in T \cup \{u_1, \ldots, u_{k-1}\}$ for $1 < k \leq l$.*

The set of terminals T satisfies $T = V \cup C$ where $V = \{x_1, \ldots, x_p\}$ is a finite set of variables and $C = \{c_1, \ldots, c_q\}$ is a finite set of constants. The number of instructions l is the length of Γ.

Usually an slp $\Gamma = \{I_1, \ldots, I_l\}$ will be identified with the set of variables u_i introduced at each instruction u_i, thus $\Gamma = \{u_1, \ldots, u_l\}$. Each of the non-terminal variables u_i can be considered as an expression over the set of terminals T constructed by a sequence of recursive compositions from the set of functions F. Following [2] we denote by $SLP(F, T)$ the set of all slp's over F and T.

Example 1. Let F be the set given by the three binary standard arithmetic operations $F = \{+, -, *\}$ and let $T = \{1, x_1, x_2, \ldots, x_n\}$ be the set of terminals. Any slp Γ in $SLP(F, T)$ represents a n-variate polynomial with integer coefficients.

An output set of a slp $\Gamma\{u_1,\ldots,u_l\}$ is any set of non-terminal variables of Γ, that is $O(\Gamma) = \{u_{i_1},\ldots,u_{i_t}\}$. The function computed by a slp $\Gamma = \{u_1,\ldots,u_l\}$ over F and T with set of terminal variables $V = \{x_1,\ldots,x_p\}$ and with output set $O(\Gamma) = \{u_{i_1},\ldots,u_{i_t}\}$, denoted by $\Phi_\Gamma : I^p \rightarrow O^t$, is defined recursively in the natural way and satisfies $\Phi_\Gamma(a_1,\ldots,a_p) = (b_1,\ldots,b_t)$, where b_j stands for the value of the expression over V of the non terminal variable u_{i_j} when we replace each variable x_k with a_k; $1 \le k \le p$.

3   Vapnik-Chervonenkis Dimension of Families of *slp*'s

In the last years GP has been applied to a range of complex learning problems including that of classification and symbolic regression in a variety of fields like quantum computing, electronic design, sorting, searching, game playing, etc. A common feature of both tasks is that they can be thought of as a supervised learning problem (see [5]) where the hypothesis class $\mathcal{C}$ is the search space described by the genotypes of the evolving structures. In the seventies the work by Vapnik and Chervonenkis ([6], [7], [8]) provided a remarkable family of bounds relating the performance of a learning machine (see [9] for a modern presentation of the theory). The Vapnik- Chervonenkis dimension (VCD) is a measure of the capacity of a family of functions (or learning machines) as classifiers. The VCD depends on the class of classifiers. Hence, it does not make sense to calculate VCD for GP in general, however it makes sense if we choose a particular class of computer programs as classifiers. Our aim is to study in depth the formal properties of GP algorithms focusing on the analysis of the classification complexity (VCD) of straight line programs.

3.1   Estimating the VC Dimension of Slp's Parameterized by Real Numbers

The following definition of VC dimension is standard. See for instance [7].

Definition 2. *Let $\mathcal{C}$ be a class of subsets of a set X. We say that $\mathcal{C}$ shatters a set $A \subset X$ if for every subset $E \subset A$ there exists $S \in \mathcal{C}$ such that $E = S \cap A$. The VC dimension of $\mathcal{C}$ is the cardinality of the largest set that is shattered by $\mathcal{C}$.*

Through this section we deal with concept classes $\mathcal{C}_{k,n}$ such that concepts are represented by k real numbers, $w = (w_1,...,w_k)$, instances are represented by n real numbers, $x = (x_1,...,x_n)$, and the membership test to the family $\mathcal{C}_{k,n}$ is expressed by a formula $\Phi_{k,n}(w,x)$ taking as inputs the pair concept/instance (w,x) and returning the value 1 if "x belongs to the concept represented by w" and 0 otherwise

We can think of $\Phi_{k,n}$ as a function from $\mathbb{R}^{k+n}$ to $\{0,1\}$. So for each concept w, define:

$$C_w := \{x \in \mathbb{R}^n : \Phi_{k,n}(w,x) = 1\}. \tag{1}$$

The goal is to obtain an upper bound on the VC dimension of the collection of sets

$$\mathcal{C}_{k,n} = \{C_w : w \in \mathbb{R}^k\}. \tag{2}$$

Now assume that the formula $\Phi_{k,n}$ is a boolean combination of s atomic formulas, each of them having the following forms:

$$\tau_i(w, x) > 0 \tag{3}$$

or

$$\tau_i(w, x) = 0 \tag{4}$$

where $\{\tau_i(w, x)\}_{1 \leq i \leq s}$ are infinitely differentiable functions from $\mathbb{R}^{k+n}$ to $\mathbb{R}$. Next, make the following assumptions about the functions τ_i. Let $\alpha_1, ..., \alpha_v \in \mathbb{R}^n$. Form the sv functions $\tau_i(w, \alpha_j)$ from $\mathbb{R}^k$ to $\mathbb{R}$. Choose $\Theta_1, ..., \Theta_r$ among these, and define

$$\Theta : \mathbb{R}^k \to \mathbb{R}^r \tag{5}$$

as

$$\Theta(w) := (\Theta_1(w), ..., \Theta_r(w)) \tag{6}$$

Assume there is a bound B independent of the α_i, r and $\epsilon_1, ..., \epsilon_r$ such that if $\Theta^{-1}(\epsilon_1, ..., \epsilon_r)$ is a $(k-r)$-dimensional $\mathcal{C}^\infty$- submanifold of $\mathbb{R}^k$ then $\Theta^{-1}(\epsilon_1, ..., \epsilon_r)$ has at most B connected components.

With the above setup, the following result is proved in [10].

Theorem 1. *The VC dimension V of a family of concepts $\mathcal{C}_{k,n}$ whose membership test can be expressed by a formula $\Phi_{k,n}$ satisfying the above conditions satisfies:*

$$V \leq 2 \, log_2 \, B + 2k \, log_2 \, (2es) \tag{7}$$

Next we state our main result concerning the VCD of a collection of subsets accepted by a family of slp's. We will say that a subset $C \subset \mathbb{R}^n$ is accepted by a slp Γ if the function computed by Γ, Φ_Γ, expresses the membership test to C. For slp's $\Gamma = (u_1, \ldots, u_l)$ of length l accepting sets we assume that the output is the last instruction u_l and takes values in $\{0, 1\}$.

Theorem 2. *Let $T = \{t_1, \ldots, t_n\}$ be a set of terminals and let $F = \{+, -*, /, sign\}$ be a set of functionals where $\{+, -, *, /\}$ denotes the set of standard arithmetic operations and $sign(x)$ is a function that outputs 1 if its input $x \in \mathbb{R}$ satisfies $x \geq 0$ and outputs 0 otherwise. Let $\Gamma_{n,L}$ be the collection of slp's Γ over F and T using at most L non-scalar operations (i.e. operations in $\{*, /, sign\}$) and a free number of scalar operations (i.e. operations in $\{+, -\}$) whose output is obtained by applying the functional $sign$ either to a previously computed result or to a terminal t_j, $1 \leq j \leq n$. Let $\mathcal{C}_{n,L}$ be the class of concepts defined by the subsets of $\mathbb{R}^n$ accepted by some slp belonging to $\Gamma_{n,L}$. Then*

$$VC - dim(\mathcal{C}_{n,L}) \leq 2(n+1)(n+L)L(2L + log_2 \, L + 9) \tag{8}$$

Sketch of the proof. The first step in the proof consist of constructing of a universal slp Γ_U, over sets F_U an T_U, that parameterizes the elements of the family $\Gamma_{n,L}$. The definition of F_U and T_U depends only on the parameters n and L and will be clear after the construction. The key idea in the definition of Γ_U is the introduction of a set of parameters α, β taking values in $\{0,1\}^k$, for a suitable natural number k, such that each specialization of this set of parameters yields a particular slp belonging to $\Gamma_{n,L}$ and conversely, each slp in $\Gamma_{n,L}$ can be obtained specializing the parameters α, β. For this purpose define $u_{-n+m} = t_m$ for $1 \leq m \leq n$. Note that any non-scalar assignment u_i, $1 \leq i \leq L$, in a slp Γ belonging to $\Gamma_{n,L}$ is a function of $t = (t_1, \ldots, t_n)$ that can be parameterized as follows.

$$u_i = U_i(\alpha, \beta)(t) = \alpha^i_{-n}(\sum_{j=-n+1}^{i-1} \alpha_j{}^i u_j) * (\sum_{j=-n+1}^{i-1} \beta_j{}^i u_j) + \tag{9}$$

$$+ (1 - \alpha^i_{-n})[\beta^i_{-n} \frac{\sum_{j=-n+1}^{i-1} \alpha_j{}^i u_j}{\sum_{j=-n+1}^{i-1} \beta_j{}^i u_j} + (1 - \beta^i_{-n})sgn(\sum_{j=-n+1}^{i-1} \beta_j{}^i u_j)], \tag{10}$$

for some suitable values $\alpha = (\alpha_j{}^i)$, $\beta = (\beta_j{}^i)$, with $\alpha_j{}^i, \beta_j{}^i \in \{0,1\}$.

Let us consider the family of parametric slp's $\{\Gamma_{(\alpha,\beta)}\}$ where for each (α, β) the slp $\Gamma_{(\alpha,\beta)} := (U_1(\alpha, \beta), \ldots, U_L(\alpha, \beta))$. Next replace the family of concepts $C_{n,L}$ with the class of subsets of $\mathbb{R}^n$ $\mathcal{C} := \{C_{(\alpha,\beta)}\}$ where for each (α, β), the set $C_{(\alpha,\beta)}$ is given as follows.

$$C_{(\alpha,\beta)} := \{t = (t_1, \ldots, t_n) \in \mathbb{R}^n : t \text{ is accepted by } \Gamma_{(\alpha,\beta)}\} \tag{11}$$

In the new class $\mathcal{C}$ parameters $\alpha_j{}^i$, $\beta_j{}^i$ are allowed to take values in $\mathbb{R}$. Since $C_{n,L} \subset \mathcal{C}$ it is enough to bound the VC dimension of $\mathcal{C}$.

Claim A. The number of parameters α, β is exactly

$$(n + 1)(n + L)L \tag{12}$$

Claim B. For each i, $1 \leq i \leq L$, the following holds:

(1) The function $U_i(\alpha, \beta)(t)$ is a piecewise rational function in the variables α, β, t of formal degree bounded by $3.2^i - 2$.
(2) U_i is defined up to a set of zero measure and there is a partition of the domain of definition of U_i by subsets $(\Omega_j^i)_{1 \leq j \leq n_i}$ with $n_i \leq 2^i$ such that each Ω_j^i is defined by a conjunction of i rational inequalities of the form $p \geq 0$ or $p < 0$ with degree $\deg p \leq 3.2^i - 2$. Moreover, the restriction of U_i to the set Ω_j^i, $U_i|_{\Omega_j^i}$, is a rational function of degree bounded by $3.2^i - 2$.
(3) Condition $U_L(\alpha, \beta)(t) = 1$ can be expressed by a boolean formula of the following form:

$$\bigvee_{1 \leq i \leq 2^L} \wedge_{1 \leq j \leq L} p_{i,j} \epsilon_{i,j} 0; \tag{13}$$

where for each i, j, $p_{i,j}$ is a rational function in the variables α, β, t of degree bounded by $3.2^L - 2$ and $\epsilon_{i,j}$ is a sign condition in $\{\geq, <\}$.

Proof. Claim A follows by counting parameters. Claim B follows by induction on i.

In order to achieve the result in Equation 8, according Theorem 1 we have to estimate the number of connected components of a set defined by r equations of the form:

$$\Theta_1(t^1, \alpha, \beta) - \epsilon_1 = 0, \ldots, \Theta_r(t^r, \alpha, \beta) - \epsilon_r = 0, \tag{14}$$

where $\epsilon_i, \in \mathbb{R}$, $t^i \in \mathbb{R}^n$, and from Claim B, item 3, the $\Theta_i(t^i, \alpha, \beta)$ are polynomials in the variables (α, β) of degree bounded by $d = 3.2^L - 2$. Fortunately the solution to this problem is given by the following result by J. Milnor ([11]).

Lemma 1. *Assume that $\Theta_1(w), \ldots, \Theta_r(w)$ are polynomials in k variables with degree at most d. Then, if $\epsilon = (\epsilon_1, \ldots, \epsilon_r)$ is a regular value of $\Theta(w) := (\Theta_1(w), \ldots, \Theta_r(w))$ as defined above, then*

$$B \leq (2d)^k \tag{15}$$

To conclude it is enough to apply Theorem 1 with $k = (n+1)(n+L)L$, $s = 6.L.2^L$ and $B \leq (2(3.2^L - 2))^k$.

3.2   Estimating the Average Error of slp's

We show how to apply the bound in Equation 8 to estimate the average error with respect to the unknown distribution from which the examples are drawn.

Let $\Gamma_{(\alpha,\beta)} := (U_1(\alpha, \beta), \ldots, U_L(\alpha, \beta))$ be a family of parametric slp's as defined in the former Section . The average error of a classifier with parameters (α, β) is

$$\varepsilon(\alpha, \beta) = \int Q(t, \alpha, \beta; y) d\mu, \tag{16}$$

where Q measures the loss between the semantic function of $\Gamma_{(\alpha,\beta)}$ and the target concept, and μ is the distribution from which examples $\{(t_i, y_i)\}_{1 \leq i \leq m}$ are drawn to the GP machine. For classification problems, the error of misclassification is given taking $Q(t, \alpha, \beta; y) = |y - \Gamma_{(\alpha,\beta)}(t)|$. Similarly, for regression tasks one takes $Q(t, \alpha, \beta; y) = (y - \Gamma_{(\alpha,\beta)}(t))^2$. Since μ is usually unknown, one replaces theoretical error $\varepsilon(\alpha, \beta)$ by empirical error $\varepsilon_m(\alpha, \beta) = \frac{1}{m} \sum_{i=1}^{m} Q(t_i, \alpha, \beta; y_i)$. Now, the results by Vapnik state that the average error $\varepsilon(\alpha, \beta)$ can be estimated independently of the distribution of $\mu(t, y)$ due to the following formula.

$$\varepsilon(\alpha, \beta) \leq \varepsilon_m(\alpha, \beta) + \sqrt{\frac{h(log(2m/h) + 1) - log(\eta/4)}{m}}, \tag{17}$$

Here h must be substituted by the upper bound given in Equation 8. The constant η is the probability that the bound is violated.

4   SLP-Based Genetic Programming

For the construction of each individual $\Gamma \in SLP(F, T)$ of the initial population we adopt the process described in [2]. For each instruction $u_k \in \Gamma$ first an element $f \in F$ is random selected and then the function arguments of f are also randomly chosen in $T \cup \{u_1, \ldots, u_{k-1}\}$, if $k > 1$ and in T if $k = 1$. We keep homogeneous populations of equal length slp's. Next, we describe the recombination operator.

Definition 3. *(slp-crossover)(see [2]) Let $\Gamma = \{u_1, \ldots, u_L\}$ and $\Gamma' = \{u'_1, \ldots, u'_L\}$ be two slp's over F and T. First, a position k in Γ is randomly selected; $1 \leq k \leq L$. Let $S_{u_k} = \{u_{j_1}, \ldots, u_{j_m}\}$ be the piece of the code of Γ related to the evaluation of u_k with the assumption that $j_1 < \ldots < j_m$. Next randomly select a position t in Γ' with $m \leq t \leq L$ and modify Γ' by making the substitution of the subset of instructions $\{u'_{t-m+1}, \ldots, u'_t\}$ in Γ', by the instructions of Γ in S_{u_k} suitably renamed. The renaming function $\mathcal{R}$ over S_{u_k} is defined as $\mathcal{R}(u_{j_i}) = u'_{t-m+i}$, for all $i \in \{1, \ldots, m\}$. With this process we obtain the first offspring from Γ and Γ'. For the second offspring we symmetrically repeat this strategy, but now we begin by randomly selecting a position k' in Γ'.*

Example 2. Let us consider the following slp's:

$$\Gamma \equiv \begin{cases} \mathbf{u_1} := \mathbf{x + y} \\ u_2 := u_1 * u_1 \\ \mathbf{u_3} := \mathbf{u_1 * x} \\ u_4 := u_3 + u_2 \\ u_5 := u_3 * u_2 \end{cases} \qquad \Gamma' \equiv \begin{cases} \mathbf{u_1} := \mathbf{x * x} \\ \mathbf{u_2} := \mathbf{u_1 + y} \\ u_3 := u_1 + x \\ \mathbf{u_4} := \mathbf{u_2 * x} \\ u_5 := u_1 + u_4 \end{cases}$$

If $k = 3$ then $S_{u_3} = \{u_1, u_3\}$, and t must be selected in $\{2, \ldots, 5\}$. Assumed that $t = 3$, the first offspring will be:

$$\Gamma_1 \equiv \begin{cases} u_1 := x * x \\ \mathbf{u_2} := \mathbf{x + y} \\ \mathbf{u_3} := \mathbf{u_2 * x} \\ u_4 := u_2 * x \\ u_5 := u_1 + u_4 \end{cases}$$

For the second offspring, if the selected position in Γ' is $k' = 4$, then $S_{u_4} = \{u_1, u_2, u_4\}$. Now if $t = 5$, the offspring will be:

$$\Gamma_2 \equiv \begin{cases} u_1 := x + y \\ u_2 := u_1 * u_1 \\ \mathbf{u_3} := \mathbf{x * x} \\ \mathbf{u_4} := \mathbf{u_3 + y} \\ \mathbf{u_5} := \mathbf{u_4 * x} \end{cases}$$

Next we describe mutation. The first step when mutation is applied to a slp Γ consists of selecting an instruction $u_i \in \Gamma$ at random. Then a new random selection is made within the arguments of the function $f \in F$ that constitutes the instruction u_i. The final step is the substitution of the selected argument for another one in $T \cup \{u_1, \ldots, u_{i-1}\}$ randomly chosen.

4.1 Fitness Based on Structural Risk Minimization

Under Structural Risk Minimization a set of possible models C forms a nested structure $C_1 \subset C_2 \subset \cdots \subset C_L \subset \ldots \subset C$. Here C_L represents the set of models of complexity L, where L is some suitable parameter depending on the problem. We require that VC-dimension V_L of model C_L is an increasing function of L. In our particular case C is the class of slp's having length bounded by some constant l with set of terminals $T =: \{t_1, \ldots, t_n\}$ and set of functionals $F = \{+, -, *, /, sign\}$ and C_L is the class of slp´s in C which use at most L non-scalar operations. In this situation one chooses the model that minimizes the right side of Equation 17.

For practical use of Equation 17 we adopt the following formula with appropriately chosen practical values of theoretical constants (see [12] for the derivation of this formula).

$$
\varepsilon_m(h) \left(1 - \sqrt{p(h) - p(h) \ln p(h) + \frac{\ln m}{2m}} \right)^{-1}, \tag{18}
$$

where $\varepsilon_m(h)$ means empirical risk, $p(h) = \frac{h}{m}$ and h is a function such that for each SLP Γ of length bounded by l with set of terminals T and set of functionals F the following holds: $h(\Gamma) = V_L$ where $L = min\{k : \Gamma \in C_k\}$, that is, h is the VC-dimension of the class of models using as many non-scalar operations as Γ.

4.2 Experimentation

We consider instances of Symbolic Regression for our experimentation. The symbolic regression problem has been approached by Genetic Programming in several contexts. Usually, in this paradigm a population of tree-like structures encoding expressions, is evolved. We adopt slp's as the structures that evolve within the process. We will keep homogeneous populations of equal length individuals along the process.

Experiment 1. We compare four crossover methods: uniform, one-point crossover, 2-point crossover and slp-crossover as defined in Definition 3. For this purpose we have performed 200 executions of the algorithm for each instance and crossover operator. We have run our implemented algorithm based on GP with straight line programs on the following three target functions:

$$
f_1(x) = x^4 + x^3 + x^2 + x; \quad f_2(x) = cos(2x); \quad f_3(x) = 2.718\, x^2 + 3.1416\, x
$$

These functions are also proposed in the book by John Koza, as illustration examples of the tree-based GP approach ([13]). The sample set contains only 20 points for the three target functions. The following values for the control parameters are the same for all tested functions: population size $M = 500$; number of generations $G = 50$; probability of crossover $p_c = 0,9$; probability of mutation $p_m = 0,01$; length of the homogeneous population of slp's $L = 20$. The set of functions is $F = \{+, -, *, //\}$ for f_1 and f_3 and $F = \{+, -, *, //, sin\}$ for

f_2. In the above sets "$//$" indicates the protected division i.e. $x//y$ returns x/y if $y \neq 0$ and 1 otherwise. The basic set of terminals is $T = \{x, 1\}$. We include in T a constant c_0 for the target function f_3. The constant c_0 takes random values in the interval range $[-1, 1]$ and is fixed before each execution. In table 2 we show the corresponding success rates for each crossover method and target function. The success rate (SR) is defined as the ratio of the successful runs with respect to the total number of runs. In our case one run will be considered successful if an individual with a empirical risk lower than 0.2 is found. The umbral of 0.2 is obtained by multiplying the number of test points, 20, by 0.01, that is the maximum error allowed to consider a point as a success point. We can see that uniform crossover and slp-crossover are the best recombination operators for the studied target functions. On the other hand the one point crossover is the worst of the studied recombination operators. Note that uniform crossover and two point crossover are not defined for the tree structure.

Experiment 2. This experiment was designed to test the effectiveness of VC regularization of the slp model using the fitness based in Structural Risk Minimization as given by formula in Equation 18 (VC-fitness) instead of fitness based on Empirical Risk Minimization (ERM-fitness). We describe first experimental comparison using the methodology and data sets from [12]. First we describe the experimental setup.

Target functions. The following target functions where used:
Discontinuous piecewise polynomial function defined as follows:

$$g_1(x) = 4(x^2(3 - 4x), \ x \in [0, 0.5],$$

$$g_1(x) = (4/3)x(4x^2 - 10x + 7) - 3/2, \ x \in (0.5, 0.75],$$

$$g_1(x) = (16/3)x(x - 1)^2, \ x \in (0.75, 1].$$

Sine-square function: $g_2(x) = sin^2(2\pi x), \ x \in [0, 1]$

Two-dimensional *sinc* function: $g_3(x) = \frac{sin\sqrt{x_1^2 + x_2^2}}{x_1^2 + x_2^2}, \ x_1, \ x_2 \in [-5, 5]$

Polynomial function: $g_4(x) = x^4 + x^3 + x^2 + x, \ x \in [0, 1]$

Estimators used. We use slp's of fixed length l with functionals in $\{+, -, *, /\}$ for target functions g_2, g_3, g_4. For target function g_1 we add the *sign* operator. In the experiments we set $l = 6, 10, 20$ and 40. The model complexity is measured by the number of non-scalar operations in the considered slp, that is, the number of instructions which do not contain $\{+, -\}$-operators. This is a measure of the non-linearity of the regressor as suggested by Theorem 2. Each estimator has being obtained running a GP algorithm with the following values for the control parameters: population size $M = 500$ or $M = 100$; number of generations $G = 50$; probability of crossover $p_c = 0, 9$; probability of mutation $p_m = 0, 01$. In all trials, slp-crossover, as described in Definition 3, was used.

Experimentation procedure. A training set of fixed size n is generated. The x-values follows from uniform distribution in the input domain. The prediction risk $\varepsilon_{n_{test}}$ of the model chosen by the GP algorithm based on SLP evolution is estimated by the mean square error (MSE) between the model (estimated from training data) and the true values of target function $g(x)$ for independently generated test inputs $(x_i, \hat{y}_i)_{1 \leq i \leq n_{test}}$, i. e.:

$$\varepsilon_{n_{test}} = \frac{1}{n_{test}} \sum_{i=1}^{n_{test}} (g(x_i) - \hat{y}_i)^2 \qquad (19)$$

Table 1. Prediction Risk: VC-fitness vs. ERM-fitness

	population size 100				population size 500			
	g_1	g_2	g_3	g_4	g_1	g_2	g_3	g_4
				length 6				
ERM-fitness	0.1559	0.1502	0.0456	0.1054	0.0670	0.1501	0.0456	0.1054
VC-fitness	0.0396	0.1502	0.0456	0.1019	0.0706	0.1443	0.0456	0.1019
Comparative	3.9368	1	1	1.0343	0.9490	1.0401	1	1.0343
				length 10				
ERM-fitness	0.1559	0.2068	0.0456	0.1054	0.0396	0.1502	0.0456	0.0377
VC-fitness	0.1559	0.1434	0.0456	0.1791	0.0396	0.1502	0.0456	0.1054
Comparative	1	1.4421	1	0.5884	1	1	1	0.3576
				length 20				
ERM-fitness	0.1566	0.1324	0.0456	0.4982	0.0396	0.1396	0.0456	0.0870
VC-fitness	0.1559	0.2068	0.0456	0.1852	0.0661	0.1827	0.0456	0.0633
Comparative	1.0004	0.6402	1	2.6900	0.5990	0.7640	1	1.3744
				length 40				
ERM-fitness	0.1029	0.1439	0.0456	0.1357	0.0745	0.1502	0.0456	0.2287
VC-fitness	0.1559	0.2068	0.0456	0.5709	0.0326	0.1502	0.0456	0.0805
Comparative	0.6600	0.6958	1	0.2376	2.2852	1	1	2.8409

The above experimental procedure is repeated 100 times using 100 different random realizations of n training samples (from the same statistical distribution). Experiments were performed using a small training sample ($n = 20$) and a large test set ($n_{test} = 200$). Table 1 shows comparison results for fitness based on ERM (ERM-fitness) and fitness based on VC-regularization (VC-fitness). Experiments for each target function $g = g_i$, $1 \leq i \leq 4$ are divided into four groups corresponding to different values of the length of the evolved slp´s ($l = 6$, $l = 10$, $l = 20$, $l = 40$). For each length two possible population sizes are considered (100 and 500 individuals). For each population size two fitness are considered: ERM-fitness and VC-fitness. The values in the fitness rows represent the estimated prediction error of the selected model. The values in the comparative rows represent the ratio between the prediction risk for the regressor

Table 2. SR over 200 independent runs for the crossover operators

Function	1-point	2-point	uniform	slp
f_1	100	100	100	100
f_2	30	35	53	53
f_3	42	71	91	93

obtained using the ERM-fitness and the corresponding value for the regressor obtained using the VC-fitness. Accordingly, the values in the comparative rows that are bigger than or equal to 1 represent a better performance of VC-fitness. If we consider an experiment successful when the comparative value is ≥ 1, then the success rate is greater than 70%. If we consider an experiment strictly successful when the the the comparative value is > 1, then the strict success rate is greater than 30%.

5 Conclusions and Future Research

We have calculated a sharp bound for the VC dimension of the GP genotype defined by computer programs using straight line code. We have used this bound to perform VC-based model selection under the GP paradigm showing that this model selection method consistently outperforms LGP algorithms based on empirical risk minimization. A next step in our research is to compare VC-regularization of slp's with other regularization methods based on asymptotical analysis like Akaike information criterion (AIC) or Bayesian information criterion (BIC). A second goal in our research on SLP-based GP is to study the experimental behavior of the straight line program computation model under Vapnik-Chervonenkis regularization but without assuming previous knowledge of the length of the structure. This investigation is crucial in practical applications for which the GP machine must be able to learn not only the shape but also the length of the evolved structures. To this end new recombination operators must be designed since the crossover procedure employed in this paper only applies to populations having fixed length chromosomes.

Acknowledgements

This work was partially supported by Spanish Grants TIN2007-67466-C02-02, MTM2007-62799 and FPU program.

References

1. Giusti, M., Heintz, J., Morais, J.E., Pardo, L.M.: Straight line programs in Geometric elimination Theory. Journal of Pure and Applied Algebra 124, 121–146 (1997)
2. Alonso, C.L., Montaña, J.L., Puente, J.: Straight line programs: a new Linear Genetic Programming Approach. In: Proc. 20th IEEE International Conference on Tools with Artificial Intelligence, ICTAI, pp. 517–524 (2008)

3. Brameier, M., Banzhaf, W.: Linear Genetic Programming. Springer, Heidelberg (2007)
4. Aldaz, M., Heintz, J., Matera, G., Montaña, J.L., Pardo, L.: Time-space tradeoffs in algebraic complexity theory. Journal of Complexity 16, 2–49 (1998)
5. Teytaud, O., Gelly, S., Bredeche, N., Schoenauer, M.A.: Statistical Learning Theory Approach of Bloat. In: Proceedings of the 2005 conference on Genetic and Evolutionary Computation, pp. 1784–1785 (2005)
6. Vapnik, V., Chervonenkis, A.: Ordered risk minimization. Automation and Remote Control 34, 1226–1235 (1974)
7. Vapnik, V.: Statistical learning theory. John Wiley & Sons, Chichester (1998)
8. Vapnik, V., Chervonenkis, A.: On the uniform convergence of relative frequencies of events to their probabilities. Theory of Probability and its Applications 16, 264–280 (1971)
9. Lugosi, G.: Pattern classification and learning theory. In: Principles of Nonparametric Learning, pp. 5–62. Springer, Heidelberg (2002)
10. Karpinski, M., Macintyre, A.: Polynomial bounds for VC dimension of sigmoidal and general Pffafian neural networks. J. Comp. Sys. Sci. 54, 169–176 (1997)
11. Milnor, J.: On the Betti Numbers of Real Varieties. Proc. Amer. Math. Soc. 15, 275–280 (1964)
12. Cherkassky, V., Yunkian, M.: Comparison of Model Selection for Regression. Neural Computation 15, 1691–1714 (2003)
13. Koza, J.R.: Genetic Programming: On the Programming of Computers by Means of Natural Selection. The MIT Press, Cambridge (1992)

A Statistical Learning Perspective of Genetic Programming

Nur Merve Amil, Nicolas Bredeche, Christian Gagné*, Sylvain Gelly,
Marc Schoenauer, and Olivier Teytaud

TAO, INRIA Saclay, LRI, Bat. 490, Université Paris-Sud, 91405 Orsay CEDEX,
France (*) LVSN, GEL-GIF, Univ. Laval, Qubec, Canada, F1V0A6

Abstract. This paper proposes a theoretical analysis of Genetic Programming (GP) from the perspective of statistical learning theory, a
well grounded mathematical toolbox for machine learning. By computing the Vapnik-Chervonenkis dimension of the family of programs that
can be inferred by a specific setting of GP, it is proved that a parsimonious fitness ensures universal consistency. This means that the empirical
error minimization allows convergence to the best possible error when
the number of test cases goes to infinity. However, it is also proved that
the standard method consisting in putting a hard limit on the program
size still results in programs of infinitely increasing size in function of
their accuracy. It is also shown that cross-validation or hold-out for
choosing the complexity level that optimizes the error rate in generalization also leads to bloat. So a more complicated modification of the
fitness is proposed in order to avoid unnecessary bloat while nevertheless
preserving universal consistency.

1 Introduction

This paper is about two important issues in Genetic Programming (GP), that
is Universal Consistency (UC) and code bloat. UC consists in the convergence
to the optimal error rate with regards to an unknown distribution of examples.
A restricted version of UC is consistency, which focus on the convergence to the
optimal error rate within a restricted search space. Both UC and consistency
are well studied in the field of statistical learning theory. Despite their possible
benefits, they have not been widely studied in the field of GP. Code bloat is the
uncontrolled growth of program size that may occur in GP when relying on a
variable length representation [10,11]. This has been identified as a key problem
in GP for which there have been several empirical studies. However, very few
theoretical studies addressed this issue directly. The work presented in this paper
is intended to provide some theoretical insights on the bloat phenomenon and
its link with UC in the context of GP-based learning taking a statistical learning
theory perspective [23].

Statistical learning theory provides several theoretical tools to analyze some
aspects of learning accuracy. Our main objective consists in performing both
an in-depth analysis of bloat as well as providing appropriate solutions to avoid

L. Vanneschi et al. (Eds.): EuroGP 2009, LNCS 5481, pp. 327–338, 2009.

it. Section 2 shortly exposes issues of code bloat with GP. Section 3 and 4 present all the aforementioned results about code bloat avoidance and UC and propose a new approach ensuring both. Then, Section 5 provides some extensions of the previous theoretical results on the use of cross-validation and hold-out methodologies. Follows some experimental results in Section 6, illustrating the accuracy of the theoretical results. Section 7 finally concludes this paper with a discussion on the consequences of those theoretical results for GP practitioners and uncover some perspectives of work.

2   Code Bloat in GP

Due to length constraints, we do not introduce here some important theories around code bloat: introns, fitness causes bloat, and removal bias. The reader is refered to [1,3,11,13,16,17,21] for more informations around that. Some common solutions against bloat rely either on specific operators (e.g. size-fair crossover [12], or different fair mutation [14]), on some parsimony-based penalization of the fitness [22] or on abrupt limitation of the program size such as the one originally used by Koza [10]. Also, some multi-objective approachs have been proposed [2,5,7,15,19]. Some other more particular solutions have been proposed but are not widely used yet [18,24]. Also, all proofs are removed due to length constraints. Readers familiar with mathematics like in e.g. [6] should however be able to guess the main ideas.

Although code bloat is not clearly understood, it is yet possible to distinguish at least two kinds of code bloat. We first define *structural bloat* as the code bloat that necessarily takes place when no optimal solution can be approximated by a set of programs with bounded length. In such a situation, optimal solutions of increasing accuracy will also exhibit an increasing complexity (larger programs), as larger and larger code will be generated in order to better approximate the target function. This extreme case of structural bloat has also been demonstrated in [9]. The authors use some polynomial functions of increasing difficulty, and demonstrate that a precise fit can only be obtained through an increased bloat (see also [4] for related issues about problem complexity in GP). Another form of bloat is the *functional bloat*, which takes place when program length keeps on growing even though an optimal solution (of known complexity) does lie in the search space. In order to clarify this point, let us use a simple symbolic regression problem defined as follow: given a set S of test cases, the goal is to find a function f (here, a GP-tree) that minimizes the Mean Square Error (or MSE). If we intend to approximate a polynomial (e.g. $14 * x^2$ with $x \in [0, 1]$), we may observe code bloat since it is possible to find arbitrarily long polynomials that gives the exact solution (e.g. $14x^2 + 0 * x^3 + \ldots$), or sequences of polynomials of length growing to ∞ and accuracy converging to the optimal accuracy (e.g. $P_n(x) = 14x^2 + \sum_{i=1}^{n} \frac{1}{n!i!} x^i$). Most of the works cited earlier are in fact concerned with functional bloat, which is the most surprising, and the most disappointing kind of bloat. We will consider various levels of functional bloat: cases where length of programs found by GP runs to infinity as the number of test cases runs to infinity whereas a bounded-length solution exists, and also cases where large programs are found with high probability by GP whereas a small program is optimal.

Another important issue is to study the convergence of the function given by GP toward the actual function used to generate the test cases, under some sufficient conditions and when the number of test cases goes to infinity. This property is known in statistical learning as *Universal Consistency* (UC). Note that this notion is slightly different from that of universal approximation, commonly referred in symbolic regression, where GP search using operators $\{+, *\}$ is assumed to be able to approximate any continuous function. UC is rather concerned with the behavior of the algorithm when the number of test cases goes to infinity: the existence of a polynomial that approximates a given function at any arbitrary precision does not imply that any polynomial approximation built from a set of sample points will converge to that given function when the number of points goes to infinity. Or more precisely, UC can be stated informally as follows (a formal definition will be given later).

A GP setting corresponds to symbolic regression from examples if it takes as inputs a finite number of examples $x_1, \ldots, x_n$ with their associated labels $y_1, \ldots, y_n$ and outputs a program P_n. Universal consistency holds if, when pairs $(x_1, y_1), \ldots, (x_n, y_n)$ (test cases) are identically independently distributed as the random variable (x, y), $L(P_n) \rightarrow L^*$ where $L(p) = Pr(y \neq p(x))$ and where $L^* = \inf_p$ measurable $L(p)$. In all of this paper, $Pr(.)$ denotes probabilities, as the traditional notation $P(.)$ is used for programs.

3 Negative Results without Regularization and Resampling

Definition 1 precisely defines the programs space under examination. Theorem 1 evaluates its VC-dimension [23]. Many theorems, in the sequel, are based only on VC-dimensions and hold for other sets of programs as well.

It should be noted the mildness of the hypothesis behind our results. We consider any programs of bounded length, working with real variables, provided that the computation time is *a priori* bounded. Usual families of programs in GP verify this hypothesis and much stronger hypothesis. For example, usual tree-based representations avoid loops and therefore all quantities that have to be bounded in lemma below (typically, number of times each operator is used) are bounded for trees of bounded depths. This is also true for direct acyclic graphs. We here deal with a very general case; much better constants can be derived for specific cases, without changing the fundamental results in the sequel of the paper.

Definition 1 (Set of programs studied). *Let* $F(n, t, q, m, z)$ *be the set of functions from* $\mathbb{R}^{z-m}$ *towards* $\{0, 1\}$ *which can be computed by a program with a maximum of n lines as follows:*

(1) A run uses at most t operations. (2) Each line contains one operation among the followings:

- *Operations* $\alpha \mapsto \exp(\alpha)$ *(at*
  most q times);
- *Operations* $+, -, \times,$ *and* $/;$
- *Jumps conditioned on* $>, \geq, <,$
  $\leq,$ *and* $=;$
- *Output 0;*
- *Output 1;*
- *Labels for jumps;*
- *Constants (at most* m *differ-*
  ent);
- *Variables (at most* z *different,*
  with $z \geq m$).

We note $F(n, t, q, m, z)$ as F for short when there is no ambiguity. The parameters n, t, q, m, z are then implicit.

The following property is central for our results.

Theorem 1 (Finite VC-dimension of the computing machine). *Consider* q', t' *and* $d' \geq 0$. *Let* $F = F(n, t, q, m, z)$ *be the set of programs described by Definition 1, where* $q \leq q'$, $T(n, t, z) \leq t'$, *and* $1 + m \leq d'$.

$$VCdim(F) \leq t'^2 d' \left(d' + 19\log_2(9d')\right)$$
$$\leq (d'(q'+1))^2 + 11d'(q'+1)(t' + \log_2(9d'(q'+1)))$$

If $q = 0$ *(no exponential) then* $VCdim(F) \leq 4d'(t'+2)$.

Interpretation. *The theorem demonstrates that interesting and natural families of programs have finite VC-dimension. Effective methods can associate a VC-dimension to these families of programs.*

We now consider how to use such results in order to ensure UC. First, we show why simple empirical risk minimization (i.e. minimizing the error observed without taking into account programs complexity) does not ensure consistency. More precisely, for some distribution of test cases and some i.i.d. (independent identically distributed) sequence of test cases $\{(x_1, y_1), \ldots, (x_n, y_n), \ldots\}$, there exists $P_1, \ldots, P_n, \ldots$ such that $\forall n \in \mathbb{N}, \forall i \in \{1, 2, \ldots, n\}$ $P_n(x_i) = y_i$, and however $\forall n \in \mathbb{N}$ $\Pr(P_n(x) = y) = 0$. This can be proved by considering that x is uniformly distributed in $[0, 1]$ and y is a constant equal to 1. Then, consider P_n, the program that compares its entry to $x_1, x_2, \ldots, x_n$, and outputs 1 if the entry is equal to x_j for some $j \leq n$, and otherwise outputs 0. With probability 1, this program output 0, whereas almost surely the desired output y is 1.

We therefore conclude that minimizing the empirical risk is not enough for ensuring any satisfactory form of consistency. Let's now show that structural risk minimization (i.e. taking into account a penalization for complex structures) can ensure UC and fast convergence when the solution can be written within finite complexity.

Theorem 2 (Universal consistency of genetic programming with structural risk minimization). *Consider* q_k, t_k, m_k, n_k, *and* z_k *increasing integer sequences. Define* $\mathcal{F}_k$ *the set of programs with at most* t_k *lines executed,* z_k *variables,* n_k *lines,* q_k *exponentials, and* m_k *constants (*$\mathcal{F}_k = F(n_k, t_k, q_k, m_k, z_k)$ *of Definition 1) and* $\mathcal{F} = \cup_k \mathcal{F}_k$. *Then with* $q'_k = q_k$, $t'_k = T(n_k, t_k, z_k)$, *and* $d'_k = 1 + m_k$, *define* V_k *as:*

- *If* $\forall k$ $q_k = 0$, *then* $V_k = 4d'_k(t'_k + 2)$.

– *Otherwise,* $V_k = (d'_k(q'_k + 1))^2 + 11d'_k(q'_k + 1)(t'_k + \log_2(9d'_k(q'_k + 1)))$.

Now given s test cases, consider $P \in \mathcal{F}$ minimizing $\hat{L}(P) + \sqrt{\frac{32}{s}V(P)\log(es)}$, where $V(P) = V_k$ where k is minimal such that $P \in \mathcal{F}_k$. Then, the generalization error, with probability 1, converges to L^*; moreover, if one optimal program belongs to $\mathcal{F}_k$, then for any s and ϵ such that $V_k \log(es) \leq s\epsilon^2/512$, the generalization error with s test cases is larger than $L^* + \epsilon$ with probability at most $\Delta \exp(-s\epsilon^2/128) + 8s^{V_k} \exp(-s\epsilon^2/512)$ where $\Delta = \sum_{j=1}^{\infty} \exp(-V_j)$.

Interpretation. *This theorem shows that genetic programming for binary classification, provided that structural risk minimization is performed (i.e. if we optimize an ad hoc compromise between complexity of programs and accuracy on empirical data), is universally consistent and verifies some convergence rate properties.*

We now prove the non-surprising fact that if it is possible to approximate the optimal function (the Bayesian classifier) without reaching it exactly, then the complexity of the program runs to infinity as soon as there is convergence of the generalization error to the optimal one.

Proposition 1 (Structural bloat in genetic programming). *Consider $\mathcal{F}_1 \subset \mathcal{F}_2 \subset \mathcal{F}_3 \subset \ldots$, where $\mathcal{F}_V$ is a set of functions from X to $\{0,1\}$ with VC-dimension bounded by V. Consider $(V(s))_{s \in \mathbb{N}}$ a non decreasing sequence of integers and $(P_s)_{s \in \mathbb{N}}$ a sequence of functions such that $P_s \in \mathcal{F}_{V(s)}$. Define $L_V = \inf_{P \in \mathcal{F}_V} L(P)$ and $V(P) = \inf\{V; P \in \mathcal{F}_V\}$ and suppose that $\forall V\ L_V > L^*$. Then, $\left(L(P_s) \xrightarrow{s \to \infty} L^*\right) \Longrightarrow \left(V(P_s) \xrightarrow{s \to \infty} \infty\right)$.*

Interpretation. *This is structural bloat: if the space of programs approximates but does not contain the optimal function and cannot approximate it within bounded size, then bloat occurs. Note that for any $\mathcal{F}_1, \mathcal{F}_2, \ldots$, the assumption $\forall V\ L_V > L^*$ holds simultaneously for all V for many distributions, as we consider countable unions of families with finite VC-dimension (e.g. see [6, chap. 18]).*

We now show that, even in cases in which an optimal short program exists, the usual procedure (known as the method of Sieves; see also [20]) defined below, consisting in defining a maximum VC-dimension depending upon the sample size and then using a family of functions accordingly, leads to bloat.

Theorem 3 (Bloat with the method of Sieves). *Let $\mathcal{F}_1, \ldots, \mathcal{F}_k, \ldots$ be nonempty sets of functions with finite VC-dimensions $V_1, \ldots, V_k, \ldots$, and let $\mathcal{F} = \cup_n \mathcal{F}_n$. Then given s i.i.d. test cases, consider $\hat{P} \in \mathcal{F}_s$ minimizing the empirical risk $\hat{L}$ in $\mathcal{F}_s$.*

From theorems about the method of Sieves, we already know that if $V_s = o(s/\log(s))$ and $V_s \to \infty$, then $\Pr\left(L(\hat{P}) \leq \hat{L}(\hat{P}) + \epsilon(s, V_s, \delta)\right) \geq 1 - \delta$ and almost surely $L(\hat{P}) \to \inf_{P \in \mathcal{F}} L(P)$.

We now state that if $V_s \to \infty$, and noting $V(P) = \min\{V_k; P \in \mathcal{F}_k\}$, then $\forall V_0,\ \delta_0 > 0,\ \exists \Pr$, a distribution of probability on X and Y, such that $\exists g \in \mathcal{F}_1$ such that $L(g) = L^*$, and for s sufficiently large $\Pr\left(V(\hat{P}) \leq V_0\right) \leq \delta_0$.

Interpretation. *The result in particular implies that for any V_0, there is a distribution of test cases such that $\exists g; V(g) = V_1$ and $L(g) = L^*$, with probability 1, $V(\hat{P}) \geq V_0$ infinitely often as s increases. This shows that bloat can occur if we use only an abrupt limit on code size, if this limit depends upon the number of test cases (a fortiori if there's no limit). Note that this result, proved thanks to a particular distribution, could indeed be proved for the whole class of classification problems for which the conditional probability of $Y = 1$ (conditionally to X) is equal to $\frac{1}{2}$ in an open subset of the domain.*

4   Universal Consistency without Bloat

In this section, we consider a more complicated case where the goal is to ensure UC, while simultaneously avoiding non-necessary bloat. This means that an optimal program does exist in a given family of functions and convergence towards the minimal error rate is performed without increasing the program complexity. This is achieved by: i) merging regularization and bounding of the VC-dimension, and ii) penalization of the complexity (i.e. length) of programs by a penalty term $R(s, P) = R(s)R'(P)$ depending upon the sample size and the program. $R(.,.)$ is user-defined and the algorithm looks for a classifier with a small value of both R' and L. In the following, we study both the UC of this algorithm (i.e. $L \to L^*$) and the no-bloat theorem (i.e. $R' \to R'(P^*)$ when P^* exists). Note that the bound $V_s = o(\log(s))$ is much stronger than the usual limit used in the method of Sieves (see Theorem 3).

Theorem 4 (No-bloat theorem). *Let $\mathcal{F}_1, \dots, \mathcal{F}_k, \dots$ with finite VC-dimensions $V_1, \dots, V_k, \dots$. Let $\mathcal{F} = \cup_n \mathcal{F}_n$. Define $V(P) = V_k$ with $k = \inf\{t | P \in \mathcal{F}_t\}$. Define $L_V = \inf_{P \in \mathcal{F}_V} L(P)$. Consider $V_s = o(\log(s))$ and $V_s \to \infty$. Consider also that $\hat{P}_s$ minimizes $\hat{\tilde{L}}(P) = \hat{L}(P) + R(s, P)$ in $\mathcal{F}_s$, and assume that $R(s, .) \geq 0$. Assume that $\sup_{P \in \mathcal{F}_{V_s}} R(s, P) = o(1)$. Then, $L(\hat{P}_s) \to \inf_{P \in \mathcal{F}} L(P)$ almost surely. Note that for well chosen family of functions, $\inf_{P \in \mathcal{F}} L(P) = L^*$. Moreover, assume that $\exists P^* \in \mathcal{F}_{V^*} \quad L(P^*) = L^*$. With $R(s, P) = R(s)R'(P)$ and with $R'(s) = \sup_{P \in \mathcal{F}_{V_s}} R'(P)$, we get the following results:*

1. **Non-asymptotic no-bloat theorem:** *For any $\delta \in]0, 1]$, $R'(\hat{P}_s) \leq R'(P^*) + (1/R(s))2\epsilon(s, V_s, \delta)$ with probability at least $1 - \delta$. This result is in particular interesting for $\epsilon(s, V_s, \delta)/R(s) \to 0$.*
2. **Almost-sure no-bloat theorem:** *If for some $\alpha > 0$, $R(s)s^{(1-\alpha)/2} = O(1)$, then almost surely $R'(\hat{P}_s) \to R'(P^*)$ and if $R'(P)$ has discrete values (such as the number of instructions in P or many complexity measures for programs) then for s sufficiently large, $R'(\hat{P}_s) = R'(P^*)$);*
3. **Convergence rate:** *For any $\delta \in]0, 1]$, with probability at least $1 - \delta$,*

$$L\left(\hat{P}_s\right) \leq \inf_{P \in \mathcal{F}_{V_s}} L(P) + \underbrace{R(s)R'(s)}_{= o(1) \text{ by hypothesis}} + 2\epsilon(s, V_s, \delta),$$

where $\epsilon(s, V, \delta) = \sqrt{\frac{4 - \log(\delta/(4s^2 V))}{2s - 4}}$.

Interpretation. *Combining a code limitation and a penalization leads to UC without bloat.*

5 Some Negative Results with Subsampling: Hold-Out or Cross-Validation

When one tries to learn a relation between x and y, the "true" cost function (typically the mean squared error of a given approximate relation, which is an expectation under some usually unknown law of probability) is generally not available. It is usually replaced by its empirical mean on a finite sample. Minimizing this empirical mean is natural, but this can be done over various families of functions (e.g. trees with depth 1, 2, 3, and so on). Choosing between these various levels is hard. Typically, the empirical mean decreases as the complexity is increased, but this decrease is not generally a decrease of the generalization error, as trees of larger depth have usually a very bad generalization error due to overfitting. Therefore, the problem is somewhat multi-objective: there is a conflict between the empirical error and the complexity level. This multi-objective optimization setting has been studied in [2,5,7].

This section is devoted to hold-out and cross-validation as tools for UC without bloat. First, let's consider hold-out for choosing the complexity level. Consider $X_0, \ldots, X_N, Y_0, \ldots, Y_N$, $2(N+1)$ samples (each of them consisting in n examples, i.e. $X_i = (X_{i,1}, X_{i,2}, \ldots, X_{i,n})$ and $Y_i = (Y_{i,1}, Y_{i,2}, \ldots, Y_{i,n})$), the X_i's being learning sets, the Y_i's being (hold-out) test sets. Consider that the function can be chosen in many complexity levels, $F_0 \subset F_1 \subset F_2 \subset F_3 \subset \ldots$, where F_0 is non-empty and $F_i \neq F_{i+1}$. Note $\hat{L}_k(f)$ the error rate of the function f in the set X_k of examples: $\hat{L}_k(f) = \frac{1}{n} \sum_{i=1}^n l(f, X_{k,i})$ where $l(f, x) = 1$ if f fails on x and 0 otherwise. Define $f_k = \arg\min_{F_k} \hat{L}_k(.)$. In hold-out, after the complete learning, the resulting classifier is $f_{k^*(n)}$, where $k^*(n) = \arg\min_{k \leq N(n)} l_k$ and $l_k = \frac{1}{n} \sum_{i=1}^n l(f_k, Y_{k,i})$. In the sequel, we assume that $f \in F_k \Rightarrow 1 - f \in F_k$ and that $VCdim(F_k) \to \infty$ as $k \to \infty$. The case with hold-out leads to different cases, namely:

Greedy case: all X_k's and Y_k's are independent; this means that we test separately each complexity level F_k with different learning sets X_k and test sets Y_k.

Case with pairing: X_0 is independent of Y_0, $\forall k, X_k = X_0$ and $\forall k, Y_k = Y_0$; this means that we use the same learning set for all complexity levels and the same test set for all complexity levels. This case is far more usual.

Theorem 5 (No bloat avoidance with greedy hold-out). *Consider greedy hold-out for choosing between complexity levels $0, 1, \ldots, N(n)$. If $N(n)$ is a constant, then for some distribution of examples $\forall k \in [0, N], P(k^*(n) = k) \to 1/(N+1)$. If $N(n) \to \infty$ as $n \to \infty$, then for some distribution of examples such that an optimal function lies in F_0, greedy hold-out leads to $k^*(n) \to \infty$ as $n \to \infty$ and therefore $\limsup_{n \to \infty} k^*(n) = \infty$.*

All the following results are in the general case of N a non decreasing function of n.

Proposition 2 (Bloat cannot be controlled by hold-out with pairing, first result). *Consider the case with pairing. For arbitrarily large v, there exists a distribution with optimal function in F_0 such that $\liminf_{n\to\infty} \Pr(k^*(n) \geq v) > 0$.*

Now, let's consider a distribution that depends on n. This is interesting, as it provides lower bounds on what can be guaranteed, for a given value of n, independently of the distribution. For technical reasons, and without loss of generality with renumbering of the F_k, we assume that F_{v+1} has a VC-dimension larger than F_v. We can show that, *with a distribution dependent on n*, $\limsup_{n\to\infty} k^*(n) \to \infty$. This leads to this other negative theorem about the control of bloat by hold-out.

Proposition 3 (Bloat can not be controlled by hold-out with pairing, second result). $\limsup_n k^*(n) = \infty$, *where the distribution depends on n but is always such that an optimal function lies in F_0.*

This result above is in the setting of a distribution which depends on n; it is of course not interesting for modelizing the evolution of one particular problem as the number of examples increases, but it shows that no bound on $k^*(n)$ for $n \geq n_0$ can be provided, whatever may be n_0, for hold-out with pairing, unless the distribution of problems is taken into account.

Cross-Validation for the Control of Bloat. We now turn our attention to the case of cross-validation. We formalize N-folds cross-validation as follows:

$$f_k^i = \arg\min_{F_k} L(., X'^i_k), \quad k^* = \arg\min \frac{1}{N} \sum_{i=1}^{N} L(f_k^i, X_k^i)$$

$$X'^i_k = (X_k^1, X_k^2, \ldots, X_k^{i-1}, X_k^{i+1}, X_k^{i+2}, \ldots, X_k^N) \text{ for } i \leq N$$

where for any i and k, X_k^i is a sample of n points.

Greedy cross-validation could be considered as in the case of hold-out above: all X_k^i could be independent. This leads to the same result (for some distribution, $k^*(n) \to \infty$) with roughly the same proof. We therefore only consider cross-validation with pairing, i.e. $\forall i, k, k'$, $X_k^i = X_{k'}^i$. For short, we note $X_k^i = X^i$.

Theorem 6 (Negative result on subsampling). *Assume that F_k has a VC-dimension going to ∞ as $k \to \infty$. One can not avoid bloat with only hold-out or cross-validation, in the sense that with paired hold-out, or greedy hold-out, or cross-validation, for any V, there exists some distribution for which almost surely, $k^*(n) > V$ infinitely often whereas an optimal function lies in F_0.*

Note that propositions above show in some cases stronger forms of bloat. If we consider greedy hold-out, hold out with pairing and cross-validation with pairing, then: (1) For some well-chosen distribution of examples, greedy hold-out almost surely leads to (i) $k^*(n) \to \infty$ if $N \to \infty$ (ii) $k^*(n)$ asymptotically uniformly distributed in $[[0, N]]$ if N finite, whereas an optimal function lies in F_0 (theorem 5). (2) Whatever may be $V = VCdim(F_v)$, for some well-chosen distribution, hold-out with pairing almost surely leads to $k^*(n) > V$ infinitely often whereas an optimal function lies in F_0 (proposition 2). (3) Whatever may be $V = VCdim(F_v)$, for some well-chosen distribution, cross-validation with pairing almost surely leads to $k^*(n) > V$ infinitely often whereas an optimal function lies in F_0.

6   Experimental Results

Some theoretical elements presented in Sections 3 and 4 are verified experimentally in this section. The experimentation are conducted using Koza-style GP [10], with a problem setup similar to the classical symbolic regression example, modified for binary classification. This is covered by theoretical results above. The GP branches used are the addition, subtraction, multiplication, protected division, and if-less-than. This last branch takes four arguments, returning the third argument if the first argument is less than the second one, otherwise returning the fourth argument. The GP terminals are the x variable, and the 0 and 1 constants. The learning task consists in minimizing the error $e(i)$ between the desired output $y_i = \{-1, 1\}$ and the obtained output $\hat{y}_i$ of the tested GP tree for the x_i input, as in the following: $e(i) = \max(1 - y_i \hat{y}_i, 0)$. The fitness measure used in the experiments consists in minimizing the sum of the errors to which is added a complexity factor that approximate the VC-dimension of the GP program: $f = \frac{1}{s} \sum_{i=1}^{s} e(i) + k\sqrt{\frac{t^2 \log_2(t)}{s}}$, where t is the number of nodes of the GP program tested, s is the number of test cases used for fitness evaluation, and k is a trade-off weight in the composition of the complexity penalization relatively to the accuracy term. The s test cases are distributed uniformly in $x_i \in [0, 1]$, with associated $y_i = \{-1, 1\}$. For $x_i < 0.4$, each y_i are equal to 1 with probability 0.25 (so $y_i = -1$ with probability 0.75), for $x_i \in [0.4, 0.6[$, $y_i = 1$ with probability 0.5, and for $x_i \geq 0.6$, $y_i = 1$ with probability 0.75. Thus, the associated classifier with best generalization capabilities would return $y_i^* = -1$ for $x_i < 0.4$, $y_i^* = 1$ for $x_i \geq 0.6$ and a random output for $x_i \in [0.4, 0.6[$, with a minimal generalization error of 0.3. After the evolutions, each best-of-run classifier is thus evaluated by a fine sampling of the input space, with the generalization error evaluated as the difference between the output given by the tested best-of-run classifier and the output obtained by a classifier with best generalization capabilities. Five types of GP evolutions have been tested: i) no limitation on the tree size (no depth limit and complexity trade-off $k = 0$), ii) depth limitation on the tree size of 17 levels (complexity trade-off $k = 0$), iii) soft complexity penalty in the fitness (complexity trade-off $k = 0.0001$), iv) medium complexity penalty in the fitness (complexity trade-off $k = 0.001$), and v) important complexity penalty in the fitness (complexity trade-off $k = 0.01$). For the three last approaches, the depth limitation of 17 levels is still maintained. The

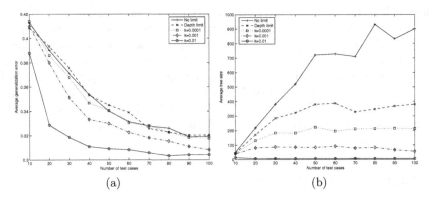

Fig. 1. Generalization errors and tree sizes observed for different size limitations. Figure (a) shows the average generalization errors observed, with apparently better results for the approaches where the fitness includes some parsimony pressure. Figure (b) shows the average tree sizes obtained, where important bloat is observed for the no limitation and maximum depth limitations.

selection method used is lexicographic parsimony pressure [15], that is regular tournament selection 4 participants, with the smallest participant taken in case of ties. Other GP parameters are: population of 1000 individuals; evolutions on 200 generations; crossover probability of 0.8; subtree, swap and shrink mutation of probability 0.05 each; and finally half-and-half initialization with maximal depth of 5. All the experiments have been implemented using the GP facilities of the Open BEAGLE (http://beagle.gel.ulaval.ca, [8]) C++ framework for evolutionary computations. The experiments have been conducted different number of test cases varying from $s = 10$ to $s = 100$ by steps of 10. One hundred evolutions is done for each combinations of approaches tested and number of test cases, for a total of 50 000 evolutions. Figure 1 shows the average generalization errors and tree size obtained for the different approaches in function of the number of test cases used for fitness evaluation. These results show that bloat occurs when no limitation of size occurs, even when lexicographic parsimony pressure is used (see curve *No limit* of Figure 1b), which validates Theorem 3. Then, as stated by Theorem 2, UC is achieved using moderate complexity penalization in the fitness measure, with a convergence toward optimal generalization error of 0.3 (see curve *k=0.001* of Figure 1a). Third, as predicted by Theorem 4, increasing the penalization leads to both UC and no bloat (see curve *k=0.01* of both Figures 1a and 1b). Note that Theorem 3 asserts that this result cannot be achieved by *a priori* scaling of the complexity, and that Section 5 shows that this can not be achieved by cross-validation.

7   Conclusion

In this paper, we have proposed a theoretical study of two important issues in Genetic Programming (GP) known as Universal Consistency (UC) and code bloat. We have shown that the understanding of the bloat phenomenon in GP could benefit from classical results from statistical learning theory. The main limit of our work is that it deals only with the statistical elements of genetic programming (effect of noise) and not with the dynamics (the effect of bounded computational power). Application of theorems from learning theory has led to two original outcomes with both positive and negative results. Firstly, results on UC of GP: there is almost sure asymptotic convergence to the optimal error rate in the context of binary classification with GP with any of the classical forms of regularizations (from learning theory): the method of Sieves, or Structural Risk Minimization. Secondly, results on code bloat: i) if the ideal target function does not have a finite description then code bloat is unavoidable (structural bloat: obviously, if there's no finite-length program with optimal error, then reaching the optimal error is, at best, only possible with an infinite growth of code), and ii) code bloat can be avoided by simultaneously bounding the length of the programs with some *ad hoc* limit and using some parsimony pressure in the fitness function (functional bloat), i.e. by combining Structural Risk Minimization and Sieves. An important point is that all methods leading to no-bloat use a regularization term; in particular, cross-validation or hold-out methods do not reach no-bloat.

Acknowledgements. This work was supported in part by the PASCAL Network of Excellence, and by postdoctoral fellowships from the ERCIM (Europe) and the FQRNT (Québec) to C. Gagné. We thank Bill Langdon for very helpful comments.

References

1. Banzhaf, W., Nordin, P., Keller, R.E., Francone, F.D.: Genetic Programming: an introduction. Morgan Kaufmann Publisher Inc., San Francisco (1998)
2. Bleuler, S., Brack, M., Thiele, L., Zitzler, E.: Multiobjective genetic programming: Reducing bloat using SPEA2. In: Proceedings of the 2001 Congress on Evolutionary Computation CEC 2001, COEX, World Trade Center, 159 Samseong-dong, Gangnam-gu, Seoul, Korea, pp. 536–543. IEEE Press, Los Alamitos (2001)
3. Blickle, T., Thiele, L.: Genetic programming and redundancy. In: Hopf, J. (ed.) Genetic Algorithms Workshop at KI 1994, pp. 33–38. Max-Planck-Institut für Informatik (1994)
4. Daida, J.M., Bertram, R.R., Stanhope, S.A., Khoo, J.C., Chaudhary, S.A., Chaudhri, O.A., Polito II, J.A.: What makes a problem GP-Hard? Analysis of a tunably difficult problem in genetic programming. Genetic Programming and Evolvable Machines 2(2), 165–191 (2001)
5. De Jong, E.D., Watson, R.A., Pollack, J.B.: Reducing bloat and promoting diversity using multi-objective methods. In: Proceedings of the Genetic and Evolutionary Computation Conference, GECCO 2001, pp. 11–18. Morgan Kaufmann Publishers, San Francisco (2001)
6. Devroye, L., Györfi, L., Lugosi, G.: A Probabilistic Theory of Pattern Recognition. Springer, Heidelberg (1997)
7. Ekart, A., Nemeth, S.: Maintaining the diversity of genetic programs. In: Foster, J.A., Lutton, E., Miller, J., Ryan, C., Tettamanzi, A.G.B. (eds.) EuroGP 2002. LNCS, vol. 2278, pp. 162–171. Springer, Heidelberg (2002)

8. Gagné, C., Parizeau, M.: Genericity in evolutionary computation software tools: Principles and case study. International Journal on Artificial Intelligence Tools 15(2), 173–194 (2006)
9. Gustafson, S., Ekart, A., Burke, E., Kendall, G.: Problem difficulty and code growth in genetic programming. Genetic Programming and Evolvable Machines 4(3), 271–290 (2004)
10. Koza, J.R.: Genetic Programming: On the Programming of Computers by Means of Natural Selection. MIT Press, Cambridge (1992)
11. Langdon, W.B.: The evolution of size in variable length representations. In: IEEE International Congress on Evolutionary Computations (ICEC 1998), pp. 633–638. IEEE Press, Los Alamitos (1998)
12. Langdon, W.B.: Size fair and homologous tree genetic programming crossovers. Genetic Programming And Evolvable Machines 1(1/2), 95–119 (2000)
13. Langdon, W.B., Poli, R.: Fitness causes bloat: Mutation. In: Late Breaking Papers at GP 1997, pp. 132–140. Stanford Bookstore (1997)
14. Langdon, W.B., Soule, T., Poli, R., Foster, J.A.: The evolution of size and shape. In: Advances in Genetic Programming III, pp. 163–190. MIT Press, Cambridge (1999)
15. Luke, S., Panait, L.: Lexicographic parsimony pressure. In: GECCO 2002: Proceedings of the Genetic and Evolutionary Computation Conference, pp. 829–836. Morgan Kaufmann Publishers, San Francisco (2002)
16. McPhee, N.F., Miller, J.D.: Accurate replication in genetic programming. In: Genetic Algorithms: Proceedings of the Sixth International Conference (ICGA 1995), Pittsburgh, PA, USA, pp. 303–309. Morgan Kaufmann, San Francisco (1995)
17. Nordin, P., Banzhaf, W.: Complexity compression and evolution. In: Genetic Algorithms: Proceedings of the Sixth International Conference (ICGA 1995), Pittsburgh, PA, USA, pp. 310–317. Morgan Kaufmann, San Francisco (1995)
18. Ratle, A., Sebag, M.: Avoiding the bloat with probabilistic grammar-guided genetic programming. In: Artificial Evolution VI. Springer, Heidelberg (2001)
19. Silva, S., Almeida, J.: Dynamic maximum tree depth: A simple technique for avoiding bloat in tree-based GP. In: Cantú-Paz, E., Foster, J.A., Deb, K., Davis, L., Roy, R., O'Reilly, U.-M., Beyer, H.-G., Kendall, G., Wilson, S.W., Harman, M., Wegener, J., Dasgupta, D., Potter, M.A., Schultz, A., Dowsland, K.A., Jonoska, N., Miller, J., Standish, R.K. (eds.) GECCO 2003. LNCS, vol. 2723, pp. 1776–1787. Springer, Heidelberg (2003)
20. Silva, S., Costa, E.: Dynamic limits for bloat control: Variations on size and depth. In: Deb, K., et al. (eds.) GECCO 2004. LNCS, vol. 3103, pp. 666–677. Springer, Heidelberg (2004)
21. Soule, T.: Exons and code growth in genetic programming. In: Foster, J.A., Lutton, E., Miller, J., Ryan, C., Tettamanzi, A.G.B. (eds.) EuroGP 2002. LNCS, vol. 2278, pp. 142–151. Springer, Heidelberg (2002)
22. Soule, T., Foster, J.A.: Effects of code growth and parsimony pressure on populations in genetic programming. Evolutionary Computation 6(4), 293–309 (1998)
23. Vapnik, V.: The Nature of Statistical Learning Theory. Springer, Heidelberg (1995)
24. Zhang, B.-T., Mühlenbein, H.: Balancing accuracy and parsimony in genetic programming. Evolutionary Computation 3(1) (1995)

Quantum Circuit Synthesis with Adaptive Parameters Control

Cristian Ruican, Mihai Udrescu, Lucian Prodan, and Mircea Vladutiu

Advanced Computing Systems and Architectures Laboratory
University "Politehnica" Timisoara, 2 V. Parvan Blvd., Timisoara 300223, Romania
{crys,mudrescu,lprodan,mvlad}@cs.upt.ro
http://www.acsa.upt.ro

Abstract. The contribution presented herein proposes an adaptive genetic algorithm applied to quantum logic circuit synthesis that dynamically adjusts its control parameters. The adaptation is based on statistical data analysis for each genetic operator type, in order to offer the appropriate exploration at algorithm runtime without user intervention. The applied performance measurement attempts to highlight the "good" parameters and to introduce an intuitive meaning for the statistical results. The experimental results indicate an important synthesis runtime speedup. Moreover, while other GA approaches can only tackle the synthesis for quantum circuits over a small number of qubits, this algorithm can be employed for circuits that process up to 5-6 qubits.

1 Introduction

The implementation of the meta-heuristic approach for quantum circuit synthesis makes use of ProGeneticAlgorithm [ProGA] [2] framework, that provides a robust and optimized C++ environment for developing genetic algorithms. The problem of setting values for different control parameters is crucial for genetic algorithm performance [7]. This paper introduces a genetic algorithm tailored for evolving quantum circuits. Our ProGA framework is used for genetic algorithm implementation, its architecture being extended in order to handle statistical information. The statistical data are analyzed on-the-fly by the adaptive algorithm and the results are used for adjusting the genetic parameters during run-time processes. The framework becomes a useful tool for the adaption behavior and it is designed to allow an easy development for this type of engineering problem, all its low-level details being implemented in a software library. It also allows for different configurations, thus making the comparison between the characteristics of the emerged solutions become straightforward; this is a fact that will be later used in the experiments section. In theory, the parameter control involved in a genetic algorithm may be related to population size, mutation probability, crossover probability, selection type, etc. Our proposal focuses on the parameter control using statistical information from the current state of the search. From the best of our knowledge, this is the first meta-heuristic approach for the GA-based quantum circuit synthesis. The experimental results prove the fact that

L. Vanneschi et al. (Eds.): EuroGP 2009, LNCS 5481, pp. 339–350, 2009.
© Springer-Verlag Berlin Heidelberg 2009

parameter control provides a higher convergence rate and therefore an important runtime speedup. Moreover, previous GA approaches for quantum circuit synthesis were effective only on a small number of qubits (up to 3-4 qubits [12]), while this solution works as well for 5-6 qubit circuits.

A quantum circuit is composed of one or more quantum gates, acting over a qubit set, according to a quantum algorithm. Designing a quantum circuit in order to implement a given function is not an easy task, because even if we know the target unitary transformation we don't know how to compose it out of elementary transformations. Even if the circuit is eventually rendered, we don't have information about its efficiency. This is the main reason why we propose a genetic algorithm approach for the synthesis task. The genetic algorithms will evolve a possible solution that will be evaluated against other previous solutions obtained, and ultimately a close-to-optimal solution will be indicated. In a circuit synthesis process the correct values for the parameter control are hard to be determined; this is the main reason why we propose an adaptive genetic algorithm that will control and change these values during run-time.

1.1   Problem Definition

Quantum circuit synthesis is considered to be the automatic combination and optimization of quantum circuits in order to implement a given function. The quantum circuit synthesis is an extensively investigated topic [5][9][10][12]. Several research groups have published results and significant progress has been reported in gate number reduction, qubit reduction, or even runtime speedup. Quantum circuit synthesis can have an important role in the development of quantum computing technology; in the last decades, the automatic classical circuit synthesis has improved the use or new circuits (in terms of development time, delay time, integration scale, cost and time to market, etc), allowing developers to be evermore creative. New complex applications are possible, while the classic physical technology limits are pushed to the edges. To shortly present the problem definition, we may consider -as requirement- the construction of a given quantum function out of a set of elementary operators (which are implemented by elementary gates). Solving the problem may be possible by following an already known path when dealing with the digital or analog circuit synthesis; however when the problem is moved in the quantum world, the situation becomes different. From examining the state-of-the-art, there is no common accepted path to follow for finding a solution [6][9][12] to be rendered.

2   Proposed Approach

The architecture is important in the realm of system development. At first glance, within our proposal different parts may be identified: a high-level description language parser used to map the quantum circuit description to a low-level representation, an algorithm responsible for the optimization of the abstract circuit, and a genetic algorithm responsible for the synthesis and optimization tasks.

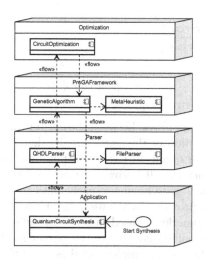

Fig. 1. Software flow proposed approach

The proposed breakdown structure indicates a layered software architecture (see Figure 1), each layer being responsible for a dedicated task. The rippled computation allows for intermediate results that can be used or optimized in the next layer. Thus, starting from a circuit description in a high-level language, after applying all the phases, the process eventually leads to the corresponding circuit. As intermediate results, we have the abstract description of the circuit, the internal data representation used for the optimization, and other program relevant information. Our Quantum Hardware Description Language (QHDL) [8] parser uses a generic implementation to create the internal data structure that is used, later on, by the genetic algorithm. The adjustment for the genetic algorithm parameters control is made by the meta-heuristic component. In the end, the evolved solution is optimized and maybe a new evolution cycle is triggered. The synthesis solution is provided as result, namely the circuit layout.

3   Genetic Algorithm Details

A dedicated genetic algorithm is used to emerge a circuit synthesis solution. The obtained solution is not necessarily the optimal one, thus providing incentives for tuning the algorithm.The terminal set that is used in the synthesis problem is composed of quantum gates (any gate from a database may be randomly used in the chromosome encoding), of the implemented methods that generate random numbers (used in the selector probabilities and in the gate selector when genetic operators are applied), and of the constant gate characteristic values (i.e. quantum circuit cost and efficiency). The function set is derived from the nature of the problem, and is composed of the mathematical functions necessary to evaluate the circuit output function (tensor product, multiplication and equality). The fitness measure specifies the solution quality within the synthesis algorithm. Therefore, the fitness assignments to a chromosome indicate how close to the algorithm target the individual output is.

Considering the discrete search space $\mathbb{X}$, the objective function $f : \mathbb{X} \to \mathbb{R}$ our scope is the find the $max_{x \in \mathbb{X}} f$ where x is a vector of decision variables, $f(x) = f(x_1, ..., x_n)$. It is a maximization problem, because the goal is to find the optimum quantum circuit that implements a given input function. The fitness function is defined as:

$$eval(x) = f(x) + W \times penalty(x) \tag{1}$$

where

$$f = \frac{f(evolved\ circuit)}{f(initial\ circuit)} \qquad (2)$$

and

$$penalty = 1 - \frac{number\ of\ evolved\ gates - number\ of\ initial\ gates}{number\ of\ initial\ gates} \qquad (3)$$

The fitness operator is implemented as a comparison between the output function of the chromosome and the output function of the given circuit, therefore revealing the differences between them. The quantum circuit output function is computed by applying the tensor product for all horizontal rows and then multiplying all the results (see Figure 2), each gate having a mathematical representation for its logic. The initial circuit is provided by the user via a high-level language hardware description. A penalty function is used in order to indicate the fact that there is a more efficient chromosome than that of the given circuit. In our optimization approach, the penalty function is implemented as the difference between the number of gates from the evolved circuit, and from the given circuit, divided by the number of given gates; it is applied only when the evolved circuit has the same functionality as the given circuit. The penalty is considered as a constraint for the algorithm and it is used to assure that a better circuit is obtained from the given one. For simplicity, we consider $W = 1$, the focus being, in this paper, on the operator performance.

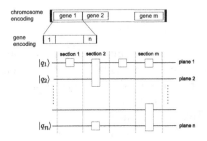

Fig. 2. Chromosome Encoding

The circuit representation is essential for the chromosome encoding (see Figure 2). Our approach is to split the circuit representation in sections and plains [1], a representation that will be used in the chromosome definition. Following Nature, where a chromosome is composed of genes, in our chromosome the genes are the circuit sections. We are able to encode the circuit within the chromosome [3], and to represent a possible candidate solution. A gene will store the specific characteristics of a particular section, and the genetic operators will be applied at the gene level or inside the gene. The mutation operator is applied inside the gene, only one locus being replaced, or at the gene level when the entire gene is randomly replaced with new locus gates (see Algorithm 1 and Algorithm 2).

The crossover operator is much more complex than mutation. In this case, the gates selected from parents are used to create offsprings, by copying their contents and properties. Thus, using a crossover probability, two genes are selected for reproduction and, by using one or two cut points, the content from between is exchanged (see Algorithm 3 and Algorithm 4).

Algorithm 1. MUTATION INSIDE THE GENE

Require: Selected gene.
Ensure: The new offspring.

1: get the number of present genes and the gene length
2: randomly select a gene and a locus
3: detect the gate corresponding to the selected locus
4: **if** empty position inside of the quantum gate **then**
5:     **repeat**
6:         move the selected locus to the right
7:     **until** a different gate id is detected (to detect the quantum gate end)
8: **end if**
9: search to the left the neighboring gate and memorize its locus
10: search to the right the neighboring gate and memorize its locus
11: generate a new random quantum gate or more quantum gates between the left and
    right locus positions

Algorithm 2. MUTATION AT THE GENE LEVEL

Require: Selected gene.
Ensure: The new offspring.

1: get the number of present genes and the gene length
2: randomly select a gene
3: perform the gene mutation by replacing the complete contents with new randomly
   generated gates

---

4   Adaptive Behavior

Almost every practical genetic algorithm is controlled by several parameters, such as population size, selector type, mutation probability, crossover probability. It is considered from a meta heuristic point of view, that genetic algorithms contain all necessary information for adaptive behavior. Nevertheless, in the following subsections, we present how the adaptive behavior optimizes the circuit synthesis search algorithm (from the users point of view, the setting of parameters is far from being a trivial task). ProGA is an object oriented framework used for genetic algorithm implementations. Software methods and design patterns are applied in order to create the necessary abstract levels. The framework is fostering for genetic algorithm implementations, through derivation of new classes from abstract ones. The type of algorithm (steady state or non-overlapping), the population structure, the encoding of the genome, and the initial settings for parameter control are made within this framework. An important framework characteristic is the possibility of extending its functionality. Thus, as is presented in Figure 3, the Adaption Control can use the framework interface, thus allowing its integration into the system.

Algorithm 3. CROSSOVER INSIDE THE GENE

Require: Selected gene.
Ensure: The new offspring.

1: create a new offspring
2: copy parent1 details into offspring
3: randomly select a cut1 point between 0 and chromosome length minus 1
4: detect the gate corresponding to the selected locus
5: **if** empty position inside a quantum gate **then**
6:    **repeat**
7:       move the selected locus to the right
8:    **until** a different gate id is detected (to detect the quantum gate end)
9: **end if**
10: search to the left the neighboring gate and memorize its index
11: search to the right the neighboring gate and memorize its index
12: **if** the crossover operator has one point **then**
13:    calculate Start index as Cut1 point
14:    calculate Stop index as the chromosome length
15: **end if**
16: **if** the crossover operator has two cut points **then**
17:    randomly select a cut2 point between right index and chromosome length minus 1
18:    calculate Start index as left index
19:    calculate Stop index as Cut2 point
20: **end if**
21: exchange the elements of offspring and parent2 between the Start and Stop indexes

---

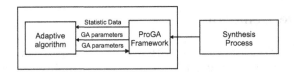

Fig. 3. Adaptive Control Integration within ProGA

The framework will provide the data necessary for statistical analysis and the actual values for parameter control, while the Adaptive Control component will return to the framework, the new adjusted values in order to determine parameters control. The Adaptive Control is considered an external tool for the genetic algorithm implementation, and it is responsible only with the appropriate update of the parameter control. Any fitness higher than or equal to 1 is considered as a solution for the synthesis problem. For each solution, statistical data is saved for later analysis: the generation number when the solution is evolved, the resulted fitness value, the chromosome values that have generated the solution, and the time required for evolution. Identical solutions are not saved into the history list, because it is not important - from an algorithmic point of view - to analyze identical data values. Thus, the history list will always contain better solutions for the given synthesis problem.

Algorithm 4. CROSSOVER AT THE GENE LEVEL

Require: Selected gene.
Ensure: The new offspring.

1: create a new offspring
2: copy parent1 details into offspring
3: random select a cut1 point between 0 and chromosome length minus 1
4: **if** the crossover operator has one cut point **then**
5:     calculate Start index as Cut1 × GeneLength
6:     calculate Stop index as the chromosome length
7: **end if**
8: **if** the crossover operator has two cut points **then**
9:     randomly select a cut2 point between 1 and number of genes
10:    **repeat**
11:        selection for cut2
12:    **until** it is different than cut1
13:    order the cut1 and cut2 points
14:    calculate Start index as Cut1 × GeneLength
15:    calculate Stop index as Cut2 × GeneLength minus 1
16: **end if**
17: exchange the elements of offspring and parent2 between the Start and Stop index

---

4.1 Operator Performance

Two types of statistical data are used as input for the adaptive algorithm. The first type is represented by the fitness results for each population, corresponding to the best, mean and worst chromosomes. The second type is represented by the operator performance. Following an idea proposed in reference [11], the performance records are essential in deciding on operator's reward.

- Absolute: when the resulted offspring has a better fitness than the best fitness from the previous generation.
- Relative: when the resulted offspring has a better fitness than its parents, but is not absolute.
- In Range: when the resulted offspring has a fitness situated between the parents fitness values.
- Worse: when the resulted offspring has a lower fitness than its parents fitness values.

For the circuit synthesis algorithm, the mutation and crossover probabilities are major parameters, that need to be controlled in a dynamic manner. Because we have defined two mutation and two crossover operators, there are four related statistical data that need to be memorized and then later analyzed for the parameter control adjustment. Each operator offspring result is important and needs to be recorded. The first type of mutation, called mutation A, is responsible with gate mutations inside of genes, while the second type of mutation, called mutation B, is applied at the chromosome level. The same rules are defined for the crossover operator (applied at the gene level and called crossover A, and at chromosome level and therefore called crossover B). The operators are implemented within the ProGA framework, and adding an extra-layer for the adaptive algorithm is sufficient for the meta-heuristic extension. From the meta-heuristic point of view, it is not important to know the operator implementation details, the only requirement is to be informed about the number of operators, because for all of them a separate statistical structure will be reserved. The algorithm will receive breeding feedback from each operator and will analyze the returned data by computing the operator performance and deciding on its adjustment rate. The change of parameter controls is made by using the feedback data from the search (stored as statistical data). The adaptive algorithm distinguishes between the qualities of evolved solutions by different operators, and then adjusts the rates based on merits. As described, the adaptive algorithm is external to the genetic algorithm framework, the only interaction consisting of the transfer of parameter rates and feedback data.

4.2   Performance Assessment

Several statistical functions may provide valuable information about data distribution; functions as Maximum, Minimum, Arithmetic Average and Standard Deviation may be applied on any kind of statistical data.

For each generation, the maximum, arithmetic average and minimum fitness values are provided by the genetic algorithm framework and then stored in the statistical data. When the genetic evolution has evolved a solution, other statistical functions are computed: maximum for all the generation maximum fitness, arithmetic average on all the maximum values, etc. Thus, we defined statistical functions on each generation, and statistical functions over all the generations, which are only used to evaluate the algorithm efficiency.

The second type of statistical data for the operator performance is computed when the operator is applied. For example, when the Crossover B result is available, the resulted offspring fitness value is compared against the previous best fitness and, if it is higher, the Absolute value is increased with one step. If it is lower, the comparison continues and if the offspring fitness is higher than the parents fitness, then the Relative value is increased, etc. After each generation, the operator performance is updated with statistical data. Following the 1/5 Rechenberg [4] rule, after five generations the analysis of the acquired data has to be made. Parameters α, β, γ and δ are introduced to rank the operator performance; they are not adjusted during the algorithm evolution. Their scope is only to statically rank the operator performance. Thus, an absolute improvement will have a higher importance in comparison with the relative improvement; a worse result will drastically decrease the operator rank. The operator reward is updated according to the following formula:

$$\sigma(op) = \alpha \times Absolute + \beta \times Relative - \gamma \times InRange - \delta \times Worse \qquad (4)$$

5 Experiments

During the performed experiments, several variables were used to measure, control and manipulate the application results. The proposed synthesis tool allows two different types of statistical data; the correlation research allows the measuring of statistical data and, at the same time, looking for possible relationship situations between some set of variables, while in the experimental research some variables are influenced in order to see their effect on other variable sets. The data analysis of the experimental results also creates correlations between the manipulated variables and those affected by the manipulation.

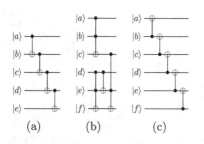

(a)   (b)   (c)

Fig. 4. Benchmark - Five-Qubit[5] (a), 2-Qubit RippleCarry Adder[13] (b) and Six-Qubit[5] (c) circuits

The experiments were conducted on a computer having the following configuration: Intel Pentium M processor at 1.86GHz, 1GB RAM memory and SuSe 10.3 as the operating system. In order to avoid lucky guesses, the experiments have been repeated for 20 times, the average result being used for comparison. The configurations for the algorithm parameters used in our experiments are shown in Table 1. Each case study is started with a benchmark quantum circuit (see Figure 4) that is used for the synthesis algorithm evaluation. For each benchmark the name of the circuit is presented along with its number of qubits (with garbage qubits if present) and the circuit cost. Two synthesis configurations (see Table 1) are used to evolve a synthesis solution, different parameters being manipulated during the test evolution (i.e. the behavior will dynamically adjust the mutation and the crossover probabilities), and the results are presented as graphs.

Table 1. Configurations Used in the Performed Experiments

Variable Name	Configuration 1	Configuration 2
Population size / Generations	50/150	50/150
Mutation type	Multiple	Single
Crossover type	Two points	One point
Selector type	Roulette Wheel	Rank
Elitism percent	0.1	0.05
Mutation probability	0.03	0.05
Crossover probability	0.3	0.4
Performance increase/decrease	0.1/0.1	0.15/0.2

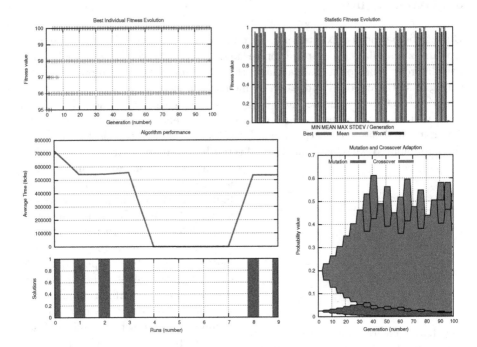

Fig. 5. Statistic Results for Configuration 2

5.1   Five-Qubit Circuit

For the Five-Qubit test circuit, we used the xor5 (see Figure 4a), its output being the EXOR of all its inputs [5]. In Figure 5 the algorithm results are presented by using plots. Using the proposed configurations, we have evolved solutions for the employed benchmark circuit and, in the right-bottom corner, the automatic adjustment for the mutation and crossover probabilities are also highlighted.

5.2   Six-Qubit Circuit

For the Six-Qubit test circuit, we used the graycode function (see Figure 4c); if the circuit for such a function is run in reverse then, the output is the ordinal number of the corresponding Gray code pattern [5]. Using the parameters set by configurations 1 and 2, several equivalent solution are emerged, differentiated by the number of gates, costs and feasibility.

A second six-qubit circuit is an add cell (see Figure 4b) taken from reference [13], and our synthesis methodology is able to evolve the same solution in less than 32 seconds.

5.3   Going beyond 6-Qubit Circuits

Other genetic algorithm based approaches have presented effective solutions only for three or four-qubit circuits. As stated in [12], the main encountered

Table 2. Test of the Approach Convergence

Number of inputs per q-gate	Mutation probability	Crossover probability	Real time (average 20 runs) as in [12]	Runtime (average 20 runs)
3-input	0.4	0.6	< 30sec	< 13sec
3-input	0.2	0.6	< 60sec	< 8sec
4-inputs	0.6	0.4	< 30sec	< 16sec
5-inputs	0.2	0.4	Not reported	< 31sec
6-inputs	0.2	0.6	Not reported	< 33sec

difficulties where: the complexity of performing tensor product for large matrixes, a high number of individuals used for the total population, and the complexity of encoding a specific quantum gate.

Our approach tackles these problems first by using an OOP (object-oriented programming) environment (backed by a framework architecture that employs optimization techniques); this improves the effectiveness of using quantum operations (including the tensor product). Second, our chromosome representation and meta-heuristic approach allows for using small populations (about 300 individuals) within the genetic evolution process. Also, another improvement comes from the fact that our approach uses a more flexible encoding scheme for the quantum gates. As a result, the experiments can be performed within our synthesis framework for 5 and 6 qubit circuits (see Table 2). Even so, attempting to perform synthesis over a larger number of qubits will also have to confront the complexity problem of matrix multiplication. However, we intend to further investigate this matter and optimize our framework, in order to extend the effectiveness of our approach for even larger quantum circuits.

6 Conclusion

The physicists still have a long way to go in order to bring the quantum circuit implementation details into a clear view, including solutions to the decoherence and gate support problems. The engineers have also a role to play in order to build a real quantum computer as a super-machine based on solid-state qubits.

This paper has presented a new approach for the automated tuning of parameters control, defined in a genetic algorithm that is used for the synthesis of quantum circuits. Statistical data are saved on each generation and analyzed by an algorithm that dynamically adjusts the parameter control values. Also, this paper offered a strategy to implement the Rechenbergs rule and the operators performance analysis in a circuit synthesis algorithm. The experiments and the source code availability prove the effectiveness of the approach for the quantum circuit synthesis task.

Future work will focus on the automatic adjustment of the population size and for the selector type, depending on the problem complexity type (direct relation with the number of qubits involved in the circuit description).

References

1. Ruican, C., Udrescu, M., Prodan, L., Vladutiu, M.: Automatic Synthesis for Quantum Circuits Using Genetic Algorithms. In: Beliczynski, B., Dzielinski, A., Iwanowski, M., Ribeiro, B. (eds.) ICANNGA 2007. LNCS, vol. 4431, pp. 174–183. Springer, Heidelberg (2007)
2. Ruican, C., Udrescu, M., Prodan, L., Vladutiu, M.: A Genetic Algorithm Framework Applied to Quantum Circuit Synthesis. Nature Inspired Cooperative Strategies for Optimization (2007)
3. Ruican, C., Udrescu, M., Prodan, L., Vladutiu, M.: Software Architecture for Quantum Circuit Synthesis. In: International Conference on Artificial Intelligence and Soft Computing (2008)
4. Eiben, E.A., Michalewicz, Z., Schoenauer, M., Smith, J.E.: Parameter Control in Evolutionary Algorithms. In: Parameter Setting in Evolutionary Algorithms (2007)
5. Maslov, D.: Reversible Logic Synthesis Benchmarks Page (2008), http://www.cs.uvic.ca/%7Edmaslov
6. Maslov, D., Dueck, G.W., Miller, M.D.: Quantum Circuit Simplification and Level Compaction. IEEE Transactions on Computer-Aided Design of Integrated Circuits and Systems (2008)
7. Herrera, F., Lozano, M.: Fuzzy adaptive genetic algorithms: design, taxonomy, and future directions. Soft Computing 7(8), 545–562 (2003)
8. Stillerman, M., Guaspari, D., Polak, W.: Final Report-A Design Language for Quantum Computing. Odyssey Research Associates, Inc., New York (2003)
9. Svore, K., Cross, A., Aho, A., Chuang, I., Markov, I.: Toward a Software Architecture for Quantum Computing Design Tools. IEEE Computer, Los Alamitos (2006)
10. Rubinstein, B.I.P.: Evolving quantum circuits using genetic programming. In: Proceedings of the 2001 Congress on Evolutionary Computation (2001)
11. Gheorghies, O., Luchian, H., Gheorghies, A.: Walking the Royal Road with Integrated-Adaptive Genetic Algorithms. University Alexandru Ioan Cuza of Iasi (2005), http://thor.info.uaic.ro/~tr/tr05-04.pdf
12. Lukac, M., Perkowski, M.: Evolving quantum circuits using genetic algorithm. In: Proceedings of the 2002 NASA/DOD Conference on Evolvable Hardware (2002)
13. Van Meter, R., Munro, V.J., Nemoto, K., Itoh, K.M.: Arithmetic on a Distributed-Memory Quantum Multicomputer. ACM Journal on Emerging Technologies in Computer Systems 3(4), A17 (2008)

Comparison of CGP and Age-Layered CGP Performance in Image Operator Evolution

Karel Slaný

Faculty of Information Technology, Brno University of Technology
Božetěchova 2, 612 66 Brno, Czech Republic
slany@fit.vutbr.cz

Abstract. This paper analyses the efficiency of the Cartesian Genetic Programming (CGP) methodology in the image operator design problem at the functional level. The CGP algorithm is compared with an age layering enhancement of the CGP algorithm by the means of achieved best results and their computational effort. Experimental results show that the Age-Layered Population Structure (ALPS) algorithm combined together with CGP can perform better in the task of image operator design in comparison with a common CGP algorithm.

1 Introduction

Cartesian Genetic Programming (CGP) was introduced by J. F. Miller and P. Thomson in 1999 [8]. When comparing with a standard genetic programming approach, CGP represents solution programs as bounded $(c \times r)$-node directed graphs. It utilizes only a mutation operator which is operating in small populations.

The influence of different aspects of the CGP algorithm have been investigated; for example the role of neutrality [2, 14], bloat [6], modularity [13] and the usage of search strategies [7]. In order to evolve more complicated digital circuits, CGP has been extended to operate at the functional level [9]. Gates in the nodes were replaced by high-level components, such adders, shifters, comparators, etc. This approach has been shown to suite well for the evolution of various image operators such as noise removal filters and edge detectors. In several papers it has been reported, that the resulting operators are human-competitive [5, 10].

The literature describes many techniques designed to preserve diversity in population as pre-selection [1], crowding [3], island models [11], etc. Escaping the local optima can only be achieved by the introduction of new, randomly generated individuals into the population. The simplest method, how to implement this technique, is restarting the evolution multiple times with different random number generator seeds. This increases the chances to find the global optima. But the evolutionary algorithm must have enough time to find a local optima. More sophisticated methods do not restart the process from scratch, but periodically introduce randomly generated individuals into the existing population. The algorithm has to ensure, that the new incomes are not easily discarded

L. Vanneschi et al. (Eds.): EuroGP 2009, LNCS 5481, pp. 351–361, 2009.

by existing better solutions and receive enough time to evolve. Such an algorithm is the Age-Layered Population Structure (ALPS) algorithm introduced by G. Hornby in 2006 [4].

This paper deals with the comparison of a standard CGP based algorithm with a modification of ALPS in the case of image operator evolution. The performance of these algorithms is measured by comparing the achieved fitness during the evolution process.

2 Image Filter Evolution

As introduced in [9, 10], an image operator can be considered as a digital circuit with nine 8-bit inputs and a single 8-bit output. This circuit can then process grey-scaled 8-bit per pixel encoded images. Every pixel value of the filtered image is calculated using a corresponding pixel and its neighbours in the processed image.

2.1 CGP at the Functional Level

In CGP a candidate graph (circuit) solution is represented by an array of c (columns) $\times$ r (rows) interconnected programmable elements (gates). The number of the circuit inputs, n_i, and outputs, n_o, is fixed through the entire evolutionary process. Each element input can be connected to the output of any other element, which is placed somewhere in the previous l columns. Feedback connections are not allowed. The l parameter defines the interconnection degree and influences the size of the total search space. In case of $l = 1$ only neighbouring columns are allowed to connect; on the other hand, if $l = c$ then a element block can connect to any element in any previous column. Each programmable element is allowed to perform one function of the functions defined in the function set Γ. The functions stored in Γ influence the design level of the circuit. The set Γ can represent a set of gates or functions defined at a higher level of abstraction. Every candidate solution is encoded into a chromosome, which is a string of $c \times r \times (e_i + 1) + n_o$ integers as shown in fig. 1. The parameter e_i is the number of inputs of the used programmable elements. If we use two-input programmable elements, then $e_i = 2$.

CGP, unlike genetic programming (GP), operates with a small population of λ elements, where usually $\lambda = 5$. The initial population is randomly generated. In each generation step the new population consists of a parent, which is the fittest individual from the previous generation, and its mutants . In case of multiple individuals sharing the best fitness, the individual, which was not selected to be the parent in the previous generation, is selected to be the parent of the next generation. This is used mainly to ensure diversity of the small population. The mutants are created by a mutation operator, which randomly modifies genes of an individual. Crossover operator is not used in CGP. In various applications, which utilize CGP, crossover has been identified to have rather destructive effect. In the particular case of image filter evolution at the functional level the crossover

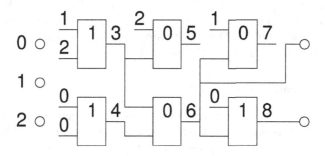

Fig. 1. An example of a 3-input circuit. Parameters are set to $l = 3$, $c = 3$, $r = 2$, $\Gamma = \{+(0), -(1)\}$. Gates 2 and 7 are not used. the chromosome looks like 1,2,1, 0,0,1, 2,3,0, 3,4,0 1,6,0, 0,6,1, 6, 8. The last two numbers represent the connection of the outputs.

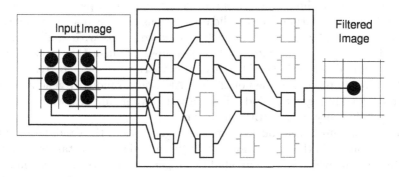

Fig. 2. Example of an image filter consisting of a 3×3 input and output matrix connected to the image operator circuit

operator does not demonstrate a very destructive behaviour [12]. However no benefits of utilizing crossover operators have been shown.

In image operator evolution the goal of CGP is to find a filter which minimizes the difference between the filtered image I_f and the reference image I_r, which must be present for a particular input image I_i. If the input and the reference image are of the size $K \times L$ pixels and a square 3×3 input and output matrix is used, then the filtered image has the size of $(K - 2) \times (L - 2)$ pixels. Because of the shape of the matrices the pixels at the edge of the filtered images can be read but they cannot be written. For an 8-bit grey-scale image the fitness value f_v can be defined by the following expression:

$$f_v = \sum_{i=1}^{K-2} \sum_{j=1}^{L-2} |I_f(i,j) - I_r(i,j)|. \tag{1}$$

The expression (1) summarizes the differences of corresponding pixels in the filtered and the reference image. When f_v drops to $f_v = 0$ then it means that the images I_f and I_r of the size $K \times L$ pixels are indistinguishable from each other (except the pixels on the edges). Papers [10, 5] show that this approach leads to good image filters. The results are satisfiable even in cases, when only a single image in the fitness function is used.

2.2   ALPS Paradigm for CGP

Premature convergence has always been a problem in genetic algorithms. One way to deal with this problem is to increase mutation probability. This will keep the diversity high. But this can also very likely destroy good alleles, which are already present in the population. When the mutation rate is set too high, then the genetic operator cannot explore narrow surroundings of a particular solution. Large population sizes can also be a solution to the diversity problem, but then more time is needed to search for a good solution.

The Age-Layered Population Structure (ALPS) [4] algorithm adds time tags into a genetic algorithm. These tags represent the age of a particular candidate solution in the population. The candidate solutions are only allowed to mutually interact and compete in groups, which contain solutions with similar age. By structuring the population into age-based groups, the algorithm ensures that a newly generated solution cannot easily be outperformed by a better solution, which is already present in the population. Also, new, randomly generated solutions are added in regular periods. These are the two main parts which maintain population diversity in the ALPS algorithm.

The age measure is the count of how many generations the candidate solution has been evolving in the population. Newly generated solution start with the age of 0. Individuals which were generated by an genetic operator such as mutation or crossover receive the age value of the oldest parent increased by 1. Every time a candidate solution is taken to be a parent, its age increases by 1. In cases a candidate solution is used multiple times to be a parent during a generation cycle, its age is still increased by 1 only once.

The population is defined as a structure of age layers, which restrict the competition and breeding among candidate solutions. Each layer, except for the last layer, has a maximum age limit. This limit allows only solutions with the age below its value to be in the layer. The last layer has no maximal age restriction, so that any best solution can stay in this layer for an unlimited time. The structure of the layers can be defined in various ways. Different systems for setting the age limits, which can be used, are shown in tab. 1. The limit values are multiplied by an *age-gap* parameter, thus obtaining maximum age of an individual in each layer.

The ALPS algorithm was designed for maintaining diversity in difficult GP based problems. Its main genetic operator is crossover with tournament selection. But crossover is not used in CGP, instead only mutation and elitism are utilized. Some modifications need to be done in order to make the ALPS algorithm work with CGP in the case of image filter evolution. These changes mainly involve removing the crossover operator from the algorithm.

During every generation cycle each layer interacts with other layers by sending individuals to the next layer or by receiving new individuals from the previous

Table 1. Various aging scheme distributions, which can be used in the ALPS algorithm

aging scheme	maximum age in layer					
	0	1	2	3	4	5
linear	1	2	3	4	5	6
Fibonacci	1	2	3	5	8	13
polynomial	1	2	4	9	16	25
exponential	1	2	4	8	16	32
factorial	1	2	6	24	120	720

layer. The original ALPS algorithm starts with randomly initialized first layer. Other layers are empty and will be filled during the process of evolution. The individuals grow older and move to next layers or are discarded. In regular intervals the bottom layer is replaced by newly generated individuals. Let us call the parameter describing this behaviour *randomize-period*. The value of the parameter stands for the number of generations between two randomization of the bottom layer.

Whenever the age of a member in a particular layer exceeds the age limit for this layer, then such a member is moved to the next layer. This formulation can cause trouble in implementation of the algorithm. Just imagine the case, when new members have to be moved into a fully occupied higher layer. In this case, the layer, which has to accept members or offspring from a previous layer, is divided into halves. The first half is used for generating new members from the original layer and the second half is populated by the newly incomes. After this step both halves behave again as a single layer.

Also elitism, similar to CGP, is used. Each layer keeps its best evolved member and only replaces it with an individual with a better or a least the same fitness. If this individual is selected to be a parent, its age is not increased in order to keep it in the current layer.

During the process of evolution the size limits of the population do not change, but the number of individuals in the layers may vary. This is caused by the fact, that in the initial phase the algorithm starts with only one populated layer. Also in certain situations a layer can lead into extinction, when current layer members and its offspring are moved into next layer and no newcomers have arrived.

3   Experimental Set-Up

In order to evolve image operators a set of function has been adopted from [10].

The CGP and modified ALPS-CGP algorithms are used to find a random shot noise filter and a Sobel filter using a set of three pictures as training data. The evolved image operators are connected to a 3×3 input and output mask. The size of the pictures is 256×256 pixels.

The experiment consists of two main groups which differ in the way how the evolved image operators is defined. In the first set of experiments the image operator shall consist of 8 columns $\times$ 6 rows of programmable elements with the

Table 2. Function set used in the experiments. All functions have 8-bit inputs and outputs.

ID	Function	Description	ID	Function	Description
0	$x \vee y$	binary or	4	$x +_{sat} y$	saturated addition
1	$x \wedge y$	binary and	5	$(x + y) >> 1$	average
2	$x \oplus y$	binary xor	6	$Max(x, y)$	maximum
3	$x + y$	addition	7	$Min(x, y)$	minimum

Fig. 3. Images used in described experiments. Top row contains input images entering the evolved image operators. Bottom row shows reference images used for fitness evaluation. The evolution searches for a Sobel operator and for a random shot noise removal filter.

interconnection parameter $l = 1$. The value $l = 1$ allows only interconnection of neighbouring columns. This is because of an easy implementation as a pipelined filter in hardware. The second set of experiments utilizes chromosomes which consist of only a single row with 48 programmable elements with the interconnection parameter set to $l = 48$. This value ensures unlimited interconnection, except that the output of an evolved circuit cannot connect directly to its input.

The CGP algorithm uses population size of 8 individuals. Mutation probability is set to 8% in both algorithms. The ALPS-CGP algorithm uses 5 layers. Each layer can hold up to 8 individuals. The polynomial aging scheme is used with *age-gap* = 20. The bottom layer is regenerated with random individuals every *randomize-period* = 5 cycles. Each evolutionary process of 10000 generations is repeated 100 times. Average data are used to compare both of the two algorithms.

The measured data are compared according to the evaluation number. That means the fitness values are plotted against the number of evaluations which the algorithm has performed rather than to the generation it has reached. This is

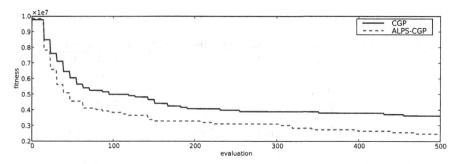

Fig. 4. The progression of the fitness value during the first 500 evaluations when evolving the Sobel filter by using the camera pictures. The image operator consists of 8×6 elements.

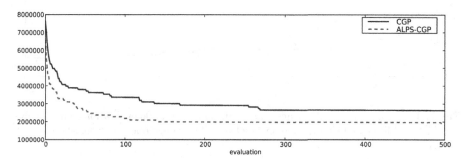

Fig. 5. The progression of the fitness value during the first 500 evaluations when evolving the Sobel filter by using the circle pictures. The image operator consists of 8×6 elements.

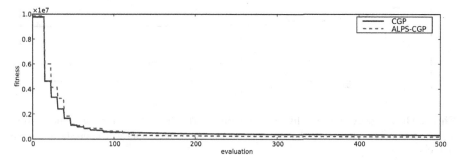

Fig. 6. The progression of the fitness value during the first 500 evaluations when evolving the random shot noise removal filter by using the Lena pictures. The image operator consists of 8×6 elements.

done because of the fact, that the ALPS-CGP algorithm uses larger populations. Thus it has a greater chance of exploring larger amounts of search-space in a single generation cycle than the CGP algorithm.

Table 3. The average achieved fitness after finishing 10000 generations

set of experiments	image set	average fitness CGP	average fitness ALPS
1	camera Sobel	2 034 740	1 371 492
1	circle Sobel	1 948 039	1 345 093
1	Lena impulse	52 980	31 681
2	camera Sobel	2 323 204	1 893 629
2	circle Sobel	2 176 023	1 580 389
2	Lena impulse	47 557	31 284

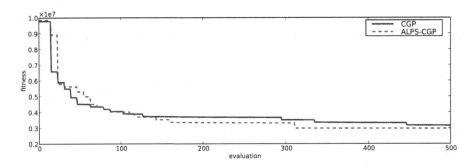

Fig. 7. The progression of the fitness value during the first 500 evaluations when evolving the Sobel filter by using the camera pictures. The image operator consists of 1×48 elements.

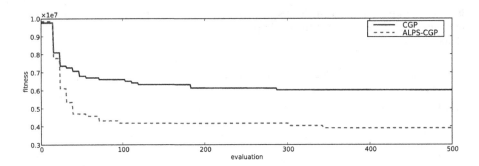

Fig. 8. The progression of the fitness value during the first 500 evaluations when evolving the Sobel filter by using the circle pictures. The image operator consists of 1×48 elements.

3.1   Results

In the first set of experiments image operators with rectangular arrangement of programmable elements with the interconnection parameter $l = 1$ were evolved.

The graphs in fig. 4, 5 and 6 show that the ALPS-CGP algorithm is behaving slightly better than the standard CGP algorithm. In average we have obtained similar or better image operators in less time than using only the CGP algorithm.

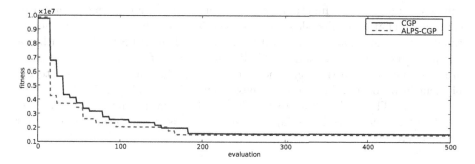

Fig. 9. The progression of the fitness value during the first 500 evaluations when evolving the random shot noise removal filter by using the Lena pictures. The image operator consists of 1×48 elements.

The average fitness values, measured after reaching final generation number, are summarized in tab. 3.

In the second set of experiments the evolutionary process searched for image operators with a linear structure with high interconnection parameter, allowing more complex structures to be designed. This also increases the search space of all possible solutions. The graphs in fig. 7, 8 and 9 show the behaviour of the algorithms in the first 500 evaluations. Again we have obtained similar results.

4   Discussion

In the first group of experiments we can observe a performance gain, when using the ALPS-CGP algorithm. In the initial phase, when the algorithms are started, the ALPS-CGP algorithm converges faster to local optima than the simple CGP algorithm. But then still keeps improving the fitness value of the evolved image operators. The ALPS algorithm in average achieves better fitness values.

In the second group of experiments the algorithms show similar behaviour as in group one. Again ALPS has achieved better fitness values than the CGP algorithm. Only in the case of the random shot noise filter the algorithms show approximately the same performance. This may be because the random shot filter is easier to construct from the function set Γ. Also the second group of experiments showed, that the evolved image operators achieve slightly worse results, as in the first case. The explanation may be the fact, that in the second case the elements, which the image operator consists of, are allowed to connect more freely in comparison with the first experimental set. The search space is much greater. Another explanation may be the fact, that the configuration of the image operator is taken from [10], and might be more optimized for the role of an image operator.

The whole system is implemented in SW. To finish all runs in the first or second set of experiments 6 days of computation time are needed when using a two 1800 MHz dual-core AMD Opteron system. This is mainly caused by the time-consuming fitness function - every pixel in the training images has to be computed separately. The slow performance is also a drawback, when the optimal performance parameters need to be found. The ALPS-CGP algorithm is not much more complex than the CGP algorithm. In fact the main differences are the time-tags and the consequent restriction. These are not difficult to implement in hardware. Therefore the next step will be implementing the system in a FPGA. This will give a greater chance of evaluating the ALPS-CGP and CGP performance. Also larger input masks can be used, which can lead to better image operators.

5   Conclusions

An analysis of the performance of a standard CGP approach and an ALPS enhanced CGP algorithm in the task of image operator evolution was performed. The performance of the algorithms was measured in six cases of system settings. Experiments have shown that the ALPS-CGP algorithm performs better than the standard CGP algorithm. However in more difficult cases the performance of the ALPS-CGP algorithm appears not to be much superior. Further experiments, including hardware implementation, are needed to receive a better comparison of the two algorithms.

Acknowledgements

This work was supported by the Grant Agency of the Czech Republic under No. 102/07/0850 *Design and hardware implementation of a patent-invention machine* and the Research intention No. MSM 0021630528 – Security-Oriented Research in Information Technology.

References

[1] Cavicchio, D.J.: Adaptive search using simulated evolution. Ph.D thesis, University of Michigan (1970)

[2] Collins, M.: Finding needles in haystacks is harder with neutrality. Genetic Programming and Evolvable Machines 7(2), 131–144 (2006)

[3] De Jong, K.A.: Analysis of the Behavior of a Class of Genetic Adaptive Systems. Ph.D thesis. University of Michigan (1975)

[4] Hornby, G.S.: Alps: the age-layered population structure for reducing the problem of premature convergence. In: GECCO 2006: Proceedings of the 8th annual conference on Genetic and evolutionary computation, pp. 815–822. ACM, New York (2006)

[5] Martínek, T., Sekanina, L.: An evolvable image filter: Experimental evaluation of a complete hardware implementation in fpga. In: Moreno, J.M., Madrenas, J., Cosp, J. (eds.) ICES 2005. LNCS, vol. 3637, pp. 76–85. Springer, Heidelberg (2005)

[6] Miller, J.F.: What bloat? cartesian genetic programming on boolean problems. In: 2001 Genetic and Evolutionary Computation Conference Late Breaking Papers, pp. 295–302 (2001)

[7] Miller, J.F., Smith, S.L.: Redundancy and computational efficiency in cartesian genetic programming. IEEE Trans. Evolutionary Computation 10(2), 167–174 (2006)

[8] Miller, J.F., Thomson, P.: Cartesian genetic programming. In: Poli, R., Banzhaf, W., Langdon, W.B., Miller, J., Nordin, P., Fogarty, T.C. (eds.) EuroGP 2000. LNCS, vol. 1802, pp. 121–132. Springer, Heidelberg (2000)

[9] Sekanina, L.: Image filter design with evolvable hardware. In: Cagnoni, S., Gottlieb, J., Hart, E., Middendorf, M., Raidl, G.R. (eds.) EvoIASP 2002, EvoWorkshops 2002, EvoSTIM 2002, EvoCOP 2002, and EvoPlan 2002. LNCS, vol. 2279, pp. 255–266. Springer, Heidelberg (2002)

[10] Sekanina, L.: Evolvable Components: From Theory to Hardware. Springer, Heidelberg (2004)

[11] Skolicki, Z., De Jong, K.A.: Improving evolutionary algorithms with multi-representation island models. In: Yao, X., Burke, E.K., Lozano, J.A., Smith, J., Merelo-Guervós, J.J., Bullinaria, J.A., Rowe, J.E., Tiňo, P., Kabán, A., Schwefel, H.-P. (eds.) PPSN 2004. LNCS, vol. 3242, pp. 420–429. Springer, Heidelberg (2004)

[12] Slaný, K., Sekanina, L.: Fitness landscape analysis and image filter evolution using functional-level cgp. In: Ebner, M., O'Neill, M., Ekárt, A., Vanneschi, L., Esparcia-Alcázar, A.I. (eds.) EuroGP 2007. LNCS, vol. 4445, pp. 311–320. Springer, Heidelberg (2007)

[13] Walker, J.A., Miller, J.F.: Investigating the performance of module acquisition in cartesian genetic programming. In: GECCO 2005: Proceedings of the 2005 conference on Genetic and evolutionary computation, pp. 1649–1656. ACM, New York (2005)

[14] Yu, T., Miller, J.F.: Neutrality and the evolvability of boolean function landscape. In: Miller, J., Tomassini, M., Lanzi, P.L., Ryan, C., Tetamanzi, A.G.B., Langdon, W.B. (eds.) EuroGP 2001. LNCS, vol. 2038, pp. 204–217. Springer, Heidelberg (2001)

Author Index